DIESEL LOCOMOTIVE R(

U. S., Canada, Mexico

BY CHARLES W. McDONALD

Editor: George H. Drury
Art Director: Lawrence Luser
Copy Editors: Marcia Stern, Bob Hayden
Art and Layout: Patti Keipe, Mark Watson

On the cover: Canadian National SD40s, running through on a Milwaukee Road freight train, and Burlington Northern SD40-2s stand at Daytons Bluff in St. Paul, Minnesota, in September 1984. Photo by Mike Danneman.

Distributed by:
Airlife Publishing Ltd.
101 Longden Road, Shrewsbury SY3 9EB, England

First edition: First printi   992. Second printing, 1993.

Printed in United States

CW00503998

DEDICATION

To the two individuals who made this book possible by their patience, understanding, and support: My wife and friend, Pam McDonald, and Dr. John H. Watson-Poodle, trusted friend.

Library of Congress Cataloging-in-Publication Data

McDonald, Charles W.
     Diesel locomotive rosters : U.S., Canada, Mexico / Charles W. McDonald. — 3rd ed.
     p. cm.
  ISBN 0-89024-112-0 : $12.95
  1. Diesel locomotives—United States—Registers. 2. Diesel locomotives—Canada—Registers. 3. Diesel locomotives—Mexico—registers. I. Title.
  TJ619.2.M33 1992   92-17907
  625.2'66'0973—dc20   CIP

To order additional copies of this book or other Kalmbach books, call toll free at (800) 533-6644

INTRODUCTION

The railroad industry has changed greatly in the 10 years since the first edition of DIESEL LOCOMOTIVE ROSTERS was published. Several major mergers have occurred, and new railroads have come into being as a result of spinoffs by the large railroads. Many of these spinoffs are short lines; others, often more interesting to the student of locomotives, are regional railroads.

The locomotive market

Regional railroads have the potential to exert considerable influence on the locomotive market, because they operate a large number of older units acquired for the most part from Class 1 railroads. Currently the Class 1 railroads have many locomotives whose leases will expire in the next few years. These units will be at least 15 years old when they come on the market. The disposition of these units is worth watching, because a strong demand for used locomotives could increase sales of new locomotives.

It has become clear that in North America there are only 11 consistent buyers of new locomotives. Some small railroads regularly buy new power a unit or two at a time, but most can't afford new power in quantity. The two principal U. S. builders are located in areas of the country with high labor costs and high taxes. It is likely that some change in the manufacture of motive power will have to occur before sales approach the levels of the 1970s.

The tax laws work positively for the locomotive builders. The great numbers of locomotives delivered in the mid to late 1970s are reaching the end of their depreciation cycle, which could result either in an upswing in deliveries of new power or in a proliferation of locomotive rebuilding programs.

Perhaps one surprise is the comparative stability of the price of new motive power. New high-tech locomotives sell for substantially the same price (about $1.25 million) that they did in 1985. This price stability is a result of both the lack of demand for new power and the intense competition between the builders for what business remains. Privately, builders will admit they are selling locomotives for less than the cost of building them in order to keep their plants in operation.

Used locomotives

The used-locomotive market has tightened up in recent years. The cheap locomotives of 1985 are gone. The mega-railroads have disposed of many of their first-generation road-switchers. As a result, larger locomotives are now in service on short lines than was the case in 1985. Years ago a Geep was remarkable in shortline service; today it is common. The continued growth of shortline railroads and the startup of new short lines will create a strong demand for used power.

The demand for used EMD locomotives should continue. Regional railroads have demonstrated a preference for rebuilt or overhauled second-generation EMD power. Used GE road units are hard to sell to anyone; few regionals or short lines use GE power. In the long run this should contribute to a resurgence of EMD in the motive power market, as should the introduction of EMD 70-series models. GE will have to work hard to maintain its market share.

Industrial locomotives

This book includes industrial locomotives for the first time. For the most part they spend their lives out of the public eye, working behind chain link fences. No public record (public in the sense that federal agencies monitor common-carrier railroads) is kept of their acquisition, sale, and scrapping. An author trying to catalog such locomotives (by conservative estimate at least 4,000 exist) may find himself presenting inaccurate data. I have no doubt that some of the information I have presented in the industrial section of this book is obsolete. It is presented as a snapshot, the best information available at the time of compilation. Readers' comments and corrections are welcome.

Industrial railroads are listed alphabetically by state (Canadian provinces follow the U. S. listings), city, and owner. Data elements include number, model, builder, previous ownership, and date built. Supplemental notes either follow the date or appear on the next line. Information on some locomotives is fragmentary; data elements for some of the locomotives may be missing.

About the information in this book

More than 550 railroads are listed in this book, but, given the rate at which short lines spring up and die, by the time you read this some of the railroads will be history and new ones will have taken their place. Roster data changes daily on the locomotive scene, and the best any author can do is provide a snapshot. The information should be useful for some time to come, and it will be historical reference when it is superseded by a new edition.

In the pages that follow, common carrier railroads of North America are listed alphabetically by name. The information provided for each road's locomotives includes:

Numbers: A plus sign indicates a range of road numbers greater than the quantity of units

Quantity: Number of units in service

Model designations sometimes have a suffix:

d	Derated: rebuilt with a lower horsepower rating
m	Modified: major changes such as chopped nose, new hoods, or new trucks
r	Rebuilt: upgraded electrically or mechanically but with no increase in the horsepower rating
u	Uprated: increased horsepower rating
L	Units built lighter than standard for special service
M	Full-width body and nose and upgraded cab interior
W	Full-width nose

Builder:

ABB   ABB Traction is a division of ASEA Brown Boveri, which is the product of a merger of two long-established European electrical manufacturers.

Alco   American Locomotive Co., Schenectady, N. Y., and its predecessors had been in the locomotive business since the 1850s. The company was renamed Alco Products in 1956. Alco produced its last diesel locomotive in 1969.

Atlas   Atlas Car & Manufacturing Co., Cleveland, Ohio, began building industrial rail equipment in 1896. During the 1930s and 1940s it also built small diesel locomotives.

ATSF   Atchison, Topeka & Santa Fe Railway's shops at Cleburne, Texas, designed and produced the CF7, essentially a cab unit rebuilt to a hood unit. The CF7 has become the shortline locomotive of the 1990s the way GE's 44-tonner and 70-tonner were in the 1950s. Cleburne also rebuilt a group of U36Cs; Santa Fe's San Bernardino, Calif., shops are best known for the switcher rebuilding program that created the SSB1200.

BBD   Bombardier Inc., Montreal, was the successor to Montreal Locomotive Works, building locomotives based on Alco designs. In July 1985 Bombardier announced it was discontinuing the production of freight diesels.

Bkvl   Brookville Locomotive Division, Pennbro Corp., Brookville, Pa., has long history of producing small industrial and mine locomotives.

BLW   Baldwin Locomotive Works, Eddystone, Pa., was the leading builder of steam locomotives and had a history reaching back to the 1830s, but it did not make the transition to diesel production altogether successfully. Baldwin merged with Lima-Hamilton in 1950.

BLH   Baldwin-Lima-Hamilton, Eddystone, Pa., was created by the merger of Baldwin and Lima-Hamilton in 1950. BLH built its last new diesel in 1956.

CLC   Canadian Locomotive Co., Kingston, Ont., was a major builder of steam locomotives. In the 1940s it marketed Baldwin and Whitcomb diesels in Canada, and later it manufactured diesels of Fairbanks-Morse design. It also built some small diesels of its own design and bought the Davenport-Porter designs when that company ceased producing locomotives.

Dav   Davenport Locomotive Works, Davenport, Iowa, produced a line of industrial locomotives and occasionally built small switchers for Class 1 railroads. In 1950 Davenport added H. K. Porter's locomotives to its own line. It ceased building locomotives in 1956.

EE   English Electric is represented by five boxcab electrics built for the National Harbours Board of Montreal in 1924 and 1926 and sold to Canadian National in the 1940s.

EMC   Electro-Motive Corporation was formed in 1924 to design and sell gasoline-powered railcars. It was purchased by General Motors in 1930 and merged with Winton Engine in 1941 to form Electro-Motive Division.

EMD   Electro-Motive Division, General Motors, La Grange, Ill., was created in 1941 by the merger of GM subsidiaries Electro-Motive Corporation and Winton Engine Co. EMD introduced mass production techniques to the manufacture of diesel locomotives and became the largest North American diesel locomotive builder.

FM   Fairbanks-Morse, Beloit, Wis., built locomotives with opposed-piston prime movers from 1944 to 1963.

GE   General Electric, Erie, Pa., produced the electrical equipment for Alco diesel locomotives and from 1940 to 1953 jointly marketed the line as Alco-GE diesels. In addition, GE built electric locomotives and an extensive line of small diesels for industrial use. In 1960 GE introduced the U25B, a 2500-h.p. hood unit, and within three years became the number-two locomotive builder in the U. S.

GMD   General Motors Diesel, Ltd., later Diesel Division, General Motors of Canada, Ltd., London, Ont., produces locomotives of EMD design. In recent years almost all EMD production has been moved from La Grange to London.

ICG   Illinois Central Gulf Railroad's, Paducah, Ky., shops are known primarily for the rebuilding program that produced the GP8, GP10, and GP11.

Lima   Lima-Hamilton, Lima Locomotive Works, Lima, Ohio, was a major steam locomotive builder. It merged with General Machinery Corp. of Hamilton, Ohio, in 1947, and from 1949 to 1951 produced a line of switchers and road-switchers under the Lima-Hamilton name. In 1950 the company merged with Baldwin.

MK   Morrison-Knudsen, Boise, Idaho, is a major rebuilder of locomotives; some of its rebuilding activity is extensive enough to result in essentially new units. Recently MK's specialty has been locomotives for commuter service.

MLW   Montreal Locomotive Works, Montreal, was Alco's Canadian affiliate, producting Alco-designed locomotives and then continuing the line after Alco's demise. Bombardier Ltd. purchased control of MLW in 1975.

Plym   Plymouth Locomotive Works, Plymouth, Ohio, was an early builder of small gasoline and diesel locomotives. Most of Plymouth's production has been for short lines and industrial railroads.

Porter   H. K. Porter, Inc., Pittsburgh, Pa., produced steam and diesel locomotives for industrial use. In 1950 it sold its diesel designs to Davenport Locomotive Works.

RLW   Republic Locomotive Works, Greenville, S. C., was established in 1980 as a rebuilder of medium-size locomotives. Recent production includes FL9s rebuilt with AC electrical systems for Metro-North and Long Island.

Vulcan   Vulcan Iron Works, Wilkes-Barre, Pa., a builder of industrial steam locomotives, introduced internal combustion locomotives in the 1920s. Production ceased in 1954.

West   Westinghouse Electric & Manufacturing Co. had provided equipment for railroad electrification since the 1890s and supplied electrical components for Baldwin and Fairbanks-Morse diesels. Between 1928 and 1936 it also built a line of diesel locomotives at South Philadelphia.

Whit   Whitcomb Locomotive Works, Rochelle, Ill., began building internal combustion and electric locomotives in 1906. In 1931 Baldwin purchased control of the company, and in 1952 production was moved to Baldwin's Eddystone plant under the Baldwin-Lima-Hamilton name.

Date built: The year or years in which the locomotives were built

Notes: Short notes appear on the same line as the entry; longer notes appear at the foot of the roster and are numbered.

Acknowledgments

The data in the book has come from numerous people who work for the railroads and the locomotive manufacturers and also from observers of the railroad industry. I thank them all for sharing the information. J. David Ingles, Senior Editor of TRAINS Magazine, read the page proofs and filled in much information of the history of the locomotives and for many rosters provided up-to-date information. I particularly appreciate his help.

The Contemporary Diesel Spotter's Guide, by Louis A. Marre and Jerry A. Pinkepank, also published by Kalmbach, is a necessary companion to this book. I recommend it highly. In addition, anyone seriously interested in locomotives will find indispensable the quarterly magazine *Extra 2200 South* (P. O. Box 8110-820, Blaine, WA 98230-2107 or P. O. Box 1004, Garibaldi Highlands, BC V0N 1T0).

Readers are invited to send comments, corrections, and information to the author in care of Kalmbach Publishing Co. The information presented in this book has been extracted from a much larger data base. Additional information is available from that data base. Write to the author in care of the publisher for price quotations.

CHARLES W. McDONALD

Minot, North Dakota
May 1992

LOCOMOTIVE MODELS

The locomotive models in this book are listed below with basic specifications, quantity in service in 1992, and production totals. Weights are not given. The weight of a locomotive depends on everything from fuel capacity (and the amount in the tank) to whether the unit was ordered ballasted for extra tractive effort or stripped down for light-rail branch lines. The nominal weight for most four-axle units is 240,000 pounds; for six-axle units, 360,000 pounds.

Model	H.P.	Wheel arrgt.	Common Carrier	Industrial	Total built	Notes
American Locomotive Co. (plus Alco designs built by Montreal Locomotive Works)						
C415	1500	B-B	2	6	26	
C420	2000	B-B	48	6	131	
C424	2400	B-B	56	4	190	
C425	2500	B-B	18	0	91	
C430	3000	B-B	6	0	16	
C628	2750	C-C	50	0	181	
C630	3000	C-C	81	0	133	
C636	3600	C-C	3	1	34	
HH600	600	B-B	0	4	78	
HH660	660	B-B	0	11	43	
FA1	1600	B-B	5	0	412	
FA2	1600	B-B	5	0	395	
MRS1	1600	C-C	23	5	83	
PA1	2000	A1A-A1A	2	0	170	
RS1	1000	B-B	39	29	417	
RS2	1500	B-B	6	10	383	
RS3	1600	B-B	43	18	1370	
RS11	1800	B-B	95	30	426	
RS27	2400	B-B	2	0	27	
RS32	2000	B-B	3	1	35	
RS36	1800	B-B	8	0	40	
RSD1	1000	C-C	1	15	150	
RSD4	1600	C-C	1	0	36	
RSD12	1800	C-C	75	3	161	
RSD15	2400	C-C	8	2	87	
S1	660	B-B	32	56	540	
S2	1000	B-B	67	210	1502	
S3	660	B-B	5	37	292	
S4	1000	B-B	34	83	797	
S6	900	B-B	13	56	126	
T6	1000	B-B	9	6	57	

Model	H.P.	Wheel arrgt.	Common Carrier	Industrial	Total built	Notes
ASEA-Brown-Boveri						
ALP44		B-B	15	0	15	NJ Transit
Atchison, Topeka & Santa Fe Railway						
CF7	1500	B-B	146	54	233	Rebuilt F7
SF30C	3000	C-C	70	0	70	Rebuilt U36C
SSB1200	1200	B-B	4	0	29	
Bombardier Inc.						
HR412W	2000	B-B	10	0	11	
HR616W	3000	C-C	20	0	20	
LRC	2700	B-B	8	0	34	
Brookville						
All types		B	4	22		
Production totals not available						
Baldwin Locomotive Works and Baldwin-Lima-Hamilton						
25 ton	145	B	0	1		
40 ton		B	0	2		
44-50 ton	370-380	B-B	4	1		
60-65 ton	550	B-B	2	3		
AS416	1600	A1A-A1A	0	1	21	
AS616	1600	C-C	4	3	1468	
DRS66-15	1500	C-C	2	1	82	
DS44-10	1000	B-B	3	33	502	
DS44-6	600	B-B	1	4	139	
DS44-7.5	750	B-B	1	6	53	
RS4TC	660	B-B	48	0		
RS12	1200	B-B	1	0	46	
S8	800	B-B	5	7	61	
S8B	800	B-B	0	4	9	
S12	1200	B-B	28	40	451	
VO1000	1000	B-B	30	22	548	
VO660	660	B-B	2	3	142	
Canadian Locomotive Co.						
20 ton		B	0	1		

Model	H.P.	Wheel arrgt.	Common Carrier	Industrial	Total built	Notes
44 ton	380	B-B	0	4		
50 ton	400	B-B	1	2		
Seaboard Coast Line						
GP16	1600	B-B	160	0	161	Rebuilt GP/9
Electro-Motive Division, General Motors and General Motors Diesel Division						
DIESEL LOCOMOTIVES						
40		B	1	6	11	
BL2	1500	B-B	3	0	58	
DD40AX	6600	D-D	2	3	47	
E8A	2250	A1A-A1A	8	0	421	
E9A	2400	A1A-A1A	30*	0	100	
* Figure includes some rebuilt E8s						
F2A	1350	B-B	2	0	74	
F3A	1500	B-B	3	0	1111	
F7A	1500	B-B	27	3	2366	
F7B	1500	B-B	15	0	1483	
F9A	1750	B-B	11	5	87	
F9B	1750	B-B	3	6	154	
F40C	3200	C-C	15	0	15	Metra
F40PH	3000	B-B	278	0	284	
F40PH-2	3000	B-B	72	0	131	
F40PH-2C	3000	B-B	25	0	25	MBTA
F40PH-2M	2000	B-B	0	4	4	Speno
F40PH-M	3000	B-B	30	0	30	Metra
F45	3600	C-C	42	0	86	
F59PH	3500	B-B	42	0	42	GO Transit
F69PHAC	3200	B-B	2	0	2	Amtrak
FL9	1750	B-A1A	41	0	60	
FP7	1500	B-B	29	1	376	
FP9	1750	B-B	17	0	79	
FP45	3600	C-C	8	0	14	

Model	H.P.	Wheel arrgt.	Common Carrier	Industrial	Total built	Notes
FTA	1350	B-B	1	0	555	
FTB	1350	B-B	1	0	541	
G8	875	B-B	0	2	11	
G12	1310	B-B	73	0	92	
G16	1600	C-C	22	22	24	
GA8	800	B-B	18	0	22	
GMD1	1200	B-B	99	0	101	
GMDH1	800	B-B	0	1	4	
GMDH3	400	B-B	0	1	1	
GP7	1500	B-B	656	43	2729	
GP9	1750	B-B	1402	41	4092	
GP9B	1750	B-B	4	0	165	
GP15-1	1500	B-B	341	0	341	
GP15T	1500	B-B	28	0	28	
GP18	1800	B-B	108	7	390	
GP20	2000	B-B	85	12	260	
GP20C	2000	B-B	10	0	10	
GP28	1800	B-B	14	0	26	
GP30	2250	B-B	331	4	908	
GP35	2500	B-B	671	7	1333	
GP38(AC)	2000	B-B	682	3	733	
GP38-2	2000	B-B	1896	16	2188	
GP38-2B	2000	B-B	1	0	1	
GP38P-2	2000	B-B	20	0	20	NdeM
GP39(DC)	2300	B-B	23	0	23	
GP39-2	2300	B-B	182	0	249	
GP39E	2300	B-B	50	0	50	
GP39M	2300	B-B	45	0	45	
GP39V	2300	B-B	25	0	25	
GP40	3000	B-B	1057	2	1243	
GP40-2	3000	B-B	879	1	892	
GP40-2LW	3000	B-B	276	0	277C	
GP40-2W	3000	B-B	11	0	11C	
GP40G	3000	B-B	10	0	10	
GP40P	3000	B-B	13	0	13	NJT
GP40P-2	3000	B-B	3	0	3	SP
GP40TC	3000	B-B	8	0	8	Amtrak
GP40X	3500	B-B	23	0	23	
GP49	2800	B-B	15	0	15	
GP50	3500	B-B	224	0	224	
GP50L	3500	B-B	48	0	50	
GP59	3000	B-B	36	0	36	
GP60	3800	B-B	265	0	270	
GP60B	3800	B-B	23	0	25	
GP60M	3800	B-B	61	0	61	
HBU4	0	B-B	23	0	23	
MP15(DC)	1500	B-B	321	21	242	
MP15AC	1500	B-B	172	7	255	
MP15T	1500	B-B	42	0	42	
MRS1	1500	C-C	5	0	13	
NW2	1000	B-B	142	146	1143	
NW5	1000	B-B	4	3	13	
RS1325	1325	B-B	2	0	2	C&IM
SC	600	B-B	0	7	44	
SD7	1500	C-C	52	5	188	
SD9	1750	C-C	339	10	471	
SD18	1800	C-C	47	7	54	
SD24	2400	C-C	12	11	179	
SD28	1800	C-C	2	4	6	
SD35	2500	C-C	65	7	360	
SD38	2000	C-C	52	0	52	
SD38-2	2000	C-C	61	13	78	
SD38AC	2000	C-C	14	0	15	
SD39	2300	C-C	35	1	54	
SD40	3000	C-C	967	9	1257	
SD40-2	3000	C-C	3672	8	3681	

Model	H.P.	Wheel arrgt.	Common Carrier	Industrial	Total built	Notes
SD40-2B	3000	C-C	3	0	3	
SD40-2F	3000	C-C	25	0	25C	
SD40-2W	3000	C-C	122	0	122C	
SD40A	3000	C-C	13	0	18	
SD40G	3000	C-C	10	0	10	
SD40T-2	3000	C-C	305	0	310	
SD40X	3000	C-C	4	0	4	
SD45	3600	C-C	158	3	1260	
SD45-2	3600	C-C	118	0	136	
SD50	3500	C-C	407	0	410	
SD50F	3500	C-C	60	0	60	
SD60	3800	C-C	46	0	46C	
SD60	3800	C-C	516	0	520	
SD60F	3800	C-C	64	0	64	
SD60M	3800	C-C	289	0	289	
SDL39	2300	C-C	9	0	10	
SDP40	3000	C-C	14	0	20	
SDP45	3600	C-C	3	0	52	
SW	600	B-B	0	6	77	
SW1	600	B-B	77	149	661	
SW7	1200	B-B	70	22	493	
SW8	800	B-B	103	82	371	
SW9	1200	B-B	165	94	815	
SW600	600	B-B	1	10	15	
SW900	900	B-B	20	117	357	
SW1000	1000	B-B	78	33	118	
SW1001	1000	B-B	92	38	151	
SW1200	1200	B-B	375	120	755	
SW1200RS	1200	B-B	184	5	259	
SW1500	1500	B-B	733	39	807	
SW1504	1500	B-B	60	0	60	NdeM
TR4A	1200	B-B	7	3	15	
TR4B	1200	B-B	7	3	15	
TR5A	1200	B-B	2	9	10	
TR6A	800	B-B	0	9	12	
TR6B	800	B-B	0	7	12	
YBU4	0	B-B	14	0	14	
ELECTRIC LOCOMOTIVES						
AEM7	7000	B-B	56	0	59	
AEM7M	7000	B-B	7	0	7	SEPTA
GF6C	6000	C-C	7	0	7	BC Rail
SW1200MG	1200	B-B	0	9	9	
GM6	6000	C-C	0	1	1	
GM10	10,000	B-B-B	0	1	1	
English Electric						
Boxcab		B+B	5	0	9	
Fairbanks-Morse						
H10-44	1000	B-B	2	0	197	
H12-44	1200	B-B	8	14	335	
H16-66	1600	C-C	0	3	58	
General Electric						
DIESEL LOCOMOTIVES						
5-8 ton		B	0	6		
11-35 ton	150	B	0	218		
40 ton	300	B	0	3		
43, 45 ton	300	B-B	30	242		
44 ton	400	B-B	55	78	348	
50-60 ton	480	B-B	3	95		
65 ton	550	B-B	34	126		
70 ton	660	B-B	59	44	238	
80 ton	550	B-B	108	179		
85 ton	550	B-B	10	5		
86-110 ton	600	B-B	3	51		
115-132 ton	1100	B-B	2	3		
Production totals for industrial units not available						
B23-7	2300	B-B	487	1	536	

Model	H.P.	Wheel arrgt.	Common Carrier	Industrial	Total built	Notes
B23-S7	2300	B-B	11	0	11	
B30-7	3000	B-B	196	0	199	
B30-7A	3000	B-B	77	0	80	
B30-7AB	3000	B-B	119	0	120	
B36-7	3600	B-B	220	0	222	
B36-7B	3600	B-B	1	0	1	
B39-8	3900	B-B	143	0	143	
BQ23-7	2300	B-B	9	0	10	
C30-7	3000	C-C	1028	0	1062	
C30-7A	3000	C-C	50	0	50	Conrail
C30-S7	3000	C-C	44	0	44	
C32-8	3200	C-C	10	0	10	Conrail
C36-7	3600	C-C	153	0	203	
C3908	3900	C-C	159	0	162	
Dash 8-32B	3200	B-B	45	0	45	
Dash 8-32B-PHDM	3200	B-B	10	0	10	
Dash 8-40B	4000	B-B	141	1	142	
Dash 8-40BPH	4000	D-D	22	0	22	Amtrak P32DH
Dash 8-40BW	4000	B-B	60	0	60	
Dash 8-40C	4000	C-C	655	0	655	
Dash 8-40CM	4000	C-C	22	0	24	
Dash 8-40CW	4000	C-C	130	0	130	

(Many roads have applied simpler or previous model designations to their Dash 8s — B40-8 instead of Dash 8-40B, for example.)

Model	H.P.	Wheel arrgt.	Common Carrier	Industrial	Total built	Notes
MATE	0	B-B	25	0	25	
P30CH	3000	C-C	10	0	25	Amtrak
SL50	600	B-B	0	21	21	
SL50E	600	B-B	0	10	10	
SL65	600	B-B	0	3	3	
SL80	600	B-B	0	1	5	
SL85	600	B-B	0	13	13	
SL100	600	B-B	0	2	2	
SL110	600	B-B	0	22	39	
SL115	1100	B-B	0	5	5	
SL120	1100	B-B	0	1	1	
SL125	1100	B-B	0	4	4	
SL136	1100	B-B	0	2	2	
SL144	1100	B-B	0	7	29	
U18B	1800	B-B	153	2	163	
U23B	2300	B-B	351	0	465	
U23C	2300	C-C	23	0	53	
U25B	2500	B-B	14	1	478	
U25C	2500	C-C	10	0	113	
U28B	2800	B-B	25	2	148	
U28C	2800	C-C	3	0	71	
U30B	3000	B-B	87	0	291	
U30C	3000	C-C	117	20	600	
U33B	3300	B-B	5	0	137	
U33C	3300	C-C	0	5	375	
U34CH	3400	C-C	32	0	32	NJ Transit
U36B	3600	B-B	66	0	125	
U36C	3600	C-C	109	0	218	
U36CG	3600	C-C	19	0	20	

ELECTRIC LOCOMOTIVES

Model	H.P.	Wheel arrgt.	Common Carrier	Industrial	Total built	Notes
Boxcab		B+B	6	0	6	CN
Centercab		B-B	3	0	3	CN
E10B	1000	B-B	3	1		Metro-North
E25B	2500	B-B	0	2	7	
E44A	5000	C-C	8	0	66	
E50C	5000	C-C	0	2	2	
E60C	6000	C-C	0	6	6	
E60C-2	5100	C-C	39	2	41	
E60CH	6000	C-C	12	0	19	
E60CP	6000	C-C	7	0	7	

Model	H.P.	Wheel arrgt.	Common Carrier	Industrial	Total built	Notes
Illinois Central Gulf Railroad						
F10A	1600	B-B	3	0	19	
GP8	1600	B-B	85	2	146	
GP10	1850	B-B	290	4	399	
GP11	1850	B-B	53	0	54	
SD20	2000	C-C	42	0	42	
SW13	1300	B-B	11	0	15	
SW14	1400	B-B	105	0	112	
Morrison-Knudsen						
GP39H-2	2300	B-B	6	0	6	MARC
F40PHL-2	3200	B-B	5	0	5	Tri-Rail
GP40FH-2	3000	B-B	21	0	21	
F40PHM-2C	3000	B-B	9	0	9	MBTA
GP40ru	3200	B-B	6	0	6	
S2rm	1000	B-B	1	0	1	
TE56	1500	B-B	5	0	5	
Montreal Locomotive Works (other than Alco models)						
DL535	1200	C-C	8	0	13	
DL535EW	1200	C-C	2	2	4	
FPA4	1800	B-B	4	0	36	
M420B	2000	B-B	8	0	8	
M420TR	2000	B-B	14	0	17	
M420R	2000	B-B	5	0	5	P&W
M420W	2000	B-B	88	0	92	
M424	2400	B-B	53	0	54	
M630	3000	C-C	29	0	45	
M630W	3000	C-C	2	0	10	
M636	3600	C-C	94	0	95	
M640	4000	C-C	1	0	1	
RS18	1800	B-B	79	4	351	
RS23	1000	B-B	36	4	40	
RSD17	2400	C-C	1	0	1	
S11	660	B-B	1	2	10	

Model	H.P.	Wheel arrgt.	Common Carrier	Industrial	Total built	Notes
S13	1000	B-B	26	6	56	
Slug	0	B-B	17	0	17	
Norfolk Southern Railway and predecessors						
Slug	0	B-B	84	0	84	
TC10	1000	B-B	4	0	4	
Plymouth						
7-18 ton		B	0	43		
20-40 ton		B	8	123		
44-50 ton	380-440	B-B	1	25		
60-65 ton	500	B-B	0	3		
100 ton	650	B-B	0	1		
DCL, DDT, DL, DLHT, FLB, MDT, ML6, TMDT		B	0	11		
Production totals not available						
H. K. Porter						
15-35 ton		B	0	8		
44-65 ton	380-440	B-B	4	21		
Production totals not available						
Union Pacific Railroad						
SW10	1200	B-B	63	0	65	
Vulcan						
4-30 ton		B	0	29		
45-50 ton	380	B-B	1	3		
65 ton	440	B-B	0	1		
Production totals not available						
Whitcomb						
4-16 ton		B	0	10		
20-35 ton		B	2	42		
40-45 ton	320-380	B-B	1	27		
50-65 ton	440	B-B	4	78		
70-80 ton	560	B-B	1	17		
CR16		B	0	1		
Production totals not available						

ABERDEEN & ROCKFISH RAILROAD

Reporting marks: AR   **Miles:** 47
Address: P. O. Box 917, Aberdeen, NC 28315
The Aberdeen & Rockfish operates between Aberdeen and Fayetteville, North Carolina.

Nos.	Qty.	Model	Builder	Date	Notes
205	1	GP7	EMD	1951	
210	1	GP7r	EMD	1953	Ex-CSX 867
300	1	GP18	EMD	1963	
400	1	GP38	EMD	1968	
2486	1	CF7	ATSF	1975	Ex-Santa Fe 2486
2594	1	CF7	ATSF	1972	Ex-Santa Fe 2594
Total	**6**				

ABERDEEN, CAROLINA & WESTERN

Reporting marks: ACWR   **Miles:** 290
Address: 115 W. Main St., Aberdeen, NC 28315
The Aberdeen, Carolina & Western operates from Aberdeen to Star, North Carolina, and from Charlotte through Star to Gulf, N.C., on what was the main line of the Norfolk Southern Railroad, the "old" Norfolk Southern.

Nos.	Qty.	Model	Builder	Date	Notes
700	1	GP7	EMD	1953	Ex-Chesapeake & Ohio 5820
701	1	GP7	EMD	1950	Ex-Seaboard System 896
702	1	GP7	EMD	1953	Ex-Chesapeake & Ohio 5886
703	1	GP7	EMD	1950	Ex-Bangor & Aroostook 71
900	1	GP9	EMD	1956	Ex-Chesapeake & Ohio
901	1	GP9	EMD	1959	Ex-Norfolk & Western 699
1132	1	SW7	EMD	1950	Ex-Southern
Total	**7**				

ACADIANA RAILWAY

Reporting marks: AKDN
Address: 12627 Revercrest Drive, Arlington, TX 76006
The Acadiana Railway has lines from Crowley to Eunice, Lousiana, and from Bunkie to Opelousas, La., both former Missouri Pacific routes. It has trackage rights on Union Pacific between Eunice and Opelousas.

Nos.	Qty.	Model	Builder	Date	Notes
101, 102	2	NW2	EMD	1948	
1610	1	GP9	EMD	1954	Ex-Union Pacific 263
Total	**3**				

AKRON & BARBERTON BELT RAILROAD

Reporting marks: ABB   **Miles:** 66
Address: P. O. Box 712, Barberton, OH 44203-0520
The Akron & Barberton Belt is a switching road between East Akron, Barberton, and Rittman, Ohio.

Nos.	Qty.	Model	Builder	Date	Notes
1201	1	SW1200	EMD	1954	Ex-Norfolk & Western 3375
1501	1	SW1500	EMD	1966	Ex-Conrail 9601
1502	1	SW1500	EMD	1967	Ex-Conrail 9602
Total	**3**				

ALABAMA & FLORIDA RAILROAD

Reporting marks: AFLR   **Miles:** 80
Address: P. O. Box 150, Opp, AL 36467
The Alabama & Florida extends from Georgiana to Geneva, Alabama, on a former Louisville & Nashville route.

Nos.	Qty.	Model	Builder	Date	Notes
3023	1	GP40	EMD	1966	1
6011, 6076, 6084	3	GP9	EMD	1956	2
Total	**4**				

Notes:
1. Ex-Illinois Central Gulf 3023
2. Ex-Baltimore & Ohio

ALAMEDA BELT LINE

Reporting marks: ABL **Miles:** 23
Address: P. O. Box 24352, Oakland, CA 94632

Alameda Belt Line is a switching road at Alameda, California. It is owned jointly by Santa Fe and Union Pacific. It is under the same management as the nearby Oakland Terminal Railway.

Nos.	Qty.	Model	Builder	Date	Notes
2144	1	GP7u	EMD	1953	Ex-Santa Fe 2144, rebuilt 1978
2197	1	GP7r	EMD	1951	Ex-Santa Fe 2197, rebuilt 1980
Total	**2**				

ALASKA RAILROAD

Reporting marks: ARR **Miles:** 1,180
Address: P. O. Box 107500, Anchorage, AK 99510

Alaska Railroad operates freight and passenger service on its main line between Anchorage and Fairbanks and on branches to Seward and Whittier. It has short, freight-only branches to Palmer, Suntrana, and Eielson Air Force Base. The shops are at Anchorage.

Nos.	Qty.	Model	Builder	Date	Notes
1503	1	F7B	EMD	1953	1
1801-1803, 1805, 1806, 1808, 1809	7	GP7Lr	EMD	1951	2
2001, 2002	2	GP38-2	EMD	1977	3
2003-2008	6	GP40d	EMD	1969	4
2501, 2502	2	GP35	EMD	1964	

Nos.	Qty.	Model	Builder	Date	Notes
2504	1	GP30u	EMD	1963	5
2801, 2802	2	GP49	EMD	1983	
2803-2809	7	GP49	EMD	1985	
3001-3015	15	GP40-2	EMD	1975-1978	
3016-3020	5	GP40	EMD	1967	
3051	1	GP35u	EMD	1964	6
Total	**49**				

Notes:
1. Has HEP
2. Ex-U. S. Army; now have Alco Type B trucks
3. Ex-Butte, Anaconda & Pacific 108, 109
4. Rebuilt by National Railway Equipment; derated to 2000 h.p.
5. Rebuilt to GP35 specifications
6. Ex-2503; rebuilt to GP40-2 specifications by ARR

Charles W. McDonald

Alaska Railroad 1808 has undergone several changes in the past four decades: it now has a low nose, four exhaust stacks, and a large winterization hatch, and Alco trucks have replaced the switcher trucks with which it was built.

ALBANY PORT RAILROAD

Reporting marks: ALBY **Miles:** 18
Address: Port of Albany, Albany, NY 12202
Albany Port Railroad is a switching line at Albany, New York. It is jointly owned by Conrail and Delaware & Hudson.

Nos.	Qty.	Model	Builder	Date	Notes
11	1	SW9	EMD	1953	Ex-Pennsylvania Railroad
12	1	SW9	GMD	1953	Ex-Pennsylvania Railroad
Total	**2**				

ALEXANDER RAILROAD

Reporting marks: ARC **Miles:** 19
Address: P. O. Box 277, Taylorsville, NC 28681
The Alexander Railroad operates between Taylorsville and Statesville, North Carolina.

Nos.	Qty.	Model	Builder	Date	Notes
3	1	44-ton	GE	1951	Ex-Hampton & Branchville 42
6	1	S3	Alco	1953	Ex-Davenport, Rock Island & Northwestern 5
7	1	S3	Alco	1950	Ex-Penn Central 9473
8	1	SW9	EMD	1952	Ex-Union Railroad 707
Total	**4**				

ALGERS, WINSLOW & WESTERN

Reporting marks: AWW **Miles:** 16
Address: P. O. Box 188, Oakland City, IN 47660
The Algers, Winslow & Western carries coal from Enosville and Algers, Indiana, to Oakland City Junction, where it connects with Norfolk Southern.

Nos.	Qty.	Model	Builder	Date	Notes
203-206	4	SD9	EMD	1955	Ex-Central of Georgia
Total	**4**				

ALGOMA CENTRAL RAILWAY

Reporting marks: AC **Miles:** 322
Address: P. O. Box 7000, Sault Ste. Marie, ON P6A 5P6, Canada
The Algoma Central operates freight and passenger service from Sault Ste. Marie, north to Hearst. A freight-only branch reaches west from Hawk Junction to Michipicoten.

Nos.	Qty.	Model	Builder	Date	Notes
100	1	GP7Lm	GMD	1951	Ex-162, rebuilt 1978
101, 102	2	GP7Lm	GMD	1952	Ex-169, 165, rebuilt 1978
103, 104	2	GP7Lm	GMD	1951	Ex-155, 156, rebuilt 1978
140, 141	2	SW8	GMD	1952	
157, 158	2	GP7m	GMD	1951	Rebuilt 1978
167, 170	2	GP7m	GMD	1952	Rebuilt 1978
180-182	3	SD40	GMD	1971	
183-188	6	SD40-2	GMD	1973	
200-205	6	GP38-2	GMD	1981	
Total	**26**				

George H. Drury

Algoma Central GP38-2 No. 203 heads up passenger train No. 2 at Hearst, Ontario. Immediately behind it is a steam generator car, which furnishes steam to heat the passenger cars.

ALIQUIPPA & SOUTHERN RAILROAD

Reporting marks: ALQS   **Miles:** 46
Address: P. O. Box 280, Aliquippa, PA 15001
The Aliquippa & Southern is a switching line at Aliquippa, Pennsylvania.

Nos.	Qty.	Model	Builder	Date	Notes
1000	1	SW1001	EMD	1973	
1200	1	SW9	EMD	1953	
1204, 1205	2	SW1200	EMD	1955	
Total	4				

ALLEGHENY RAILROAD

Reporting marks: ALY   **Miles:** 154
Address: 316 Pine St., Warren, PA 16365
The Allegheny Railroad operates a former Pennsylvania Railroad route from Emporium to Erie, Pennsylvania.

Nos.	Qty.	Model	Builder	Date	Notes
101, 102	2	GP40	EMD	1968	Ex-Conrail 3242, 3251
103	1	CF7	ATSF	1972	Ex-Santa Fe 2590
104	1	CF7	ATSF	1974	Ex-Santa Fe 2520
105, 106	2	GP35	EMD	1964	Ex-Union Pacific 743, 748
Total	6				

ALTON & SOUTHERN

Reporting marks: ALS   **Miles:** 178
Address: 1000 22nd St., East St. Louis, IL 62207
The Alton & Southern is a belt line at East St. Louis, Illinois. It is owned jointly by St. Louis Southwestern and Union Pacific.

Nos.	Qty.	Model	Builder	Date	Notes
1500-1517	18	SW1500	EMD	1969-1971	
1522	1	MP15	EMD	1980	
Total	19				

AMADOR CENTRAL RAILROAD

Reporting marks: AMC   **Miles:** 12
Address: Highway 49, Martell, CA 95654
The Amador Central operates from Martell, California, to a connection with Southern Pacific at Ione.

Nos.	Qty.	Model	Builder	Date	Notes
9	1	S12	BLW	1951	Ex-Sharon Steel
10	1	S12	BLW	1952	Ex-Southern Pacific
Total	2				

AMTRAK
(National Railroad Passenger Corporation)

Reporting marks: AMTK   Miles:24,800
Address: 60 Massachusetts Ave., N.E., Washington, DC 20007

Nos.	Qty.	Model	Builder	Date	Notes
DIESEL LOCOMOTIVES					
7	1	45-ton	GE	1941	
9	1	65-ton	GE	1942	
104, 106, 107	3	RS3m	Alco	1951	1
192-199	8	GP40TC	GMD	1966	
200, 201	2	F40PH	EMD	1976	
202	1	F40AC	EMD	1976	2
203-235, 237-245, 247-271, 273-365, 367-409	205	F40PH	EMD	1976-1988	
410-415	6	F40PH	GMD	1978	
450, 451	2	F69PHAC	EMD	1988	
484-489	6	FL9	EMD	1957	
500-519	20	P32BH	GE	1991	
550-567	18	SSB1200	ATSF		
575-599	25	CF7	ATSF		
650-664	15	GP40r	EMD	1966-1970	

Nos.	Qty.	Model	Builder	Date	Notes
736, 738, 742-744	5	SW1	EMD	1942-1950	
747-750	4	SW8	EMD	1951-1953	
760-762	3	GP7	EMD	1951-1952	
764-770	7	GP9	EMD	1954-1957	
771-784	14	GP7	EMD	1950-1953	

ELECTRIC LOCOMOTIVES

Nos.	Qty.	Model	Builder	Date	Notes
540-547	8	E44A	GE	1961	
600-610	11	E60MA	GE	1975	3
620, 621	2	E60CP	GE	1975	
901, 902, 905-953	52	AEM7	EMD	1980-1982	
954	1	E60CP	GE	1975	

Charles W. McDonald

F40PHs like 359 and 365 make up more than half of Amtrak's diesel fleet. The 3000-h.p. machines not only haul the trains but provide electric power for light, heat, and air conditioning.

Nos.	Qty.	Model	Builder	Date	Notes
TURBINE UNITS					
58-69	12	RTG	ANF	1974	4
150-163	14	RTG	Rohr	1976	5
Total	**446**				

Notes:
1. Repowered with EMD engine
2. AC electrical equipment by ASEA Brown Boveri
3. HEP-equipped
4. Power units of ANF-Frangeco trainsets
5. Power units for Rohr trainsets

Amtrak number	Previous identity
7	U. S. Army 7078
9	U. S. Army
192-199	GO Transit 500-507
410-415	GO Transit 510-515
550-552	Santa Fe 1215-1217
553-564	Santa Fe 1220-1226, 1228, 1230-1233
565-567	Santa Fe 1239-1241
575-598	Santa Fe 2418, 2430, 2433, 2437-2440, 2445, 2453, 2454, 2456, 2457, 2459-2463, 2513, 2562, 2584, 2587, 2588, 2592, 2595
650-655	Conrail 3104, 3088, 3089, 3095, 3083, 3090
656	Illinois Central Gulf 3072
657-661	Boston & Maine 320, 321, 323, 324, 341
662-664	Milwaukee Road 2007, 2020, 2042
736, 738	Penn Central 8479, 8522
742-744	Penn Central 8543, 8559, 8575
747, 748	New York Central 8623, 8625
749	Lehigh Valley 267
750	Lehigh Valley 275
760, 761	Frisco 610, 621
762	Norfolk & Western 3452, previously Wabash 452
763-768	Union Pacific 208, 185, 241, 207, 242, 234

Amtrak number	Previous identity
770	Chicago & North Western 1726
771	Louisville & Nashville 432
772	Louisville & Nashville 1710
773	Quebec North Shore & Labrador 100
774, 775	Wabash 451, 461
776-779	Union Pacific 102, 104, 110, 129
780	Chicago & Eastern Illinois 210
781-783	Louisville & Nashville 394, 493, 495
784	Rock Island 433

Charles W. McDonald

On its electrified line between Washington and New Haven, Amtrak's passenger trains move behind AEM7s. The locomotive is based on a Swedish design; similar locomotives are found in Austria.

ANGELINA & NECHES RIVER RAILROAD

Reporting marks: ANR   **Miles:** 21
Address: P. O. Box 1328, Lufkin, TX 75902
The Angelina & Neches River operates from Lufkin to Keltys and Buck Creek, Texas.

Nos.	Qty.	Model	Builder	Date	Notes
12	1	S4	Alco	1958	
1500	1	SW1500	EMD	1972	
2000	1	GP38-2	EMD	1980	
Total	**3**				

ANN ARBOR RAILROAD

Reporting marks: AA   **Miles:** 54
Address: 121 S. Walnut St., Howell, MI 48844
The Ann Arbor, a remnant of the "old" Ann Arbor, operates from Ann Arbor, Michigan, to Toledo, Ohio.

Nos.	Qty.	Model	Builder	Date	Notes
7771, 7791, 7802	3	GP38	EMD	1969	Ex-Conrail, same numbers
Total	**3**				

APACHE RAILWAY

Reporting marks: APA   **Miles:** 54
Address: P. O. Drawer E, Snowflake, AZ 85937
The Apache Railway extends from Snowflake, Arizona, to a connection with the Santa Fe at Holbrook.

Nos.	Qty.	Model	Builder	Date	Notes
81-84	4	C420	Alco	1966	1
700, 800, 900	3	RS36	Alco	1962	2
Total	**7**				

Notes:
1. Ex-Louisville & Nashville
2. Ex-Southern Pacific; stored serviceable

APALACHICOLA NORTHERN RAILROAD

Reporting marks: AN   **Miles:** 99
Address: 300 First St., Port St. Joe, FL 32456
The Apalachicola Northern operates from Port St. Joe, Florida, north to a connection with CSX at Chattahoochee, Fla.

Nos.	Qty.	Model	Builder	Date	Notes
709-711	3	SW9	EMD	1952-1953	
712-719	8	SW1500	EMD	1969-1970	
720-722	3	GP15T	EMD	1983	
Total	14				

Charles W. McDonald

Only 28 GP15Ts were built, Chessie System 1500-1524 and Apalachicola Northern 720-722. Inside the hood is an 8-cylinder turbocharged engine.

APPANOOSE COUNTY COMMUNITY RAILROAD

Reporting marks: ANPC   **Miles:** 10
Address: P. O. Box 321, Centerville, IA 52544
The Appanoose County operates on former Burlington and Rock Island track between Centerville and Moulton, Iowa.

Nos.	Qty.	Model	Builder	Date	Notes
101	1	GP7	EMD	1951	Ex-Missouri-Kansas-Texas
973	1	GP7	EMD	1953	Ex-Detroit, Toledo & Ironton
Total	2				

ARCADE & ATTICA RAILROAD CORP.

Reporting marks: ARA   **Miles:** 15
Address: 278 Main St., Arcade, NY 14009
The Arcade & Attica extends from Arcade Junction to North Java, New York. In addition to freight service it operates steam-powered excursion trains.

Nos.	Qty.	Model	Builder	Date	Notes
110	1	44-ton	GE	1941	Stored unserviceable
111	1	44-ton	GE	1947	
112	1	65-ton	GE	1945	Ex-U. S. Navy
Total	3				

ARIZONA & CALIFORNIA RAILROAD

Reporting marks: AZCR   **Miles:** 240
Address: Parker, AZ 85344
The Arizona & California operates a former Santa Fe route from Cadiz, California, in the Mojave Desert, southeast to Matthie, Arizona, then to Phoenix by trackage rights on Santa Fe, plus a branch from Rice to Ripley, Calif.

Nos.	Qty.	Model	Builder	Date	Notes
2001-2005	5	GP20	EMD	1960	1
3801-3803	3	GP40d	EMD		2
3804	1	GP38AC	EMD		3
7405, 7424	2	GP38	EMD		
8305	1	GP38	EMD		4
9623, 9628	2	MP15DC	EMD		5
Total	14				

Notes:
1. Ex-Southeast Coal; previously Union Pacific
2. Ex-Baltimore & Ohio 3205, 3730, 3724
3. Ex-Illinois Central 9508
4. Ex-Buffalo & Pittsburgh
5. Ex-Conrail 9623, 9628; out of service

ARIZONA EASTERN RAILWAY

Reporting marks: AZER   **Miles:** 135
Address: P. O. Box Y, Claypool, AZ 85532
The Arizona Eastern extends from Miami to Bowie, Arizona, where it connects with Southern Pacific.

Nos.	Qty.	Model	Builder	Date	Notes
22, 23	2	SW9	EMD	1951	
24, 25	2	SW1200	EMD	1964	
1751-1755	5	GP9	EMD	1954-1956	
2170, 2171	2	GP20	EMD		Ex-Kyle
4112, 4137, 4148	3	GP20	EMD		Ex-SP
Total	14				

ARKANSAS & LOUISIANA MISSOURI RAILWAY

Reporting marks: ALM   **Miles:** 53
Address: P. O. Box 1653, Monroe, LA 71201
The Arkansas & Louisiana Missouri extends from Monroe, Louisiana, to Crossett, Arkansas.

Nos.	Qty.	Model	Builder	Date	Notes
10	1	NW2	EMD	1949	
11	1	SW7	EMD	1950	
12	1	SW9	EMD	1952	
14	1	SW9	EMD	1953	
Total	4				

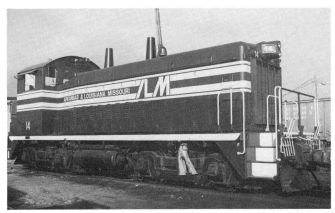

Charles W. McDonald

Arkansas & Louisiana Missouri 14 is a well-maintained SW9. The object hanging from a hook on the frame of the front truck is a rerailing frog.

ARKANSAS & MISSOURI RAILROAD

Reporting marks: AM   **Miles:** 138
Address: 107 N. Commercial St., Springdale, AR 72764
The Arkansas & Missouri operates a former Frisco route between Fort Smith, Arkansas, and Monett, Missouri.

Nos.	Qty.	Model	Builder	Date	Notes
10, 12, 14, 16, 18	5	T6	Alco	1959	1
20	1	RS1	Alco	1951	2
22	1	RS1	Alco	1943	3
42	1	RS32	Alco	1961	4
44, 46, 48	3	C420	Alco	1965	5
50	1	C420	Alco	1963	6
52, 54	2	C420	Alco	1965	7

Nos.	Qty.	Model	Builder	Date	Notes
58	1	C420	Alco	1966	8
60, 62	2	C420	Alco	1964	9
412, 417	2	C420	Alco	1964	10
1324	1	C420	Alco	1967	11
4240, 4243	2	C424	Alco	1964	12
4258, 4264	2	C425	Alco	1966	13
Total	24				

Notes:
1. Ex-Norfolk & Western 14, 46, 43, 34, 19; 10 held for disposal
2. Ex-Rutland 400
3. Ex-Atlanta & St. Andrews Bay 905; oldest Alco road-switcher in service
4. Ex-Conrail 2031, Penn Central 2031, NYC 8031
5. Ex-Louisville & Nashville 1353, 1354, 1366; Seaboard Coast Line 1214, 1215, 1227; Seaboard Air Line 112, 113, 125
6. Ex-Essex Terminal 106, Lehigh & Hudson River 22
7. Ex-Conrail 2073, 2074; Lehigh & Hudson River 23, 24
8. Ex-Delaware & Hudson 420, Conrail 2077, Lehigh & Hudson River 29
9. Ex-Delaware & Hudson 413, 416; Lehigh Valley 413, 406
10. Ex-Delaware & Hudson 412, 417; Lehigh Valley 412, 407
11. Ex-Louisville & Nashville 1324, Monon 507; for disposal
12. Ex-Kyle 4240, 4243; Burlington Northern 4240, 4243; SPS 300, 303; for sale
13. Ex-Kyle 4258, 4264; Burlington Northern 4258, 4264; SPS 320, 326; for sale

ARKANSAS CENTRAL RAILWAY

Reporting marks: ACRY   **Miles:** 1
Address: 600 S. Riverside Road, St. Joseph, MO 64502
The Arkansas Central provides switching service at Hatton, Arkansas, for its owner, Herzog Stone Products. Kansas City Southern operates the road.

Nos.	Qty.	Model	Builder	Date	Notes
4151	1	GP7	EMD	1953	Ex-Kansas City Southern 4151
Total	1				

ARNAUD RAILWAY

Reporting marks: ARND   **Miles:** 23
The Arnaud Railway runs from Arnaud Junction, 8 miles north of Sept-Iles, Quebec, on the Quebec North Shore & Labrador, to Pointe-Noire, 21 miles west of Sept-Iles.

Nos.	Qty.	Model	Builder	Date	Notes
901, 902	2	RS18	MLW	1961	
903	1	RS18	MLW	1962	
904-911	8	RS18	MLW	1964	
Total	11				

AROOSTOOK VALLEY RAILROAD

Reporting marks: AVL   **Miles:** 4
Address: P. O. Drawer 509, Presque Isle, ME 04760
The Aroostook Valley is a former interurban line operating between Presque Isle and Washburn Junction, Maine.

Nos.	Qty.	Model	Builder	Date	Notes
10, 11	2	44-ton	GE	1945	11 stored unserviceable
12	1	44-ton	GE	1949	
Total	3				

ASHLAND RAILWAY

Reporting marks: ASRY   **Miles:** 45
Address: P. O. Box 479, Stockton, NJ 08559
The Ashland Railway has two routes: Willard through Mansfield and Ashland to West Salem, Ohio, and Lakehurst to Woodmansie, New Jersey. The enginehouse for the Ohio operation is at Ashland.

Nos.	Qty.	Model	Builder	Date	Notes
32, 33	2	GP9	GMD	1956	Ex-Conrail
65, 66	2	NW2	EMD	1947, 1950	Ex-Conrail
Total	**4**				

ASHLEY, DREW & NORTHERN

Reporting marks: ADN   **Miles:** 41
Address: P. O. Box 757, Crossett, AR 71615

The Ashley, Drew & Northern extends from Crossett to Monticello, Arkansas. It is operated in conjunction with the Fordyce & Princeton Railroad. Engine facilities are at Crossett.

Nos.	Qty.	Model	Builder	Date	Notes
1513	1	CF7	ATSF	1974	
1514	1	CF7	ATSF	1974	
1815	1	GP28	EMD	1964	Ex-Missouri Pacific 2000
1816	1	GP28	EMD	1964	Ex-Illinois Central Gulf 9441
Total	**4**				

ASHTABULA, CARSON & JEFFERSON RAILROAD

Reporting marks: ACJR   **Miles:** 6
Address: 160 E. Walnut St., Jefferson, OH 44047

The Ashtabula, Carson & Jefferson operates between Carson Yard and Jefferson, Ohio. The enginehouse is at Jefferson.

Nos.	Qty.	Model	Builder	Date	Notes
107	1	S2	Alco	1950	
Total	**1**				

AT&L RAILROAD

Reporting marks: ATL   **Miles:** 49
Address: P. O. Box 29, Watonga, OK 73772

The AT&L extends from Watonga to El Reno, Oklahoma, and has a branch from Geary to Bridgeport. The enginehouse is at Watonga.

Nos.	Qty.	Model	Builder	Date	Notes
814	1	F9A	EMD	1954	1
1127	1	SW1200	EMD	1964	
1948	1	GP9	EMD	1958	2
2165, 2169	2	GP7r	EMD	1952	3
Total	**5**				

Notes:
1. Owned by Central Oklahoma Railfan Club; stand-by power
2. Ex-Burlington Northern 1948
3. Ex-Santa Fe, rebuilt 1979

ATCHISON, TOPEKA & SANTA FE RAILWAY

Reporting marks: ATSF   **Miles:** 11,700
Address: 80 E. Jackson Blvd., Chicago, IL 60604

Nos.	Qty.	Model	Builder	Date	Notes
90-93, 95-98	8	F45	EMD	1967	
100-147, 149-151, 153-162	61	GP60M	EMD	1990	
325-347	23	GP60B	EMD	1991	1
500-559	60	B40-8W	GE	1990	
1101-1109, 1115-1119, 1123-1124, 1126-1129, 1140-1146	28	Slug	ATSF	1972-1981	
1310-1325, 1327-1329	19	GP7r	EMD	1950-1953	2
1460	1	VO1000r	BLW	1943	3
1556-1575	20	SD39u	EMD	1969	4
2000, 2009-2012, 2014, 2016, 2017, 2019, 2020, 2026, 2052, 2055, 2075, 2079, 2080, 2083, 2084, 2087, 2089, 2091, 2092, 2095, 2096, 2098, 2102-2104, 2107-2109, 2113, 2115, 2117, 2119, 2120, 2122, 2123-2125, 2131, 2132, 2135-2138, 2140-2143, 2145, 2147-2149, 2159, 2172, 2181, 2183, 2192-2195, 2199, 2200, 2201, 2204, 2208, 2209, 2212, 2213, 2215-2218, 2220-2222, 2224, 2228-2233, 2235-2243	93	GP7r	EMD	1950-1953	5

GP7 No. 1315 is a product of the rebuilding program at Santa Fe's Cleburne, Texas, shops. Obvious features are the low nose, four exhaust stacks, and an angular cab roof that accommodates an air conditioning unit. Not so visible are 1315's slug controls.

Nos.	Qty.	Model	Builder	Date	Notes
2244-2247, 2250-2263, 2265-2283, 2285-2289, 2291-2299	51	GP9r	EMD	1953-1957	6
2300-2322, 2324-2351, 2353-2360	59	GP38r	EMD	1970	7
2700-2705, 2707-2729, 2731, 2732, 2734-2748, 2750-2757, 2759, 2760, 2762-2765, 2767-2785	79	GP30r	EMD	1962-1963	8
2801, 2802, 2804-2816, 2818, 2819, 2821-2845, 2847-2855, 2857-2878, 2880, 2882-2894, 2896-2899, 2901-2925, 2927-2938, 2940-2944, 2946-2949, 2951-2963	150	GP35r	EMD	1964-1965	9

Santa Fe 5062 is one of a large group of SD40-2s on the road's roster.

Nos.	Qty.	Model	Builder	Date	Notes
3400-3449	50	GP39-2r	EMD	1975	10
3000-3010, 3040, 3669-3671, 3673-3690, 3693-3705	52	GP39-2	EMD	1974-1980	
3800-3809	10	GP40X	EMD	1978	
3810-3854	45	GP50	EMD	1981, 1985	
4000-4039	40	GP60	EMD	1988-1989	
5000-5010, 5012-5014, 5016-5019	18	SD40r	EMD	1966	11
5020-5036, 5038-5057	37	SD40-2r	EMD	1977	12
5058-5192, 5200-5207, 5209-5213	148	SD40-2	EMD	1978-1981	
5255-5267	13	SDP40Fr	EMD	1974	13
5300, 5301, 5303-5308, 5310-5312, 5315, 5317-5321, 5323-5339, 5341-5362, 5364-5408, 5434-5437, 5502, 5503	107	SD45r	EMD	1966-1970	14

Nos.	Qty.	Model	Builder	Date	Notes	Nos.	Qty.	Model	Builder	Date	Notes
5705, 5707-5714	9	SD45-2	EMD	1974		7484-7499	16	B36-7	GE	1980	
5800-5868	69	SD45-2	EMD	1972-1973	15	8010-8049, 8051-					
5950-5989	40	F45	EMD	1968	16	8066, 8068-8166	155	C30-7	GE	1977-1983	
6350-6419	70	B23-7	GE	1978-1984		8736-8762	27	U36C	GE	1974	
7400-7402	3	B39-8	GE	1984		9500-9569	51	SF30C	ATSF	1985-1986	17
7410-7449	40	Dash 8-40B	GE	1988-1989		**Total**	**1,652**				

THE CF7

At the end of the 1960s the Atchison, Topeka & Santa Fe found itself with several hundred F7 freight diesels that were out of a job. They had been bumped from mainline freight duties by high-horse-power hood units, but their carbody configuration made them unsuitable for branchline and local work, where they were needed. Santa Fe couldn't afford to buy several hundred new units for that service, but the mechanical components of the Fs still had some useful life remaining. Santa Fe decided to convert the F7s to hood units. Two factors doubtless influenced that decision: the desire to keep the shop at Cleburne, Texas, busy, and the financial advantages of a capital rebuild program, which lets the railroad treat rebuilt locomotives as new ones and depreciate them over a period of several years.

The rebuilding process was complicated by the basic difference between a cab unit, such as the F7, and a hood unit. A hood unit is essentially a flat car carrying a diesel engine, a generator, and controls, with sheet-metal hoods protecting them and the crew from the weather. A cab unit is designed so that the sides carry part of the load of the machinery. Remove the sides and the frame will sag. Fabricating new frames was a major part of the conversion of the F7s to CF7s.

At first glance, the CF7 looks like a GP7 that has had its short hood lowered. It differs from a GP7 in having a shorter short hood

and a longer cab; the side members of the frame are quite different. Most of the CF7s were built with a curved cab roof; the last 54 have an angled cab roof that can accommodate an air conditioner. Mechanically, the CF7 is the same as a GP7

Santa Fe built 233 CF7s between 1970 and 1978. In the early 1980s the road began to phase them out and discovered a ready market for the CF7 in the short lines.

Charles W. McDonald

Viewed from the side, the CF7's distinguishing features are apparent: long cab, short nose, and fishbelly side sills.

Notes:
1. Cabless booster units
2. Slug control units, rebuilt 1973-1981
3. Rebuilt 1970 with EMD 1500 h.p. engine
4. Rebuilt 1985-86; 2500 h.p.
5. Rebuilt 1973-1981
6. Rebuilt 1978-1980
7. Rebuilt 1982-1985
8. Rebuilt 1981-1984
9. Rebuilt 1979-1985
10. Rebuilt 1986-1988
11. Rebuilt 1981
12. Rebuilt 1988
13. Ex-Amtrak, rebuilt 1985
14. Rebuilt 1981-1984
15. Rebuilt 1986-1988
16. Rebuilt 1982-1983
17. Rebuilt U36Cs, 3160 h.p., built 1972-75

ATLANTA & ST. ANDREWS BAY RAILWAY

Reporting marks: ASAB   **Miles:** 89
Address: P. O. Box 2775, Panama City, FL 32402
The Atlanta & St. Andrews Bay extends from Dothan, Alabama, south to Panama City, Florida, with a branch from Campbellton to Graceville, Fla. It operates and supplies motive power for the affiliated Abbeville-Grimes Railway, which operates between Dothan and Grimes, Ala. A&StAB's enginehouse is at Panama City.

Nos.	Qty.	Model	Builder	Date	Notes
500-507	8	GP38	EMD	1969	Ex-Conrail
508, 509	2	GP38-2	EMD	1973	
510	1	GP38-2	EMD	1975	
511	1	GP38	EMD	1969	Ex-Conrail
Total	12				

ATLANTIC & GULF RAILROAD

Reporting marks: AGLF   **Miles:** 77
Address: P. O. Box 1446, Clarksdale, MS 38614
The Atlantic & Gulf operates from Thomasville through Albany to Sylvester, Georgia. Engine facilities are at Albany.

Nos.	Qty.	Model	Builder	Date	Notes
11	1	GP7	EMD	1953	Ex-Union Pacific 1110
121	1	GP7	EMD	1953	Ex-Union Pacific 1121
1026	1	GP7	EMD	1950	Ex-North Carolina Port Authority 1026
2391	1	GP9	EMD	1957	Ex-Louisville & Nashville 2391
7720	1	GP8	ICG	1977	Ex-Illinois Central Gulf 7720
8047	1	GP10	ICG	1970	Ex-Illinois Central Gulf 8047
Total	6				

Charles W. McDonald

Atlanta & St. Andrews Bay 508 is a GP38-2. The road's profile doesn't call for dynamic braking.

ATLANTIC & WESTERN RAILWAY

Reporting marks: ATW **Miles:** 3
Address: P. O. Box 1208, Sanford, NC 27330
The Atlantic & Western operates on 3 miles of track between Sanford and Jonesboro, North Carolina. The enginehouse is at Sanford.

Nos.	Qty.	Model	Builder	Date	Notes
100	1	70-ton	GE	1950	
101	1	70-ton	GE	1948	
Total	2				

AUSTIN & NORTHWESTERN RAILROAD

Reporting marks: AUNW **Miles:** 163
Address: 500 Robert Martinez Jr. St., Austin, TX 78702
The Austin & Northwestern extends from Giddings, Texas, through Austin to Llano, with a branch from Fairland to Marble Falls. The enginehouse is at Austin.

Nos.	Qty.	Model	Builder	Date	Notes
11	1	GP9	EMD	1956	
14-16	3	SW9	EMD	1952	
22, 23, 44, 55	4	GP9	EMD	1955-1956	
66, 67	2	GP35	EMD	1963, 1964	
272	1	GP9	EMD	1954	
Total	11				

BALTIMORE & ANNAPOLIS RAILROAD

Reporting marks: BLA **Miles:** 6
Address: 100 W. Maple Road, Linthicum, MD 21090
The Baltimore & Annapolis operates 6 miles of track between Glen Burnie and Baltimore, Maryland. Engine facilities are at Glen Burnie.

Nos.	Qty.	Model	Builder	Date	Notes
87	1	SW9	EMD	1953	
Total	1				

BANGOR & AROOSTOOK RAILROAD

Reporting marks: BAR **Miles:** 436
Address: Northern Maine Jct. Park, RR 2, Bangor, ME 04401
The Bangor & Aroostook extends from Searsport, Maine, north to St. Leonard, New Brunswick, across the St. John River from Van Buren, Maine. The railroad has several branches, chief among them a cluster of former main lines serving Presque Isle and Caribou. The principal shops are at Northern Maine Junction, just west of Bangor.

Nos.	Qty.	Model	Builder	Date	Notes
20-24	5	GP7r	EMD	1952	1
51, 56	2	BL2	EMD	1949	3
60-63, 66-71, 73, 75	12	GP7	EMD	1950, 1952	4
78-80	3	GP9	EMD	1954	5

Charles W. McDonald

Bangor & Aroostook 73, a GP7, wears BAR's latest livery: black, orange, and gray with white trim.

Nos.	Qty.	Model	Builder	Date	Notes
81-88, 90-98	17	GP38	EMD	1966-1969	6
502	1	F3A	EMD	1947	2
Total	**40**				

Notes:
1. Ex-Santa Fe, acquired 1989
2. Ex-42; restored for service and eventually preservation
3. Stored unserviceable
4. 63, 66-69 stored unserviceable
5. 80 stored unserviceable
6. 90-94 are ex-Conrail 7662-7666, formerly Pennsylvania-Reading Seashore Lines 2002-2006; 95-98 are ex-Missouri Pacific 2006, 2003, 2007, 2004

BATH & HAMMONDSPORT RAILROAD

Reporting marks: BH   **Miles:** 9
Address: Water St., Hammondsport, NY 14840
The Bath & Hammondsport operates between Bath and Hammondsport, New York. The enginehouse is at Hammondsport.

Nos.	Qty.	Model	Builder	Date	Notes
4, 5	2	S1	Alco	1950	
Total	**2**				

BATTEN KILL RAILROAD

Reporting marks: BKRR   **Miles:** 35
Address: 1 Elbow St., Greenwich, NY 12834
The Batten Kill Railroad operates between Eagle Bridge and Thompson, New York. The enginehouse is at Greenwich.

Nos.	Qty.	Model	Builder	Date	Notes
605	1	RS3	Alco	1950	1
4116	1	RS3	Alco	1952	2
Total	**2**				

Notes:
1. Ex-Vermont Railway 605, Lehigh & Hudson River 10
2. Ex-Greenwich & Johnsonville; ex-Delaware & Hudson 4116

BAUXITE & NORTHERN RAILWAY

Reporting marks: BXN   **Track miles:** 19
Address: P. O. Box 138, Bauxite, AR 72011
The Bauxite & Northern extends from Bauxite, Arkansas, to a connection with Union Pacific at Bauxite Junction, 3 miles away. The enginehouse is at Bauxite.

Nos.	Qty.	Model	Builder	Date	Notes
15, 16	2	MP15DC	EMD	1974	
Total	**2**				

BAY COLONY RAILROAD

Reporting marks: BCLR   **Miles:** 122
Address: 4 Freight House Road, Wareham, MA 02571
Bay Colony operates several former New Haven branches in southeastern Massachusetts. The two longest segments are from Middleboro to Hyannis, South Dennis, and Falmouth, and from Braintree to North Plymouth. The engine facilities are at Buzzards Bay.

Nos.	Qty.	Model	Builder	Date	Notes
30	1	RS1	Alco	1955	1
151	1	25-ton	GE	1942	2
410, 411	2	44-ton	GE	1949, 1956	3
1052	1	S2	Alco	1943	4
1058, 1061	2	S4	Alco	1950	5
1062	1	S4	Alco	1954	6
1064	1	RS1	Alco	1952	7
1501	1	GP7	EMD	1953	8
1751	1	GP9	EMD	1955	9
Total	**11**				

Tom Nelligan

Bay Colony 1052, an Alco S2, leads a local freight along the waterfront at Buzzards Bay. The unit is light gray with yellow and orange stripes.

Notes:
1. Ex-Genesee & Wyoming 30; leased to Cape Cod Railroad
2. Ex-New York Air Brake
3. Ex-Dansville & Mount Morris; 411 is last 44-tonner built
4. Ex-Portland Terminal 1052
5. Ex-Portland Terminal; 1058 leased to Cape Cod Railroad
6. Ex-Maine Central
7. Ex-Genesee & Wyoming 25
8. Ex-Detroit & Toledo Shore Line 50
9. Ex-Conrail 7803, ex-New York Central

BC RAIL

Reporting marks: BCOL   **Miles:** 3,010
Address: P. O. Box 8770, Vancouver, BC V6B 4X6, Canada
BC Rail extends north from North Vancouver, British Columbia, to Dawson Creek, Fort Nelson, and Driftwood, B. C. A recently constructed branch to serve coal mines near Tumbler Ridge and Quintette is electrified. BC Rail operates passenger service between North Vancouver and Prince George.

Leonard G. Thompson

GF6Cs 6004 and 6005 shove hard on the rear of a coal train on BC Rail's Tumbler Subdivision.

27

Nos.	Qty.	Model	Builder	Date	Notes
DIESEL LOCOMOTIVES					
401-410	10	Slug	BC Rail	1981-1987	1
502	1	S13	MLW	1959	
601-613, 617,					
619-624, 626-630	25	RS18	MLW	1957-1966	2
631, 632	2	C420	Alco	1966	3
640-647	8	M420W	MLW	1973	
681-688	8	M420B	MLW	1975	
736-742	7	SD40-2	EMD	1978	4
743-750	8	SD40-2	EMD	1979	5
751-754, 756-759,					
761, 762	10	SD40-2	GMD	1980	
763-767	5	SD40-2	GMD	1985	
802, 803, 811	3	C425	Alco	1964	6
4601-4611	11	C40-8M	GE	1990	
ELECTRIC LOCOMOTIVES					
6001-6007	7	GF6C	GMD	1983-1984	
Total	**105**				

Notes:
1. Built from R33s
2. Being rebuilt with Caterpillar engines
3. Ex-Lehigh & Hudson River 25, 26
4. Ex-Kennecott Copper 101-107
5. Ex-Oneida & Western 9950-9957
6. Ex-Erie Lackawanna

BEAUFORT & MOREHEAD RAILROAD

Reporting marks: BMH   **Miles:** 3
Address: 16 Broad St., Beaufort, NC 28516
The Beaufort & Morehead operates between Beaufort and Morehead City, North Carolina. The enginehouse is at Beaufort.

Nos.	Qty.	Model	Builder	Date	Notes
75	1	80-ton	Whitcomb	1947	
1203	1	SW1200	EMD	1955	Ex-Missouri Pacific 1203
Total	**2**				

BEECH MOUNTAIN RAILROAD

Reporting marks: BEEM   **Miles:** 8
Address: P. O. Box 2327, Elkins, WV 26241
The Beech Mountain Railroad carries coal from Star Bridge to a connection with CSX at Alexander, West Virginia. The enginehouse is at Alexander.

Nos.	Qty.	Model	Builder	Date	Notes
113	1	S4	Alco	1946	
Total	**1**				

BELFAST & MOOSEHEAD LAKE RAILROAD

Reporting marks: BML   **Miles:** 33
Address: 11 Water St., Belfast, ME 04915
The Belfast & Moosehead Lake extends from Belfast, Maine, on the coast, inland to Burnham Junction, where it connects with Maine Central. B&ML's enginehouses are at Belfast and Thorndike.

Nos.	Qty.	Model	Builder	Date	Notes
50, 52-55	5	70-ton	GE	1946-1951	
Total	**5**				

BELT RAILWAY OF CHICAGO

Reporting marks: BRC   **Miles:** 25
Address: 6900 Central Ave., Chicago, IL 60638

Nos.	Qty.	Model	Builder	Date	Notes
420	1	S6	Alco	1957	
470-473, 477	5	GP7	EMD	1951-1952	1
480, 481	2	GP9	EMD	1956	
490-495	6	GP38-2	EMD	1972	

Nos.	Qty.	Model	Builder	Date	Notes
500, 501, 502-506	7	TR4A	EMD	1949-1950	
510, 511, 512-516	7	TR4B	EMD	1949-1950	
520-523	4	SW9	EMD	1951	
524-526	3	SW1200	EMD	1963	
530-532	3	SW1500	EMD	1967-1968	
533-536	4	MP15	EMD	1975, 1980	
600-605	6	C424	Alco	1965-1966	
Total	**48**				

Notes:
1. 471 has been rebuilt on a GP9 frame

R. B. Olson

Two Belt Railway of Chicago GP38-2s bring a freight through Blue Island, Illinois, on Baltimore & Ohio Chicago Terminal rails.

BELTON RAILROAD

Reporting marks: BRR   **Miles:** 7
Address: P. O. Box 836, Denison, TX 75020
The Belton Railroad operates between Smith and Belton, Texas. The enginehouse is at Belton.

Nos.	Qty.	Model	Builder	Date	Notes
504	1	SW1	EMD	1940	Ex-Terminal Railroad Association of St. Louis 504
Total	**1**				

BERLIN MILLS RAILWAY

Reporting marks: BMS   **Miles:** 12
Address: 650 Main St., Berlin, NH 03570
Berlin Mills Railway is a terminal road at Berlin, New Hampshire, serving the mills of its owner, James River Paper Company.

Nos.	Qty.	Model	Builder	Date	Notes
731	1	SW1	EMD	1949	1
741	1	SW1	EMD	1948	2
745	1	SW1	EMD	1948	3
Total	**3**				

Notes:
1. Ex-Amtrak 731, Penn Central 8428, New York Central 602
2. Ex-Amtrak 741, Penn Central 8530, Pennsylvania 9150
3. Ex-Amtrak 745, Penn Central 8580, Pennsylvania 9200

BESSEMER & LAKE ERIE RAILROAD

Reporting marks: BLE   **Track miles:** 640
Address: P. O. Box 68, Monroeville, PA 15146
The Bessemer & Lake Erie extends from North Bessemer, Pennsylvania, near Pittsburgh, north to Lake Erie at Erie, Pa., and Conneaut, Ohio.

Nos.	Qty.	Model	Builder	Date	Notes
210	1	SD38-2	EMD	1975	1
451, 454	2	SD7	EMD	1953	
658	1	SD38-2	EMD	1974	2
715A, 718A, 719A, 722A	4	F7A	EMD	1952	
715B, 716B, 719B, 721B	4	F7B	EMD	1952	
725A, 727A, 728A	3	F7A	EMD	1953	
803	1	SD7	EMD	1952	
821+847	15	SD9	EMD	1956-1958	3
851-856	6	SD18	EMD	1962	
861-863	3	SD38	EMD	1967	
865-869	5	SD38AC	EMD	1971	
870-878, 890, 892	8	SD38-2	EMD	1973-1976	
Total	**53**				

Notes:
1. Ex-Duluth, Missabe & Iron Range 210
2. Ex-Elgin, Joliet & Eastern 658
3. Some units retired

Nate Clark Jr.

An SD38, an SD9, and an SD18 lead a Bessemer & Lake Erie ore train across the Conrail diamonds at Shenango, Pa.

BIRMINGHAM SOUTHERN RAILROAD

Reporting marks: BS   **Track miles:** 84
Address: P. O. Box 579, Fairfield, AL 35064
Birmingham Southern is a switching road at Birmingham, Alabama.

Nos.	Qty.	Model	Builder	Date	Notes
200	1	SW1200	EMD	1957	
210, 212, 213	3	SW1000	EMD	1967-1969	
218-225	8	SW1001	EMD	1973-1975	
226	1	SW1001	EMD	1972	1
260, 261	2	MP15	EMD	1976	
302	1	SW7r	EMD	1950	2
316	1	NW2	EMD	1946	
370	1	MP15	EMD	1976	
383, 384	2	SW8	EMD	1956	
574	1	SW1200	EMD	1981	3
575	1	SW1200	EMD	1954	3
588	1	SW1200	EMD	1953	3
601-603	3	SW1001	EMD	1976	4
630-633	4	SD9	EMD	1957	
700-704	5	GP38-2	EMD	1972-1973	5
Total	**35**				

Notes:
1. Ex-Newburgh & South Shore 1020
2. Rebuilt from SW7 202
3. Ex-Union Railroad, same numbers
4. Ex-Union Railroad 101-103
5. Ex-Elgin, Joliet & Eastern 700-704

BLACK RIVER & WESTERN RAILROAD

Reporting marks: BRW   **Miles:** 19

Address: P. O. Box 200, Ringoes, NJ 08551

The Black River & Western operates freight and excursion service between Three Bridges and Lambertville, New Jersey. The enginehouse is at Ringoes.

Nos.	Qty.	Model	Builder	Date	Notes
41, 42	2	CF7	ATSF	1977	Ex-Santa Fe
56	1	T6	Alco	1953	Ex-Pennsylvania Railroad
57	1	RS1	Alco	1948	Ex-Washington Terminal
Total	4				

BLOOMER SHIPPERS CONNECTING RAILROAD

Reporting marks: BLOL   **Miles:** 51

Address: 100 E. Locust St., Chatsworth, IL 60921

The "Bloomer Line" extends from Kempton, Illinois, to Colfax and Gibson City. The Kempton-Colfax route is a segment of the former Illinois Central Bloomington District; the Gibson City line is part of the Norfolk & Western (Wabash) Chicago-St. Louis line. The enginehouse is at Chatsworth.

Nos.	Qty.	Model	Builder	Date	Notes
55	1	SW1200	EMD	1965	Ex-Missouri Pacific 1155
91	1	GP9	EMD	1958	Ex-Burlington Northern 1949
92	1	GP9	EMD	1956	Ex-Conrail 7368
Total	3				

BLUE MOUNTAIN & READING RAILROAD

Reporting marks: BMRG   **Miles:** 30

READING, BLUE MOUNTAIN & NORTHERN RAILROAD

Reporting marks: RBMN   **Miles:** 124

Address: P. O. Box 433, Hamburg, PA 19526

The Blue Mountain & Reading and Reading, Blue Mountain & Northern are under the same management. Blue Mountain & Reading operates freight and excursion service between Temple and Hamburg, Pennsylvania, and freight from Topton to Kutztown, Pottstown to Boyertown, and Emmaus to Pennsburg. Reading, Blue Mountain & Northern operates freight from Reading to Pottsville, Good Spring, and Locust Summit, Pa. Engine facilities are at South Hamburg and Port Clinton.

Nos.	Qty.	Model	Builder	Date	Notes
103	1	NW2	EMD	1947	1
600	1	SW1	EMD	1953	2
1000	1	NW2	EMD	1948	3
1200	1	SW7m	EMD	1951	4
1201	1	SW7	EMD	1951	5
1501, 1502	2	CF7	ATSF	1977	
3300-3304	5	U33B	GE	1968	6
5706, 5898	2	E8A	EMD	1952, 1950	7
5898	1	E8A	EMD	1950	8
Total	14				

Notes:
1. Ex-Reading 103, leased from Reading Technical & Historical Society
2. Ex-Warner Co. 15
3. Ex-Conrail 9220, Erie Lackawanna 413
4. Ex-Conrail 8905, Penn Central, New York Central
5. Ex-Penn Central 8917
6. Ex-Conrail, Penn Central 2895, 2896, 2902, 2914, 2930
7. Ex-Pennsylvania 5706 and 5898, leased from private owner

BORDER PACIFIC RAILROAD

Reporting marks: BOP   **Miles:** 32

Address: P. O. Drawer 156, Rio Grande City, TX 78582-0156

The Border Pacific operates between Rio Grande City and Mission, Texas. The enginehouse is at Rio Grande City.

Nos.	Qty.	Model	Builder	Date	Notes
98	1	GP7	EMD	1953	Ex-Missouri-Kansas-Texas
Total	1				

BRANDON CORPORATION

Reporting marks: BRAN   **Track miles:** 17

Address: 28th & N Streets, Omaha, NE 68107

The Brandon Corporation operates a switching railroad at South Omaha, Nebraska.

Nos.	Qty.	Model	Builder	Date	Notes
3	1	S1	Alco	1945	Ex-South Omaha Terminal
71	1	S3	Alco	1950	
423	1	C415	Alco	1966	Ex-Rock Island
Total	3				

BRANDYWINE VALLEY RAILROAD

Reporting marks: BVRY   **Miles:** 4

Address: 50 S. First Ave., Coatesville, PA 19320

The Brandywine Valley Railroad operates between Coatesville and Valley Township, Pennsylvania, connecting Lukens Steel, its owner, with Conrail and the Octoraro Railway.

Nos.	Qty.	Model	Builder	Date	Notes
8201	1	NW2	EMD	1949	Ex-Conrail 9230
8202	1	NW2	EMD	1942	Ex-Conrail 9228
8203	1	NW2	EMD	1942	Ex-Conrail 9259
8204	1	SW1200	EMD	1964	Ex-Missouri Pacific 1184
8205	1	SW1200	EMD	1960	Ex-Southern Pacific 2315
8206	1	SW9	EMD	1950	Ex-Indiana Harbor Belt
Total	6				

BROWNSVILLE & RIO GRANDE INTERNATIONAL RAILROAD

Reporting marks: BRG   **Miles:** 33

Address: P. O. Box 3818, Brownsville, TX 78523-3818

The Brownsville & Rio Grande International Railroad is a terminal railroad at Brownsville, Texas.

Nos.	Qty.	Model	Builder	Date	Notes
147	1	SW9	EMD	1951	Ex-Seaboard Coast Line 147
198	1	SW9	EMD	1951	Ex-Seaboard Coast Line 198
Total	2				

BSDA RAILROAD

Reporting marks: BSDA   **Miles:** 1

Address: 720 Olive St., Suite 2800, St. Louis, MO 63101

BSDA is a switching railroad in St. Louis.

Nos.	Qty.	Model	Builder	Date	Notes
1209	1	SW1200	EMD	1955	Ex-Missouri Pacific 1209
Total	1				

BUCKINGHAM BRANCH RAILROAD

Reporting marks: BB   **Miles:** 19

Address: P. O. Box 336, Dillwyn, VA 23936

The Buckingham Branch Railroad operates between Bremo and Dillwyn, Virginia. The enginehouse is at Dillwyn.

Nos.	Qty.	Model	Builder	Date	Notes
8851	1	GP7	EMD	1951	Ex-CSX
Total	1				

BUFFALO & PITTSBURGH RAILROAD

Reporting marks: BPRR   **Miles:** 369

Address: 201 N. Penn St., Punxsutawney, PA 15767

The Buffalo & Pittsburgh consists of ex-CSX lines (formerly Baltimore & Ohio) extending from Buffalo, New York, south to Eidenau, Pennsylvania, with trackage rights on CSX from Eidenau to New Castle. The major branches are to Bruin, Clearfield, and Lucerne Junction, Pa. The principal engine facilities are at Butler, Pa. The Buffalo & Pittsburgh and the Rochester & Southern are both part of Genesee & Wyoming Industries, and locomotives are pooled with other GWI railroads.

Nos.	Qty.	Model	Builder	Date	Notes
101-106	6	GP40	EMD	1967	1
201-204, 206-209	9	GP9	EMD	1956	2
626, 874, 879, 886, 887	5	GP9	EMD	1959	3
922, 926	2	GP18	EMD	1960	3
3000, 3001	2	GP40	EMD	1971	4
3100, 3102, 3106, 3107, 3111, 3119	6	GP40	EMD	1967	5
6673	1	GP40	EMD	1966	6
7803, 7822	2	GP38	EMD	1969	7
Total	33				

Notes:
1. Ex-Helm Leasing
2. Ex-Chesapeake & Ohio; 208 built 1954
3. Ex-Norfolk & Western
4. Ex-CSX
5. Ex-Kyle Railroad
6. Ex-Seaboard System
7. Ex-Conrail

BUFFALO SOUTHERN RAILROAD

Reporting marks: BUFF　　　　　**Miles:** 31

Address: 8600 Depot St., Eden, NY 14057

The Buffalo Southern operates between Buffalo and Gowanda, New York. The enginehouse is at Hamburg.

Nos.	Qty.	Model	Builder	Date	Notes
28	1	RS3	Alco	1953	1
29	1	25-ton	Whitcomb	1942	
30	1	44-ton	GE	1946	1
81	1	S2	Alco	1943	1
107	1	S2	Alco	1951	1
291, 298	2	GP9	EMD	1959	2
5010	1	RS11	Alco	1961	
Total	8				

Notes:
1. Stored serviceable
2. Ex-Milwaukee Road, leased from MNVA

BURLINGTON JUNCTION RAILWAY

Reporting marks: BJRY　　　　　**Miles:** 2

Address: P. O. Box 37, Burlington, IA 52601

The Burlington Junction is a switching railroad at Burlington, Iowa.

Nos.	Qty.	Model	Builder	Date	Notes
44	1	44-ton	GE	1941	Ex-Washington & Old Dominion; Fonda, Johnstown & Gloversville
Total	1				

BURLINGTON NORTHERN RAILROAD

Reporting marks: BN　　　　　**Miles:** 28,937

Address: 9401 Indian Creek Parkway, Overland Park, KS 66210

Nos.	Qty.	Model	Builder	Date	Notes
1	1	F9Ar	EMD	1954	1
2	1	F9Br	EMD	1954	1
5, 14	2	NW2u	EMD	1949	2
20-65	46	SW1500	EMD	1968-1973	
70	1	SW1	EMD	1941	
162-164, 166	4	SW1200	EMD	1957	
169	1	SW9	EMD	1951	3
180	1	F40PH	EMD	1987	4
170-174, 176-177, 179-181, 183, 184, 188, 189, 193-199, 201, 205-207, 209-214, 222-226, 228, 231, 233-255	61	SW1200	EMD	1955-1965	
300-324	25	SW1500	EMD	1967, 1973	
325-394, 427-434, 436-449, 574-577, 579-585	53	SW1000	EMD	1966-1972	
600-602, 604	4	GP9B	EMD	1954-1957	

Nos.	Qty.	Model	Builder	Date	Notes
741, 743, 745, 748, 752, 753, 756, 758-763, 767, 770, 771, 775, 776, 778-783, 785, 787, 790-793, 797-800, 802, 804-806, 808-810, 813, 816, 822, 824, 826, 827, 834, 838					
	49	GP39d	EMD	1987	5
1000-1004	5	MP15	EMD	1975	
1237-1245	9	SD38-2	EMD		6
1355, 1356, 1358, 1364, 1365	5	GP5	EMD	1959	7
1376-1399	24	GP15-1	EMD	1977	
1400-1403, 1406-1409, 1411, 1413, 1414, 1418-1420, 1422, 1426, 1436, 1438					
	18	GP7ru	EMD	1974-1976	8
1703, 1706, 1707, 1709, 1711, 1726, 1728, 1734, 1741, 1742, 1745-1749, 1751, 1752, 1758-1760, 1763, 1764, 1769, 1774, 1777, 1782, 1783, 1787, 1799, 1800, 1802, 1804, 1812, 1813, 1816, 1819, 1821, 1829, 1836, 1839, 1841, 1851, 1854, 1858, 1860, 1861, 1863, 1868, 1869, 1875, 1877, 1878, 1883, 1889, 1896, 1898, 1900, 1910, 1913, 1914, 1916, 1917, 1920, 1922, 1923, 1938, 1942, 1951, 1954, 1956, 1958, 1960, 1961, 1965, 1966, 1969, 1970, 1971, 1977, 1980					
	66	GP9	EMD	1954-1958	
1991	1	SD60M	EMD	1991	
1993-1997	5	GP18	EMD	1960	
2000-2009	10	GP20C	EMD	1989-1990	9
2030, 2031, 2044, 2048, 2049, 2054, 2057, 2058, 2063					
	9	GP20	EMD	1960-1961	
2072-2087	16	GP38	EMD	1970, 1972	
2088-2109	22	GP38-2	EMD	1974	
2110-2125, 2127-2138	28	GP38AC	EMD	1971	
2150-2154	5	GP38-2	EMD	1980	
2155-2189	35	GP38	EMD	1970	
2255-2314, 2316-2369	114	GP38-2	EMD	1972-1976	

Nos.	Qty.	Model	Builder	Date	Notes
2543	1	GP35	EMD	1964	
2601	1	GP38-2B	EMD	1976	10
2700-2739	40	GP39-2	EMD	1981	
2750-2758	9	GP39E	EMD	1989	11
2800-2834, 2875-2884	45	GP39M	M-K	1988-1990	11
2900-2940	41	GP39E	EMD	1989-1990	11
2960-2984	25	GP39V	VMV	1990-1991	11
3040-3064	25	GP40-2	EMD	1979	
3075-3084	10	GP40G	EMD	1989	12
3100-3109	10	GP50	EMD	1980	13
3110-3157	48	GP50L	EMD	1985	
3158-3162	5	GP50	EMD	1985	13a
3500-3555	56	GP40m	MK	1988-1989	14
4000-4050, 4052-4119	119	B30-7AB	GE	1982-1983	15

Charles W. McDonald

Five Burlington Northern GP50s were built with large cabs that could accommodate the entire train crew. No. 3162 was the last GP50 built.

Nos.	Qty.	Model	Builder	Date	Notes
5000-5135, 5137-5141	141	C30-7	GE	1979-1981	
5326	1	U30C	GE	1972	
5365-5367, 5369, 5371-5394	28	U30C	GE	1974	16
5485-5492	8	B30-7	GE	1977	16
5500-5599	100	C30-7	GE	1976-1979	
5782, 5783, 5785, 5787, 5788, 5790	6	U30B	GE	1973	16
6100-6103, 6107-6110, 6113-6123, 6126, 6127, 6131, 6133-6135, 6139, 6141-6143, 6146, 6147, 6150, 6152, 6153, 6154, 6156-6164, 6166, 6167, 6174, 6176-6179, 6181, 6183-6185, 6190-6200, 6202, 6204, 6217-6219, 6221, 6223-6225, 6227, 6228, 6230-6237	89	SD9	EMD	1954-1959	
6240-6246	7	SD9u	EMD	1989-1990	17
6260-6263	4	SD38-2	EMD	1979	
6303, 6304, 6309, 6313, 6318, 6321, 6323	7	SD40	EMD	1971	
6325-6329, 63316334	9	SD40-2	EMD	1972	
6339	1	SD40	EMD	1968	18
6348-6366, 6368-6373, 6378-6385	23	SD40-2	EMD	1972, 1974	
6394, 6397-6399	4	SD40	EMD	1966	
6700-6713, 6715-6759, 6761-6764, 6766-6798, 6801-6811, 6813-6836, 6840-6847, 6850, 6900-6913, 6915-6921, 6923-6928, 6950, 7000-7018, 7020-7220, 7222-7240, 7242-7291	459	SD40-2	EMD	1972-1980	
7300-7309	10	SD40G	EMD	1989	19
7350-7358	10	SD40-2	EMD		20
7500-7502	3	SD40-2B	EMD	1972-1980	21
7800-7819, 7821-7898, 7900-7923, 7924-7940	139	SD40-2	EMD	1974-1980	
8000, 8001	2	B39-8	GE	1987	22
8000-8042, 8044-8181	181	SD40-2	EMD	1977-1980	

Nos.	Qty.	Model	Builder	Date	Notes
8300	1	SD60	EMD	1990	23
8500-8518, 8520-9599	99	B39-8	GE	1987	24
9000-9099	100	SD60	EMD	1986	25
9200-9298	99	SD60M	EMD	1990-1991	
9900-9908, 9910-9925	25	E9Ar	EMD		26
Total	**2,622**				

Notes:
1. Ex-Northern Pacific, rebuilt for executive train service; BN calls these F9-2; mechanically they are the same as GP38-2s.
2. Rebuilt to SW1200 specifications
3. Stored serviceable
4. Leased from Metra
5. GP39 rebuilt by EMD; leased from EMD
6. Leased from GATX
7. GP9 carbody with FT components, 1350 h.p.
8. GP7s rebuilt to 1800 h.p.; classified GP10

Charles W. McDonald

Burlington Northern, America's number-one coal hauler, has more than 800 SD40-2s on its roster.

9. Rebuilt with Caterpillar engines
10. No cab; rebuilt from wrecked GP38-2
11. Rebuilt from GP30s and GP35s
12. Rebuilt by BN from GP40s
13. Last units purchased by the Frisco
13a. Equipped with large cabs
14. Rebuilt by M-K from GP40s
15. Cabless units
16. Most stored
17. Rebuilt by BN

18. Stored
19. Rebuilt SD40s
20. Ex-Reserve Mining, leased from GATX
21. Cabless; rebuilt from wrecked units
22. Leased from GECX
23. Leased from EMD
24. Leased from LMX
25. Leased from Oakway
26. E8s and E9s rebuilt in 1973-1974 by Morrison-Knudsen for Chicago commuter service

CAIRO TERMINAL RAILROAD

Reporting marks: CTML   **Miles:** 28
Address: 204 Eighth St., Cairo, IL 62914
Cairo Terminal operates between Cairo and Elco, Illinois. Engine facilities are at Cairo.

Nos.	Qty.	Model	Builder	Date	Notes
101, 102	2	SW1	EMD	1950	Ex-Conrail; 101 stored
100	1	GP0	EMD	1966	Ex Illinois Central Gulf
Total	**3**				

CALIFORNIA WESTERN RAILROAD

Reporting marks: CWR   **Miles:** 40
Address: Foot of Laurel St., Fort Bragg, CA 95437
California Western operates freight and passenger service between Willits and Fort Bragg, California. The enginehouse is at Fort Bragg.

Nos.	Qty.	Model	Builder	Date	Notes
56	1	RS12	BLW	1955	
62	1	RS11	Alco	1959	Ex-Southern Pacific
64, 65	2	GP9	EMD	1955, 1954	Ex-Southern Pacific
Total	**4**				

CalTrans

With the skyline of San Francisco in the background, CalTrans F40PH-2 No. 910, named Millbrae, sets out for San Jose with three new gallery cars in tow.

CALTRANS
(California Department of Transportation)

Reporting marks: CALT   **Miles:** 47
Address: P. O. Box 7310, San Francisco, CA 94120
CalTrans operates commuter service between San Francisco and San Jose. Extension of the service to Gilroy is planned for 1992.

Nos.	Qty.	Model	Builder	Date	Notes
900-919	20	F40PH-2	EMD	1985, 1987	
Total	**20**				

CAMBRIA & INDIANA RAILROAD

Reporting marks: CI   **Miles:** 33
Address: P. O. Box 530, Colver, PA 15927
The Cambria & Indiana operates between Colver and Revloc, Pennsylvania. The enginehouse is at Colver.

Nos.	Qty.	Model	Builder	Date	Notes
15, 16	2	SW1500	EMD	1968	
17	1	SW1500	EMD	1969	
19, 20	2	MP15	EMD	1975	
32	1	SW7	EMD	1952	
40, 41, 44	3	SW9	EMD	1956-1957	
Total	**9**				

CANADIAN NATIONAL RAILWAYS

Reporting marks: CN   **Miles:** 22,518
Address: P. O. Box 8100, Montreal PQ H3C 3N4, Canada

Nos.	Qty.	Model	Builder	Date	Notes
106, 108, 110-115, 117, 119	10	S13	MLW	1959	
160-168	9	Slug	CN	1964-1965	1
200-209	10	YBU4	GMD	1980	2
211-263	53	Slug	CN	1986-1990	3
264-270	7	Slug	CN	1990-1991	4
301, 304, 309	3	S13	MLW	1959	
356	1	Slug	CN	1966	5
400, 402-405	5	SW900r	GMD	1958	
500-522	23	HBU4	GMD	1978, 1980	6
523-526	4	YBU4m	GMD	1980	6

Nos.	Qty.	Model	Builder	Date	Notes
1101, 1105, 1106, 1113, 1115-1118, 1120, 1121, 1123, 1124, 1127, 1129, 1130, 1133, 1134, 1139-1141, 1143, 1144, 1147-1151, 1153-1156, 1159, 1160, 1163, 1166, 1167, 1169-1172, 1177-1182	46	GMD1m	GMD	1958-1960	7

Charles W. McDonald

Canadian National Railways 5354 is one of 122 SD40-2Ws on the road's roster. Both major Canadian railways have specified full-width noses on recent orders for locomotives.

Nos.	Qty.	Model	Builder	Date	Notes
1205, 1206, 1210, 1211, 1213, 1215, 1217, 1218, 1227-1229, 1231, 1232, 1234, 1236, 1242, 1244, 1245, 1247, 1251, 1252, 1254, 1256, 1259-1262, 1264, 1265, 1267, 1268, 1287-1289, 1291, 1295, 1296, 1298, 1300-1303, 1305, 1308, 1310, 1311, 1314, 1315, 1317, 1318, 1320-1324, 1326-1330, 1334, 1335, 1337-1339, 1341-1344, 1346, 1348-1350, 1352, 1353, 1355-1357, 1359-1364, 1366, 1367, 1369, 1371, 1374, 1375, 1377, 1379, 1381, 1383-1389, 1391, 1392, 1394-1396	105	SW1200RS	GMD	1956-1960	
1400-1423	24	GMD1r	GMD	1958-1960	8
1600-1614	15	GMD1r	GMD	1958-1959	8, 9

Nos.	Qty.	Model	Builder	Date	Notes
1750-1752, 1754, 1757-1761, 1764, 1765, 1768, 1775, 1782, 1786					
	15	RS18d	MLW	1959-1960	10
1900-1905, 1907-1912,					
1914, 1915	14	GMD1	GMD	1958-1959	11
2000-2043	44	C630	MLW	1967-1968	
2100-2119	20	HR616	BBD	1982	12
2305-2310, 2313-2320,					
2322-2329, 2332-2339	30	M636	MLW	1970-1971	
2400-2429	30	B40-8W	GE	1990	12
3100	1	RS18	MLW	1959	
3500-3510, 3512-3525, 3527-3534, 3536-3551, 3553-3564, 3566-3579					
	75	M420W	MLW	1973-1975	13
3580-3589	10	HR412W	BBD	1981	13
3624, 3625, 3627, 3640, 3642, 3644, 3646, 3661, 3668, 3673, 3675, 3681, 3682, 3684, 3832, 3842					
	16	RS18	MLW	1957-1959	
4000-4036, 4100-4140	78	GP9r	GMD		8
4560, 4566, 4571, 4572,					
4585, 4589, 4590, 4595	8	GP9	GMD	1957-1958	
4700-4732	33	GP38-2	GMD	1972-1973	
4760-4785, 4787-4800,					
4802-4810	49	GP38-2W	GMD	1973-1974	13
5000-5010, 5012-5017, 5019-5061, 5063-5102, 5105-5139, 5141-5150, 5152-5240					
	234	SD40	GMD	1967-1971	
5241-5252, 5254-5363	122	SD40-2W	GMD	1975-1980	13
5400-5459	60	SD50F	GMD	1985, 1987	12
5500-5563	64	SD60F	GMD	1985, 1989	12
5700-5703	4	SD38-2	GMD	1975	14

Nos.	Qty.	Model	Builder	Date	Notes
ELECTRIC LOCOMOTIVES					
6710-6715	6		GE	1914-1917	15
6716, 6717, 6722-6724	5		EE	1924	15
6725-6727	3		GE	1950	16

All electric units are leased to Montreal Urban Community Transit Commission for commuter service. They also appear on the MUCTC roster.

Nos.	Qty.	Model	Builder	Date	Notes
DIESEL LOCOMOTIVES					
7000-7028	29	GP9r	GMD	1985	8
7100-7107	8	SW1200RSu			
			GMD	1956-1960	17
7200-7270	71	GP9r	GMD	1985	8
7500-7518, 7520-7522,					
7524, 7526, 7528, 7530	26	GP38r	GMD	1978	18
8700-8711	12	S13r	MLW		19
9302-9310, 9312-					
9314, 9316, 9317	14	GP40	GMD	1966-1967	
9400-9434, 9436-9486,					
9488-9667	266	GP40-2W	GMD	1974-1976	13
9668-9677	10	GP40-2W	GMD	1973-1975	20
Total	1,672				

Notes:
1. Rebuilt by CN from MLW S3s
2. Yard slugs
3. Rebuilt by CN from GP9s for yard and transfer use
4. Rebuilt by CN from MLW S3s for yard and transfer use
5. Rebuilt from MLW S3
6. Slugs for hump service
7. Rebuilt with B trucks
8. Rebuilt by CN with 645 engine components
9. Lightweight for branchline service
10. Derated to 1400 h.p.; A1A trucks from GMD1s
11. Steam generators removed

12. Full-width body, full-width nose
13. Full-width nose
14. Ex-Northern Alberta 401-404
15. Boxcab B+B units
16. Center-cab B-B units
17. Rebuilt by CN in 1985 and 1987 with GP9 components and trucks
18. Rebuilt by CN 1978-1990 for use with slugs (formerly numbered 200-226)
19. Rebuilt by CN1984-1985
20. Ex-GO Transit 700-702, 704-710; wide nose

C&S RAILROAD

Reporting marks: CSKR   **Miles:** 35
Address: Center Ave., Jim Thorpe, PA 18229
C&S operates a former Central Railroad of New Jersey line from Packerton Junction, near Jim Thorpe, in Carbon County, Pennsylvania, west into Schuylkill County.

Nos.	Qty.	Model	Builder	Date	Notes
5771	1	U36B	GE	1971	Ex-Seaboard System
Total	1				

CANEY FORK & WESTERN RAILROAD

Reporting marks: CFWR   **Miles:** 61
Address: 401 Depot St., McMinnville, TN 37110
Caney Fork & Western extends from Tullahoma to Sparta, Tennessee. The enginehouse is at McMinnville, about halfway along the route.

Nos.	Qty.	Model	Builder	Date	Notes
531	1	GP9	EMD	1954	Ex-Western Railway of Alabama
979	1	GP7	EMD	1952	Ex-Seaboard Air Line
2345	1	GP7	EMD	1951	Ex-Nashville, Chattanooga & St. Louis
Total	3				

CANTON RAILROAD

Reporting marks: CTN   **Miles:** 5
Address: 4201 Boston St., Baltimore, MD 21224
Canton Railroad is a switching line in the Canton section of Baltimore, Maryland.

Nos.	Qty.	Model	Builder	Date	Notes
1201	1	SW1200	EMD	1954	
1501	1	SW1500	EMD	1967	
Total	2				

CAPE FEAR RAILWAYS

Reporting marks: CFR   **Miles:** 10
Address: P. O. Box 70090, Fort Bragg, NC 28307
Cape Fear Railways extends from Fort Junction, at Fort Bragg, North Carolina, to Skibo, on the west side of Fayetteville. Most of its operations are within the Fort Bragg Military Reservation.

Nos.	Qty.	Model	Builder	Date	Notes
2083, 2084, 2087, 2088	4	MRS1	Alco	1953	1
Total	4				

Notes:
1. Ex-U. S. Army, same numbers; 2083, 2084 stored unserviceable

CAROLINA COASTAL RAILWAY

Reporting marks: CLNA   **Miles:** 17
Address: 1 Park West Circle, Suite 201, Midlothian, VA 23113
Carolina Coastal Railway operates from Pinetown to Belhaven, North Carolina. Enginehouses are at Pinetown and Belhaven.

Nos.	Qty.	Model	Builder	Date	Notes
95	1	45-ton	GE	1948	
127	1	S12r	BLW	1947	Ex-Atlantic Coast Line
Total	2				

CAROLINA RAIL SERVICES

Reporting marks: CRIJ
Address: P.O. Box 3629, Morehead City, NC 28557
Carolina Rail Services is a switching railroad at Morehead City, North Carolina.

Nos.	Qty.	Model	Builder	Date	Notes
1201, 1202	2	SW1200	EMD	1966	
Total	**2**				

CARTHAGE, KNIGHTSTOWN & SHIRLEY RAILROAD

Reporting marks: CKSI   **Miles:** 34
Address: P.O. Box 379, Shirley, IN 47348
The Carthage, Knightstown & Shirley extends from Emporia, Indiana, south of Anderson, to Carthage. The enginehouse is at Shirley.

Nos.	Qty.	Model	Builder	Date	Notes
215	1	44-ton	GE	1951	
Total	**1**				

CARTIER RAILWAY

Reporting marks: CART   **Miles:** 260
Address: Port Cartier, PQ G5B 2H3, Canada
The Cartier Railway extends from Port Cartier, Quebec, on the north shore of the St. Lawrence River west of Sept-Iles, north to Gagnon. The road's shops are at Port Cartier.

Nos.	Qty.	Model	Builder	Date	Notes
30-39	10	C630	Alco	1966	1
41-49	9	M636	MLW	1970-1971	2
62-68	7	RS18	MLW	1959-1960	
71, 73-76	5	M636	MLW	1972-1973	
77-79	3	C636	Alco	1968	3

Greg McDonnell

The sole purpose of the Cartier Railway is to haul iron ore from the wilds of eastern Quebec to ship-loading facilities on the shore of the St. Lawrence River.

Nos.	Qty.	Model	Builder	Date	Notes
81-85	5	M636	MLW	1975	
86, 87	2	M636	MLW	1971	4
Total	**41**				

Notes:
1. Ex-Duluth, Missabe & Iron Range 905, 906, 908, 900-904, 907, 909
2. Ex-Canadian National 2300-2304, 2311, 2312, 2318, 2321
3. Ex-Alco demonstrators 636-3, 636-2, 636-1
4. Ex-Canadian National 2330, 2331

CEDAR RAPIDS & IOWA CITY RAILWAY

Reporting marks: CIC   **Miles:** 55
Address: P. O. Box 2951, Cedar Rapids, IA 52406

The Crandic — "CR and IC" — operates its original line between the cities of its name, plus branches from Iowa City to Hills and from Cedar Rapids to Amana. The enginehouse is at Cedar Rapids.

Nos.	Qty.	Model	Builder	Date	Notes
97, 99	2	SW1200	EMD	1966	1
100-103	4	GP9	EMD	1959	2
Total	6				

Notes:
1. 97 is ex-Fort Worth Railway, previously Missouri Pacific 1146
2. Ex-Milwaukee Road 305, 308, 304, 302

CENTRAL CALIFORNIA TRACTION COMPANY

Reporting marks: CCT   **Miles:** 52
Address: 1645 N. Cherokee Road, Stockton, CA 95205

Central California Traction operates between Sacramento and Stockton, California. Engine facilities are at Stockton.

Nos.	Qty.	Model	Builder	Date	Notes
60	1	GP7	EMD	1952	Ex-Reading
1790, 1795	2	GP18	EMD	1961	Ex-Rock Island
Total	3				

CENTRAL INDIANA & WESTERN RAILROAD

Reporting marks: CEIW   **Miles:** 9
Address: P. O. Box 456, Lapel, IN 46051

Central Indiana & Western operates from Lapel to Anderson, Indiana. The enginehouse is at Lapel.

Nos.	Qty.	Model	Builder	Date	Notes
88	1	SW7	EMD	1950	Ex-Conrail
Total	1				

CENTRAL MICHIGAN RAILWAY

Reporting marks: CMGN   **Miles:** 214
Address: 4901 Towne Centre Road, Saginaw, MI 48604

Central Michigan has two separate routes: from Owosso to Durand and Bay City, Michigan, with a branch to Essexville, and from Ionia through Grand Rapids to Ravenna, Michigan — all former Grand Trunk Western lines.

Nos.	Qty.	Model	Builder	Date	Notes
4502, 4504	2	U23B	GE	1973	Ex-Missouri Pacific
8801-8804	4	GP40	EMD	1969	
8902-8904	3	U23B	GE	1973	Ex-Missouri Pacific
Total	9				

Byron C. Babbish

Central Michigan 8803, a GP40 (which appears to have only two radiator fans) is on the point of a freight heading west from Durand.

CENTRAL MONTANA RAIL

Reporting marks: CM **Miles:** 66
Address: P. O. Box 928, Denton, MT 59430

Central Montana Rail operates between Geraldine and Moccasin Junction, Montana. The enginehouse is at Denton.

Nos.	Qty.	Model	Builder	Date	Notes
1809, 1810, 1814,					
1817, 1824, 1838	6	GP9	EMD	1954-1956	1
Total	6				

Notes:
1. Ex-Burlington Northern, same numbers

CENTRAL RAILROAD CO. OF INDIANAPOLIS

Reporting marks: CERA **Miles:** 122
Address: P. O. Box 554, Kokomo, IN 46903-0554

Central Railroad of Indianapolis operates from Frankfort, Indiana, through Kokomo to Marion and from Kokomo to Peru.

Nos.	Qty.	Model	Builder	Date	Notes
83	1	F7A	EMD	1949	1
819	1	NW2	EMD	1949	2
1201	1	SW9	EMD	1951	3
1202	1	SW9	EMD	1951	4
1750	1	GP7m	EMD	1952	5
1751	1	GP7m	EMD	1952	5
Total	6				

Notes:
1. Ex-Milwaukee Road; leased from Indiana Transportation Museum; painted Monon colors
2. Ex-Indiana Harbor Belt
3. Ex-Wabash
4. Ex-Pennsylvania Railroad
5. Ex-Missouri-Kansas-Texas

CENTRAL VERMONT RAILWAY

Reporting marks: CV **Miles:** 366
Address: 2 Federal St., St. Albans, VT 05478

Central Vermont's main line extends from tidewater at New London, Connecticut, north to the Canadian border at East Alburg, Vermont. It has several short branches in Vermont. CV's principal shops are at St. Albans, Vt.

Nos.	Qty.	Model	Builder	Date	Notes
4605-4610	6	GP9r	EMD	1957	1
4917-4920, 4926, 4959	6	GP9	EMD	1957	
5800, 5801, 5804,					
5806-5811	9	GP38AC	EMD	1971	2
Total	21				

Notes:
1. Rebuilt by Grand Trunk Western
2. Ex-Grand Trunk Western

CENTRAL WESTERN RAILWAY

Reporting marks: CWRL **Miles:** 105
Address: P. O. Box 2520, Stettler, AB T0C 2L0, Canada

Central Western operates a former Canadian National route between Ferlow Junction, south of Camrose, Alberta, and Dinosaur (Munson), east of Drumheller.

Nos.	Qty.	Model	Builder	Date	Notes
4301, 4302	2	GP7u	EMD	1951	1
7438	1	GP9	GMD	1957	2
Total	3				

Notes:
1. Ex-Wabash Valley 4101, 4102
2. Ex-Conrail-Penn Central 7438, New York Central 6038

CHAPARRAL RAILROAD

Reporting marks: CHRC   **Miles:** 85
Address: P. O. Box 370, Wolf City, TX 75496
Chaparral Railroad operates from Paris, Texas, to Farmersville, then to Garland by trackage rights on Santa Fe. The enginehouse is at Wolf, Texas.

Nos.	Qty.	Model	Builder	Date	Notes
703, 704	2	GP7	EMD	1952	Ex-Florida East Coast
Total	2				

CHARLES CITY RAIL LINE

Reporting marks: CCRY   **Miles:** 4
Address: P. O. Box 51, Osage, IA 50461
Charles City Rail Line is a switching operation at Charles City, Iowa.

Nos.	Qty.	Model	Builder	Date	Notes
9086, 9363	2	GP10	ICG	1969, 1970	
Total	2				

CHATTAHOOCHEE INDUSTRIAL RAILROAD

Reporting marks: CIRR   **Miles:** 15
Address: P. O. Box 253, Cedar Springs, GA 31732
Chattahoochee Industrial extends from Hilton to Saffold, Georgia. The enginehouse is at Cedar Springs.

Nos.	Qty.	Model	Builder	Date	Notes
1	1	GP10	ICG		1
38, 97	2	RS1	Alco	1947	2
89, 90	2	Slug			
1500	1	SW1500	EMD	1969	
1505	1	SW1500	EMD	1973	3
1830	1	GP7	EMD	1954	4
Total	8				

Notes:

1. Ex-Illinois Central Gulf; named City of New Orleans
2. Ex-Washington Terminal 52 and 54; named B. W. Moore and Old 97
3. Ex-Pittsburgh & Lake Erie 1542
4. Ex-Southern 8038; named Best Friend

CHATTAHOOCHEE VALLEY RAILWAY

Reporting marks: CHV   **Miles:** 10
Address: 305 W. 5th St., West Point, GA 31833
Chattahoochee Valley Railway extends from West Point, Georgia, to McGinty, Alabama. The enginehouse is at West Point.

Nos.	Qty.	Model	Builder	Date	Notes
101	1	SW1500	EMD	1966	
102	1	SW1001	EMD	1971	Ex-New Orleans Public Belt
Total	2				

CHATTOOGA & CHICKAMAUGA RAILWAY

Reporting marks: CCKY   **Miles:** 68
Address: P. O. Box 2385, Columbus, MS 39704
The Chattooga & Chicamauga has two routes extending south from Alton Park Spur, Tennessee, to Hedges and to Lyerly, Georgia. The enginehouse is at Lafayette, Ga.

Nos.	Qty.	Model	Builder	Date	Notes
101, 102	2	CF7	ATSF	1973	
Total	2				

CHENEY RAILROAD

Reporting marks: CHNY   **Miles:** 55
Address: P. O. Box 309, Allgood, AL 35013
The Cheney Railroad operates between Greens and Carnes, Alabama, on a former CSX (ex-Louisville & Nashville) route from Birmingham to Attalla. The enginehouse is at Graystone.

Nos.	Qty.	Model	Builder	Date	Notes
65	1	65-ton	GE	1942	
Total	1				

Robert C. Del Grosso

Six-motor units of relatively low horsepower, such as the SD38-2, are usually purchased for low-speed heavy hauling. Chicago & Illinois Midland 72 is a case in point; it spends most of its life on coal trains.

CHICAGO & ILLINOIS MIDLAND RAILWAY

Reporting marks: CIM   **Miles:** 118
Address: P. O. Box 139, Springfield, IL 62705
The Chicago & Illinois Midland extends from East Peoria to Taylorville, Illinois. The road's shops are at Springfield.

Nos.	Qty.	Model	Builder	Date	Notes
18, 20-23	5	SW1200	EMD	1955	
30, 31	2	RS1325	EMD	1960	Unique to C&IM
50-54	5	SD9	EMD	1955	
60, 61	2	SD18	EMD	1961, 1962	
70-75	6	SD38-2	EMD	1974	
Total	20				

CHESAPEAKE & ALBEMARLE RAILROAD

Reporting marks: CA   **Miles:** 73
Address: 1500 Lexington Drive, Elizabeth City, NC 27909
The Chesapeake & Albermarle extends from Chesapeake, Virginia, to Edenton, North Carolina. The engineinehouse is at Elizabeth City, N.C.

Nos.	Qty.	Model	Builder	Date	Notes
2158, 2190	2	GP7	EMD	1952	Ex-Santa Fe
Total	2				

CHESTNUT RIDGE RAILWAY

Reporting marks: CHR   **Miles:** 7
Address: Palmerton, PA 18071
The Chestnut Ridge Railway operates from Palmerton to Little Gap, Pennsylvania. The enginehouse is at Palmerton.

Nos.	Qty.	Model	Builder	Date	Notes
11	1	S2	Alco	1946	
Total	1				

CHICAGO & NORTH WESTERN TRANSPORTATION CO.

Reporting marks: CNW   **Miles:** 6,450
Address: One North Western Center, Chicago, IL 60606

Nos.	Qty.	Model	Builder	Date	Notes
217	1	F7A	EMD	1951	1
315	1	F7B	EMD	1950	1
400-403	4	F7A	EMD	1949-1950	1
410, 411	2	F7B	EMD	1950	1
423	1	F7A	EMD	1950	1
903, 904, 912, 913, 948, 950, 954, 957, 964, 974	10	SD45	EMD	1967, 1969	1
1302-1316	15	MP15	EMD	1975	
4100-4111, 4113-4118, 4120-4132, 4134-4138, 4140-4145, 4147-4150, 4152-4157, 4160-4165, 4167-4170, 4172-4176, 4178-4209, 4279-4299, 4307, 4314, 4322-4324, 4328, 4331, 4333, 4341, 4344, 4356, 4357, 4360	133	GP7	EMD	1951-1958	2

The 1500-h.p. GP15-1 was designed to utilize components of traded-in GP7s and GP9s — the units it was designed to replace. Chicago & North Western's GP15-1s had conventional air filters, hence the extra louvers behind the cab.

Charles W. McDonald

Nos.	Qty.	Model	Builder	Date	Notes
4400-4406, 4408-4424	24	GP15-1	EMD	1976	
4431-4436, 4438-4442, 4445, 4447, 4453, 4455, 4457, 4459, 4460, 4463-4486	42	GP7	EMD	1951-1959	2
4527, 4537-4554, 4556-4562	26	GP9	EMD	1955-1959	
4600-4634	35	GP38-2	EMD	1979	3
4701-4706, 4711	7	GP38r	EMD	1967	4
5050, 5052-5095, 5097-5099	48	GP50	EMD	1980	
5500-5513, 5515-5518, 5520-5522, 5524-5529, 5531-5535, 5537	33	GP40	EMD	1965-1966	5
6600	1	SDCAT	EMD	1968	6
6474, 6481, 6482, 6485, 6500, 6503, 6504, 6507, 6514, 6515, 6523, 6524, 6527-6529, 6532, 6536, 6537, 6539, 6540, 6542-6544, 6549, 6556	25	SD45	EMD	1967-1970	
6564	1	SD40R	EMD	1968	

Nos.	Qty.	Model	Builder	Date	Notes
6566, 6567, 6576, 6582, 6584, 6585	6	SD45	EMD	1968-1969	
6622-6647	26	SD24d	EMD	1959	7
6650-6659	10	SD38-2	EMD	1975	
6801-6821, 6823-6909, 6911, 6913-6935	132	SD40-2	EMD	1973-1976	
7000-7035	36	SD50	EMD	1985	
8001-8055	55	SD60	EMD	1986	
8501-8577	77	C40-8	GE	1989-1991	
Total	751				

Notes:
1. Stored
2. Includes GP9s downrated to 1500 h.p. and classed as GP7s; 4100-series units are ex-Rock Island
3. Ordered by Rock Island, which declared bankruptcy before delivery; C&NW accepted them, and they were delivered in C&NW colors
4. Ex-CSX
5. Ex-Conrail
6. SD45 repowered with Caterpillar engine
7. Ex-Southern; rebuilt by C&NW and classified as SD18

CHICAGO, CENTRAL & PACIFIC RAILROAD

Reporting marks: CC **Miles:** 798
Address: 1006 E. 4th St., Waterloo, IA 50704
Chicago Central's main route extends from Chicago west to Tara, Iowa, where lines diverge to Omaha, Nebraska, and Sioux City, Iowa, on former Illinois Central track. The road's main shops are at Waterloo, Iowa.

Nos.	Qty.	Model	Builder	Date	Notes
964, 966-977, 979-981	16	GP9u	EMD	1969-1970	1
1200	1	SW9	EMD	1967	2
1300, 1301, 1328	3	SW1300	ICG	1971, 1980	2
1504, 1505, 1585, 1591-1593	6	GP8	ICG	1970-1972	2
1600-1602	3	GP7u	EMD	1952	3
1669, 1705, 1719	3	GP10	ICG	1971-1972	2, 4
1742	1	GP9	EMD	1957	2
1743-1745, 1749, 1750, 1765, 1775, 1777	8	GP10	ICG	1969-1971	2
1806	1	GP18	EMD	1960	2
1878	1	GP7u	EMD	1954	1

Nos.	Qty.	Model	Builder	Date	Notes
2000-2005	6	GP38	EMD	1967-1969	5
7983, 7984, 7989, 7990	4	GP8	ICG	1972	2
8002, 8012, 8023, 8032, 8033, 8035, 8039, 8055, 8056, 8059, 8063, 8079, 8093, 8111, 8114, 8121, 8134, 8169-8171, 8179, 8181, 8188, 8190, 8199, 8211, 8234, 8258, 8260	31	GP10	ICG	1970-1972	2
9376	1	GP9	EMD	1959	2
9400, 9402	2	GP18	EMD	1960	2
9405	1	GP9	EMD	1959	2
9408, 9413, 9414, 9420, 9426-9428	7	GP18	EMD	1960, 1963	2
9438-9440	3	GP28	EMD	1964	2
Total	**98**				

Notes:
1. Rebuilt to 2000 h.p. by Milwaukee Road; classified GP20
2. Ex-Illinois Central Gulf
3. Ex-Missouri-Kansas-Texas 95, 97, 119; 1750 h.p.
4. 1850 h.p.
5. Ex-Monongahela (2000-2004); ex-Chesapeake & Ohio (2005)

CHICAGO-CHEMUNG RAILROAD

Reporting marks: CCMG **Miles:** 4
Address: 8550 Richfield Road, Crystal Lake, IL 60014
The Chicago-Chemung Railroad operates between Harvard and Chemung, Illinois.

Nos.	Qty.	Model	Builder	Date	Notes
202	1	SW7	EMD	1950	Ex-Chesapeake & Ohio 5236
Total	**1**				

CHICAGO RAIL LINK

Reporting marks: CRL **Miles:** 30
Address: 2728 E. 104th St., Chicago, IL 60617-5766

Chicago Rail Link operates switching service on former Rock Island track at Blue Island, Kensington, and Chicago.

Nos.	Qty.	Model	Builder	Date	Notes
6	1	SW1200	EMD	1964	1
14, 15	2	GP18	EMD	1960	2
18, 19	2	SW1500	EMD	1972	3
51	1	SW9	EMD	1953	4
Total	**6**				

Notes:
1. Ex-Terminal Railroad Association of St. Louis 1227
2. Ex-Rock Island 1339, 1343
3. Ex-Missouri Pacific 1518, 1520
4. Ex-Mississippi Central 210

CHICAGO SHORT LINE RAILWAY

Reporting marks: CSL   **Track miles:** 28
Address: 9746 S. Avenue N, Chicago, IL 60617
Chicago Short Line is a switching road operating between South Chicago, Illinois, and Indiana Harbor, Indiana.

Nos.	Qty.	Model	Builder	Date	Notes
28, 29	2	SW1001	EMD	1974	
30, 31	2	SW1500	EMD	1968, 1971	
Total	**4**				

CHICAGO SOUTHSHORE & SOUTH BEND RAILROAD

Reporting marks: CSS   **Miles:** 129
Address: 505 N. Carroll Ave., Michigan City, IN 56360-5082
Chicago SouthShore & South Bend operates freight service with diesel locomotives between Kensington, Illinois, and South Bend, Indiana, on the electrified line owned by Northern Indiana Commuter Transportation District — the former Chicago South Shore & South Bend Railroad.

Nos.	Qty.	Model	Builder	Date	Notes
206	1	SW8	EMD	1951	Ex-Colorado & Wyoming
2000-2009	10	GP38-2	EMD	1981	
Total	**11**				

CLAREMONT CONCORD RAILROAD

Reporting marks: CLCO   **Miles:** 5
Address: P. O. Box 1598, Claremont, NH 03743
Claremont Concord operates between Claremont Junction and Claremont, New Hampshire. The enginehouse is at Claremont Junction.

Nos.	Qty.	Model	Builder	Date	Notes
30	1	44-ton	GE	1942	
119	1	44-ton	GE	1944	
Total	**2**				

CHICAGO, WEST PULLMAN & SOUTHERN RAILROAD

Reporting marks: CWP   **Miles:** 32
Address: 2728 E. 104th St., Chicago, IL 60617-5766
Chicago, West Pullman & Southern is a switching road between West Pullman and Irondale, Illinois.

Nos.	Qty.	Model	Builder	Date	Notes
41-46	6	SW8	EMD	1951-1952	
47, 48, 51, 52	4	SW9	EMD	1952-1953	
Total	**10**				

CITY OF PRINEVILLE RAILWAY

Reporting marks: COP   **Miles:** 19
Address: 185 E. 10th St., Prineville, OR 97754
The City of Prineville Railway connects Prineville, Oregon, with the Burlington Northern and Union Pacific at Prineville Junction, north of Bend. The enginehouse is at Prineville.

Nos.	Qty.	Model	Builder	Date	Notes
985, 989	2	GP9u	EMD	1954	1
Total	**2**				

Notes:
1. Ex-Milwaukee Road 985, 989; 2000 h.p.; classified GP20

CLARENDON & PITTSFORD RAILROAD

Reporting marks: CLP   **Miles:** 29
Address: 1 Railway Lane, Burlington, VT 05401
The Clarendon & Pittsford extends from Whitehall, New York, to Rutland, Vermont, and also operates a short stretch of track at Florence, Vt. It is affiliated with the Vermont Railway. The enginehouse is at Rutland.

Nos.	Qty.	Model	Builder	Date	Notes
203	1	GP38	EMD	1966	Ex-Maine Central 255
502	1	SW1500	EMD	1968	Ex-Toledo, Peoria & Western 304
752	1	GP9	EMD	1957	Ex-Burlington Northern 1879
Total	**3**				

William J. Husa Jr.

Colorado & Wyoming's two GP38-2s were intended for coal-hauling duty on the road's Southern Division.

COLORADO & WYOMING RAILWAY

Reporting marks: CW   **Miles:** 31
Address: P. O. Box 316, Pueblo, CO 81002

The Colorado & Wyoming operates switching service at Pueblo, Colorado, for its owner, CF&I Steel, and also carries coal from New Elk Mine to Jansen, Colorado, near Trinidad. Engine facilities are at Pueblo and New Elk Mine.

Nos.	Qty.	Model	Builder	Date	Notes
102-104	3	GP7	EMD	1951	
201-205, 207-209	8	SW8	EMD	1951	
2001, 2002	2	GP38-2	EMD	1973	
Total	**13**				

COE RAIL

Reporting marks: CRLE   **Miles:** 9
Address: 840 N. Pontiac Trail, Walled Lake, MI 48088

Coe Rail operates between Wixom and Walled Lake, Michigan. The enginehouse is at Walled Lake.

Nos.	Qty.	Model	Builder	Date	Notes
52	1	S2	Alco		Ex-Port Huron & Detroit
70	1	S1	Alco	1945	Ex-Delray Connecting
105	1	20-ton	Whitcomb	1942	
Total	**3**				

COLONEL'S ISLAND RAILROAD

Reporting marks: CISD   **Miles:** 31
Address: P. O. Box 2406, Savannah, GA 31402

Colonel's Island Railroad extends from Anguilla, Georgia, to Mydharris, south of Brunswick. The enginehouse is at Mydharris.

Nos.	Qty.	Model	Builder	Date	Notes
8236, 8237, 8230	3	GP7	EMD	1061, 1063	1
Total	**3**				

Notes:
1. Ex-Southern, same numbers

COLUMBIA & COWLITZ RAILWAY

Reporting marks: CLC   **Miles:** 8
Address: P. O. Box 209, Longview, WA 98632

The Columbia & Cowlitz extends from Columbia Junction to Ostrander Junction, Washington. The enginehouse is at Longview.

Nos.	Qty.	Model	Builder	Date	Notes
700	1	GP7r	EMD	1953	Rebuilt by M-K 1977
701	1	GP7r	EMD	1953	Rebuilt by M-K 1976
Total	**2**				

COLUMBUS & GREENVILLE RAILWAY

Reporting marks: CAGY   **Miles:** 242
Address: 201 19th St. North, Columbus, MS 39703
The Columbus & Greenville extends across the state of Mississippi from Columbus on the east to Greenville on the west. It also operates a former Illinois Central route from Cleveland to Hollandale, and a line from Greenville to Leland, to connect with that route. The enginehouse is at Columbus.

Nos.	Qty.	Model	Builder	Date	Notes
601	1	DRS6-4-1500	BLW	1951	1
608, 614, 615, 618, 619, 621	6	GP7	EMD	1952	2
801-810	10	CF7	ATSF	1977	
Total	17				

Notes:
1. On display, unserviceable
2. Ex-Florida East Coast, same numbers

Charles W. McDonald

Columbus & Greenville 614 exemplifies a locomotive model that so far shows no sign of impending extinction: the GP7.

COLUMBIA & SILVER CREEK RAILROAD

Reporting marks: CLSL   **Miles:** 11
Address: P. O. Box 1317, Shelbyville, TN 37160-1317
The Columbia & Silver Creek operates from Taylorsville to Soso, Mississippi. The enginehouse is at Soso.

Nos.	Qty.	Model	Builder	Date	Notes
9	1	SW600	EMD	1954	Ex-McLouth Steel
Total	1				

COLUMBIA TERMINAL

Reporting marks: CT   **Miles:** 24
Address: P. O. Box N, Columbia, MO 65205
Columbia Terminal operates on a former Wabash branch from Centralia to Columbia, Missouri. The enginehouse is at Columbia.

Nos.	Qty.	Model	Builder	Date	Notes
1	1	SW1200	EMD	1965	Ex-Missouri Pacific 1260
Total	1				

COMMONWEALTH RAILWAY

Reporting marks: CWRY   **Miles:** 17
Address: 1 Park Circle West, Suite 201, Midlothian, VA 23113
Commonwealth Railway operates between Suffolk and West Norfolk, Virginia. The enginehouse is at Suffolk.

Nos.	Qty.	Model	Builder	Date	Notes
517	1	CF7	ATSF	1974	
Total	1				

CONEMAUGH & BLACK LICK RAILROAD

Reporting marks: CBL **Miles:** 9
Address: 1 Locust St., Johnstown, PA 15909
Conemaugh & Black Lick is a switching road at Johnstown, Pennsylvania. It is owned by Bethlehem Steel.

Nos.	Qty.	Model	Builder	Date	Notes
100-102	3	NW2	EMD	1949	
104, 110-112, 114, 116	6	SW7	EMD	1949-1950	
122, 125	2	NW2	EMD	1948, 1945	
127, 128	2	SW1200	EMD	1964, 1965	
Total	**13**				

CONNECTICUT CENTRAL RAILROAD

Reporting marks: CCCL **Miles:** 17
Address: P. O. Box 1022, Middletown, CT 06475
Connecticut Central operates from Middlefield to Middletown, Connecticut, plus branches to Portland, Cromwell, Laurel, and North Middletown. The enginehouse is at Middletown.

Nos.	Qty.	Model	Builder	Date	Notes
35, 36	2	S4	Alco	1959, 1953	Ex-Genesee & Wyoming
Total	**2**				

CONNECTICUT DEPARTMENT OF TRANSPORTATION

Address: Union Station, Third Floor West, New Haven, CT 06511
Connecticut Department of Transportation operates "Shore Line East" diesel-powered commuter service between New Haven and Old Saybrook, Connecticut, 36 miles, on Amtrak's Northeast Corridor route. Conn DOT subsidizes Metro-North commuter service, electric and diesel, west of New Haven.

Nos.	Qty.	Model	Builder	Date	Notes
471	1	GP7u	EMD		1, 4
253, 257	2	GP38	EMD	1966	2, 4
6690, 6691	2	F7A	EMD		3, 4
Total	**5**				

Notes:
1. Ex-Maine Central 471, rebuilt 1985, has large cab
2. Ex-Maine Central 253, 257
3. Ex-Port Authority of Allegheny County; rebuilt by GE, 1981
4. Painted and lettered New Haven

CONRAIL (Consolidated Rail Corporation)

Reporting marks: CR **Miles:** 17,368
Address: 6 Penn Center Plaza, Philadelphia, PA 19103

Nos.	Qty.	Model	Builder	Date	Notes
1000-1023	24	MT4	CR	1979	1
1100-1118	19	MT6	GE	1978-1979	1
1119-1128	10	MT6	CR	1979	1
1600-1699	100	GP15-1	EMD	1979	
1900-1902, 1910-2023	117	B23-7	GE	1978-1979	

Nos.	Qty.	Model	Builder	Date	Notes
2169, 2172, 2174, 2175, 2177, 2179, 2182, 2185-2189, 2191, 2194, 2198, 2200-2202, 2204-2206, 2208-2213, 2215-2218, 2220-2223, 2225, 2228, 2230, 2232, 2234-2236, 2238, 2241, 2242, 2244-2249	51	GP30	EMD	1962-1963	2
2251, 2252, 2255-2259, 2261, 2262, 2264, 2266, 2268, 2271, 2274, 2275, 2277, 2288, 2290, 2300, 2304, 2305, 2308, 2314, 2316, 2317, 2321, 2327, 2332, 2333, 2339-2342, 2346, 2348-2350, 2353, 2356-2359, 2364-2366, 2389, 2394	47	GP35	EMD	1963-1965	

Nos.	Qty.	Model	Builder	Date	Notes
2700-2731, 2733-2738, 2740-2798	97	U23B	GE	1972-1975	2
2800-2816	17	B23-7	GE	1977	2
2971-2974	4	U36B	GE	1974	2
3050-3053, 3171, 3179, 3182, 3188, 3189, 3192-3195, 3197-3199, 3201-3206, 3208-3210, 3212-3219, 3221, 3223, 3224, 3226	37	GP40	EMD	1967-1969	2
3275-3316, 3318, 3320-3404	128	GP40-2	EMD	1973-1980	
3620, 3624, 3626, 3637, 3640-3642, 3644, 3645, 3649-3651, 3653, 3657, 3661, 3662, 3669, 3673, 3682, 3684, 3688, 3691	22	GP35	EMD	1964	3
4020-4022	3	E8A	EMD	1950-1954	4
5000-5016, 5018-5044, 5046-5059	58	B36-7	GE	1983	
5060-5089	30	B40-8	GE	1988	
5400, 5401, 5404-5413, 5428-5431, 5433-5462	46	GP7u	EMD	1950	5
6000-6021	22	C39-8	GE	1986	
6025-6049	25	C40-8	GE	1989	
6050-6149	100	C40-8W	GE	1990-1991	
6240-6244, 6246-6249, 6251, 6252, 6254-6264, 6266, 6268-6275, 6277-6297, 6299-6304, 6306-6308, 6314-6320, 6322, 6323, 6326-6329, 6331-6334, 6336, 6338-6341, 6343-6346	87	SD40	EMD	1970, 1966	
6350, 6351, 6354, 6355, 6357-6370, 6372-6376, 6378-6406, 6408-6524	170	SD40-2	EMD	1977-1979	
6550-6599	50	C30-7A	GE	1984	
6605-6609	5	C30-7	GE	1977	
6610-6619	10	C32-8	GE	1984	
6620-6644	25	C36-7	GE	1985	

Joseph R. Snopek

Conrail 6611, one of ten C32-8s on Conrail's roster, illustrates the look of General Electric's Dash 8 line. The most obvious distinguishing feature is the dynamic brake and equipment blower hump behind the cab.

Nos.	Qty.	Model	Builder	Date	Notes
6654-6666	13	SD45-2	EMD	1972	
6700-6784, 6805-6834	115	SD50	EMD	1983-1986	
6843-6867	25	SD60	EMD	1988	
6900-6912, 6914-6918	18	U23C	GE	1970	6
6925-6959	35	SD38	EMD	1970	6
7513-7531, 7533-7535, 7537, 7545-7597	76	GP9u	EMD	1954-1959	7
7635	1	GP38	EMD	1969	
7656-7659	4	GP38AC	EMD	1971	
7670-7746, 7868-7871, 7873-7915, 7917-7927, 7929, 7931-7939	145	GP38	EMD	1969-1971	

51

Nos.	Qty.	Model	Builder	Date	Notes
8040-8079, 8081-8100, 8102-8113, 8115-8166, 8168-8180, 8182-8201, 8203-8263, 8265-8271, 8273-8281					
	234	GP38-2	EMD	1973-1980	
8600, 8604, 8606, 8609, 8612, 8615, 8620, 8621	8	SW8	EMD	1950	
8632, 8634, 8641, 8646	4	SW900	EMD	1953	
8666-8668, 8670, 8673, 8682, 8684-8686	9	SW8m	EMD	1953	
8690, 8693, 8698	3	SW8	EMD	1950	
8838, 8841, 8851, 8865, 8896, 8898	6	SW7	EMD	1950-1951	
8922, 8935, 8939, 8942, 8954, 8959, 8970, 8989, 9016, 9017, 9022	11	SW9	EMD	1952-1953	
9037, 9045, 9047	3	SW7	EMD	1951-1952	
9095	1	SW9	EMD	1951	
9097	1	SW7	EMD	1950	
9121, 9128, 9130, 9140	4	SW9	EMD	1953	
9315-9326, 9328, 9330-9338, 9340-9347, 9349, 9350, 9352, 9354, 9355, 9357-9360, 9362, 9363, 9365, 9366, 9369, 9370, 9372, 9373, 9375-9378, 9380, 9381					
	53	SW1200	EMD	1956	
9400-9424	25	SW1001	EMD	1972	
9500-9600, 9609-9620	113	SW1500	EMD	1966-1973	
Total	**2,211**				

Notes:
1. Slugs rebuilt from Alco hood units
2. Retired or held for disposal
3. Many unserviceable, being retired
4. For executive train service
5. Classified GP8 by Conrail
6. For hump service
7. Rebuilt 1976-1979 by Illinois Central Gulf, Precision National, and Morrison-Knudsen; classified GP10

COPPER BASIN RAILWAY

Reporting marks: CBRY   **Miles:** 70
Address: P. O. Drawer I, Hayden, AZ 85235

Copper Basin Railway extends from Magma to Winkelman, Arizona, with a branch from Ray Junction to Ray. The enginehouse is at Hayden Junction.

Nos.	Qty.	Model	Builder	Date	Notes
201-203	3	GP18	EMD	1960	
204-208	5	GP9	EMD	1954-1956	
301-304	4	SD39	EMD	1970	Ex-Southern Pacific
401, 402	2	GP39	EMD	1970	Ex-Kennecott Copper
Total	**14**				

CORINTH & COUNCE RAILROAD

Reporting marks: CCR   **Miles:** 26
Address: P. O. Box 128, Counce, TN 38326

The Corinth & Counce operates between Corinth, Mississippi, and Counce, Tennessee, and from Sharp to Yellow Creek, Miss. It is a subsidiary of MidSouth Rail Corporation. Enginehouses are at Corinth and Counce.

Nos.	Qty.	Model	Builder	Date	Notes
901	1	SW900	EMD	1960	
1003	1	SW1000	EMD	1974	
1004, 1005	2	SW1001	EMD	1980, 1981	
Total	**4**				

CP RAIL

Reporting marks: CP   **Miles:** 13,677
Address: P. O. Box 6042, Station A, Montreal, PQ H3C 3E4, Canada

Nos.	Qty.	Model	Builder	Date	Notes
1200-1205	6	SW9r	GMD		1
1206-1214, 1237-1251, 1268-1276	33	SW1200RSr	GMD		2
1500-1511	12	GP7r	GMD		3
1512-1652, 1682-1697	157	GP9r	GMD	1980	4
3000-3020	21	GP38AC	GMD	1970-1971	
3021-3135	115	GP38-2	GMD	1983-1986	
3779	1	S11d	MLW	1953	5
4200-4250	51	C424	MLW	1963-1966	
4500-4505, 4507	7	C630M	MLW	1968	
4508-4512, 4550, 4551, 4553, 4556-4573	26	M630	MLW	1969-1970	
4700-4743	44	M636d	MLW	1969-1970	6
4744	1	M640m	MLW	1971	7
5000, 5001	2	GP30	GMD	1963	8
5002-5017, 5019-5025	23	GP35d	GMD	1964-1966	9
5400-5414	15	SD40	GMD	1968, 1971	10
5500-5559, 5561-5564	64	SD40	GMD	1966-1967	

5565-5585, 5587, 5589-5597, 5599-5633, 5635-5836, 5838, 5839, 5841, 5843, 5844, 5846-5849, 5853, 5855, 5857, 5860-5879, 5900-6080

	481	SD40-2	GMD	1972-1985	
8013-8046	34	RS23	MLW	1959	

8100, 8105-8107, 8109-8111, 8113-8115, 8119, 8120, 8122-8124, 8128, 8129, 8131-8134, 8136, 138, 8142, 8147, 8153, 8155, 8156, 8158-8162, 8165-8167, 8171

	37	SW1200RS	GMD	1958-1960	
8200, 8201, 8203-8220, 8222-8249	48	GP9u	GMD	1956	11
8921	1	RSD17	MLW	1957	12

Charles W. McDonald

CP Rail 5011 is a GP35 that has been derated to 2250 h.p. Derating is more often a matter of fuel rack settings — invisible externally — than of major surgery.

9000-9024	25	SD40-2F	GMD	1988	13
Total	**1,204**				

Notes:
1. Rebuilt 1982-1983
2. Rebuilt by CP 1981-1985 with 645 engine components
3. Rebuilt by CP 1980-1984 with 645 engine components
4. Rebuilt by CP 1980-1989 with 645 engine components
5. Rebuilt with Cummins engine
6. Derated to 3000 h.p.
7. Has A1A trucks for testing AC traction motors
8. Only Canadian GP30s
9. Derated to 2250 h.p.
10. Ex-Quebec North Shore & Labrador 204-218
11. Rebuilt by CP with 645C engine components, 1975-1978 and 1988-1990
12. Rebuilt with low nose
13. Full-width body and nose

COUNCIL BLUFFS & OTTUMWA RAILWAY

Reporting marks: COBA   **Track miles:** 27
Address: 29th Ave. and High Street, Council Bluffs, IA 51501
Council Bluffs & Ottumwa is a switching operation at Council Bluffs, Iowa.

Nos.	Qty.	Model	Builder	Date	Notes
992, 994	2	NW5	EMD	1947	Ex-Burlington Northern 992, 994
Total	**2**				

CRAB ORCHARD & EGYPTIAN RAILROAD

Reporting marks: COER   **Miles:** 15
Address: 514 N. Market St., Marion, IL 62959
The Crab Orchard & Egyptian operates from Marion to Mande, Ordill, and Herrin, Illinois. Enginehouses are at Marion and Herrin.

Nos.	Qty.	Model	Builder	Date	Notes
1	1	35-ton	Davenport	1942	
1136	1	SW1200	EMD	1966	Ex-Missouri Pacific
1161	1	SW1200	EMD	1965	Ex-Missouri Pacific
Total	**3**				

CRYSTAL CITY RAILROAD

Reporting marks: CYCY   **Miles:** 55
Address: 1003 N. First St., Crystal City, TN 78839-0672
The Crystal City Railroad extends from Gardendale, Texas, to Crystal City and Carrizo Springs. The enginehouse is at Crystal City.

Nos.	Qty.	Model	Builder	Date	Notes
1233	1	SW1200	EMD	1964	1
4159	1	GP7	EMD	1953	2
Total	**2**				

Notes:
1. Ex-Terminal Railroad Association of St. Louis 1233
2. Ex-Kansas City Southern 4159

CSX TRANSPORTATION INC.

Reporting marks: CSX   **Miles:** 19,414
Address: 500 Water St., Jacksonville, FL 32202

Nos.	Qty.	Model	Builder	Date	Notes
116	1	F7A	EMD	1948	1
117	1	F7B	EMD	1952	1
118	1	F7A	EMD	1952	1
119	1	F7B	EMD	1950	1
382	1	GP39-2	EMD	1972	
883	1	GP7	EMD	1950	
1004, 1008	2	SD35d	EMD	1964	2
1010-1029	20	Slug	EMD	1988	3
1038-1050	13	Slug	EMD	1984-1986	
1066	1	GP18	EMD	1960	
1100-1139	40	SW1500	EMD	1970-1973	
1140-1149	10	MP15	EMD	1975	
1150-1194	45	MP15AC	EMD	1977-1978	
1200-1241	42	MP15T	EMD	1984-1985	
1500-1524	25	MP15T	EMD	1982	
1700-1800, 1802-1860	160	GP16	SBD	1979-1982	4
1888-1990	103	U18B	GE	1973-1974	
2001-2004, 2006-2015, 2017-2052, 2054-2069, 2071-2080, 2082-2086, 2090-2094, 2096-2098, 2100-2104, 2106-2109, 2111-2125, 2127-2129, 2131-2134, 2136-2148, 2150-2183, 2185, 2187-2189	171	GP38	EMD	1967-1971	
2200-2330	131	Slug	CSX	1988-1991	5
2400-2404	5	SD40-2	EMD	1979-1980	
2420, 2421	2	SD40	EMD	1964	
2451-2454	4	SD40-2	EMD	1975	
2500-2716	217	GP38-2	EMD	1972-1980	
3000-3007, 3009	9	BQ23-7	GE	1978-1979	
3100-3143	44	B23-7	GE	1978	

Nos.	Qty.	Model	Builder	Date	Notes
3209-3211, 3218, 3225, 3229, 3231-3234, 3236, 3238, 3239, 3241-3244, 3246, 3247, 3249-3255, 3257-3270, 3272-3297, 3299-3303, 3305-3324	91	U23B	GE	1969-1975	
4018, 4024, 4044, 4046	4	Dash 8-40B	GE	1989	
4200-4204, 4206-4246, 4248-4266	65	GP30d	EMD	1962-1963	6
4280-4299	20	GP39	EMD	1969	
4300-4302, 4304-4308, 4310	9	GP39-2	EMD	1974	
4447	1	GP40-2	EMD	1981	
4505, 4507, 4512, 4518, 4519, 4526, 4527, 4529-4533, 4535, 4537-4541, 4543, 4550, 4552, 4559, 4562, 4565, 4567, 4572, 4573, 4575, 4577, 4584, 4586, 4589, 4590, 4594, 4598, 4599	36	SD35	EMD	1964-1965	
4600, 4601, 4606	3	SD40	EMD	1966-1971	
5200-5224	25	MATE	GE	1971-1972	
5500-5580	81	B30-7	GE	1978-1981	
5700, 5704, 5706, 5708, 5710, 5717, 5721, 5725, 5727, 5729, 5730, 5732-5735, 5739, 5740, 5745, 5746, 5750, 5751, 5753, 5754, 5757-5760, 5762, 5763, 5764, 5767, 5770, 5772-5776, 5779-5781, 5784-5786, 5788-5805	61	U36B	GE	1970-1972	
5806-5925	120	B36-7	GE	1985	
5930-5943, 5946	15	B40-8	GE	1989	
6001-6020, 6022-6035, 6037-6048, 6050-6068, 6070-6160, 6201, 6203-6249, 6276, 6279, 6280, 6295-6297, 6300, 6301, 6318, 6346-6365, 6388-6392, 6400-6499	338	GP40-2	EMD	1972-1981	
6500, 6501, 6504-6510, 6512-6524, 6526-6529, 6531-6533, 6535-6537, 6539-6545, 6547-6550, 6553-6557, 6560-6568, 6570, 6571, (continued) 6573-6582, 6584, 6585, 6587-6590, 6592-6598, 6601-6605, 6608-6612, 6614-6616, 6619-6623, 6625-6639, 6641-6645, 6647, 6649-6652, 6654, 6655, 6657, 6659, 6661-6664, 6666, 6668, 6671, 6675, 6677, 6679-6682, 6685, 6686, 6688, 6694-6697, 6701, 6703-6706, 6709, 6711, 6715, 6717, 6718, 6720, 6724-6729, 6731-6733, 6735-6742, 6744, 6748, 6749, 6752-6757, 6759-6764, 6766, 6768-6776, 6778-6780, 6782-6788, 6791-6794, 6796, 6798-6802, 6804-6807, 6809, 6810, 6815-6819, 6824, 6825, 6828-6830, 6832, 6834, 6835, 6839, 6842-6854	255	GP40	EMD	1966-1971	
6900-6943, 6947	45	GP40-2	EMD	1979-1980	
7000-7094	95	C30-7	GE	19791981	
7200, 7201, 7206-7208, 7210-7212, 7216-7220, 7222, 7224, 7226-7228, 7229, 7230, 7233, 7242, 7245, 7246, 7248-7252, 7254-7256, 7258-7260, 7265-7269, 7272, 7274-7276	44	U30C	GE	1969-1972	
7401, 7403, 7406-7410, 7412	8	GP39-2	EMD	1974	
7500-7646, 7650-7668, 7670-7675	172	C40-8	GE	1989-1991	
8000-8244, 8246-8261	261	SD40-2	EMD	1974-1981	
8302, 8304, 8305	3	SD40r	EMD	1969-1971	
8302, 8304, 8305, 8310, 8312, 8314, 8315, 8317-8324, 8328, 8329, 8331	18	SD40e	EMD	1966-1971	7
8332, 8334-8338, 8340, 8342-8344, 8347-8354, 8356, 8357, 8360-8362, 8364, 8365, 8367-8370, 8372, 8373, 8375, 8376, 8380, 8383, 8386-8389, 8391-8396, 8400, 8404-8410, 8413-8415, 8417-8421, 8423-8427, 8429-8433, 8435-8488	125	SD40e	EMD	1966-1971	7
8500-8643	144	SD50	EMD	1983-1985	

Nos.	Qty.	Model	Builder	Date	Notes
8950, 8951, 8954-8958,					
8960, 8962-8964, 8966,					
8968-8973	18	SD45-2	EMD	1972-1974	
9565, 9575	2	NW2	EMD	1949	
Total	3,109				

Notes:
1. Executive train service
2. 1500 h.p., classified H15
3. Rebuilt by CSX
4. Rebuilt GP7s and GP9s
5. Rebuilt GPs
6. 2000 h.p.; rebuilt 1981-1983
7. Rebuilt 1989-1991; classified SD40-2 by CSX

NEW AND OLD NUMBERS OF CSX LOCOMOTIVES

Tracing the ancestry of CSX locomotives is a difficult job. Both components, Chessie System and Seaboard System, renumbered their locomotives in recent years as part of merger processes, and Seaboard System's components had done the same previously — those renumberings were not completely carried out, and rebuilding programs further confuse the matter. The list that follows is admittedly incomplete; it represents the best information that was available at press time. The CSX numbering system is based on that of Seaboard System. Most SBD numbers remain unchanged.

Initials	Railroad	Initials	Railroad
AWP	Atlanta & West Point	L&N	Louisville & Nashville
B&O	Baltimore & Ohio	NYS&W	N.Y., Susquehanna & Western
C&O	Chespeake & Ohio	RDG	Reading
CHSY	Chessie System	SBD	Seaboard System
CRR	Clinchfield	SCL	Seaboard Coast Line
D&H	Delaware & Hudson	WM	Western Maryland
GARR	Georgia Railroad	WRA	Western Railway of Alabama

CSX Nos.	Model	Previous Nos.
116	F7A	CRR 800
117	F7B	CRR 869
118	F7A	CRR 200
119	F7B	CRR 250
382	GP39-2	D&H 382, D&H 7413
883	GP7	L&N 2368
1004, 1008	SD35d	SCL 4, 8
1038, 1039	Slug	B&O 138, 139; WM 138T, 139T
1040-1050	Slug	C&O/B&O 140-150
1066	GP18	SCL 1066
1100-1129	SW1500	L&N 5000-5029
1130-1139	MP15AC	L&N 4225-4234
1140-1149	MP15	L&N 5030-5039
1150-1194	MP15AC	SCL 4000-4019, 4200-4224
1500-1524	GP15T	C&O 1500-1524
1700-1745	GP16	SCL 4600-4645
1746-1800	GP16	SCL 4700-4754
1802-1855	GP16	SCL 4756-4809
1856-1860	GP16	SCL 4975-4979
1888-1899	U18B	SCL 250-261
1900-1990	U18B	SCL 300-392
2001-2049	GP38	B&O 3801-3849
2050-2098	GP38	C&O 3850-3898
2100-2119	GP38	B&O 4800-4819
2120-2129	GP38	C&O 4820-4829
2131-2148	GP38	SBD 6222-6239
2150-2179	GP38	SBD 6241-6270
2180+2189	GP38	SBD 6271+6280
2400-2404	SD40-2	B&O 7700-7704
2420, 2421	SD40	SBD 4509, 4510
2451-2454	SD40-2	SBD 4501-4504; L&N 4501-4504
2500-2559	GP38-2	SCL 500-559
2560-2608	GP38-2	SBD 4050+4099
2609-2650	GP38-2	SBD 4100+4144
2651-2716	GP38-2	SBD 6000-6065

After much experimentation, CSX settled on a diesel livery of gray, dark blue, and yellow. No. 5895 is a B36-7, one of 120 such units on the CSX roster.

CSX Nos.	Model	Previous Nos.
3000-3009	BQ23-7	SCL 5130-5138
3100-3143	B23-7	SCL 5100-5114, 5140-5154, 5500-5516
3209-3229	U23B	C&O 2309-2329
3231+3324	U23B	L&N 2700+2272, 2800+2824
4018, 4024, 4044, 4046	B40-8	NYS&W 4018, 4024, 4044, 4046
4200-4233	GP30d	C&O 3000+3046
4234-4266	GP30d	B&O 6904+6974
4280-4299	GP39	C&O 3900-19
4300+4310	GP39-2	D&H 373+389; D&H 7404+7420; RDG 7404+7420

CSX Nos.	Model	Previous Nos.
4447	GP40-2	B&O 4447
4505-4599	SD35	L&N 4505-4599
4600	SD40	SBD 8300
4601	SD40	SBD 8309
4606	SD40	SBD 8326
5200-5224	MATE	SCL 3200-3224
5500-5516	B30-7	SCL 5500-5516
5517-5580	B30-7	C&O 8235-8298
5700+5805	U36B	SCL 1748+1854
5930-5943, 5946	B40-8	NYS&W 4010+4048

CSX Nos.	Model	Previous Nos.
6001-6063	GP40-2	B&O 4100-4162
6064-6083	GP40-2	C&O 4165-4184
6084-6155	GP40-2	B&O 4185-4256
6156-6160	GP40-2	WM 4257-4261
6201-6210	GP40-2	B&O 4302-4311
6211-6220	GP40-2	WM 6211-6220
6221-6249	GP40-2	B&O 4322-4350
6276-6318	GP40-2	C&O 4378-4420
6388-6390	GP40-2	SBD 6642-6644
6391	GP40-2	SBD 6645
6392	GP40-2	B&O 4163
6400-6435	GP40-2	CSX 6161-6196
6436-6456	GP40-2	CSX 6320-6340
6457	GP40-2	CSX 6199
6458-6460	GP40-2	CSX 6342-6344
6461	GP40-2	CSX 6200
6500-6514	GP40	B&O 3684+3698
6515-6523	GP40	B&O 3740-3747
6524+6550	GP40	B&O 3747+3775
6553+6590	GP40	B&O 3778+4015
6592+	GP40	B&O 4017+4064
6641-6645	GP40	C&O 4066-4070
6647-6652	GP40	A&WP 727-732
6654, 6655	GP40	GARR 752, 753
6657+6788	GP40	SCL 1500+1634

CSX Nos.	Model	Previous Nos.	
6791-6796	GP40	WRA 701-706	
6798+6825	GP40	L&N 3001+3029	
6828+6854	GP40	C&O 4073+4099	
6900-6923	GP40-2	CSX 6290-6319	
6924-6943	GP40-2	CSX 6251-6269, 6250	
6947	GP40-2	CSX 6285	
7000-7015	C30-7	SBD 7000-7015	L&N 7000-7015
7016-7031	C30-7	SBD 7016-7031	SCL 7016-7031
7032-7051	C30-7	SBD 7032-7051	L&N 7032-7051
7052-7061	C30-7	SBD 7052-7061	SCL 7052-7061
7062-7069	C30-7	SBD 7062-7069	L&N 7062-7069
7070-7094	C30-7	SBD 7070-7094	SCL 7070-7094
7200+7276	U30C	L&N 1534-1582, 1470-1498	
7401+7412	GP39-2	D&H 7401+7412, RDG 7401+7412	
8000-8039	SD40-2	L&N 8000-8039	
8040-8066	SD40-2	SCL 8040-8066	
8067-8086	SD40-2	L&N 8067-8086	
8087-8130	SD40-2	SCL 8087-8130	
8116-8126	SD40-2	L&N 8116-8126	
8500-8643	SD50	SBD 8500-8643	
8950-8964	SD45-2	SCL 2045-2059	
8966-8973	SD45-2	CRR 3608-3615	
9565	NW2	B&O 5065	
9575	NW2	C&O 5075	

CUYAHOGA VALLEY RAILWAY

Reporting marks: CUVA   **Track miles:** 14

Address: P. O. Box 6073, Cleveland, OH 44101

The Cuyahoga Valley Railway is a switching line at Cleveland, Ohio.

Nos.	Qty.	Model	Builder	Date	Notes
110, 421, 424, 425	4	SW1001	EMD	1973-1976	
802, 855	2	SW8	EMD	1953, 1952	
960-962	3	SW900	EMD	1954-1956	
1001, 1003, 1050, 1051	4	SW1001	EMD	1966-1979	
1201, 1209, 1212, 1213	4	SW1200	EMD	1954-1957	
1214, 1215	2	SW9	EMD	1953	
1280-1282, 1284, 1286	5	SW1200	EMD	1956-1957	
Total	**24**				

DAKOTA, MINNESOTA & EASTERN RAILROAD

Reporting marks: DME   **Miles:** 965

Address: 337 22nd Ave., Brookings, SD 57006

The Dakota, Minnesota & Eastern is a former Chicago & North Western route from Winona, Minnesota, west to Rapid City, South Dakota, plus branches, also ex-C&NW, to Plainview, Minn., Mason City, Iowa, Comfrey, Minn., Watertown, S.D., Oakes, North Dakota, Mansfield, S.D., and Onida, S.D. DM&E's shops are at Huron, S.D.

Nos.	Qty.	Model	Builder	Date	Notes
544-551, 553-560	16	SD7u	EMD	1952-1954	1
1463, 1471, 1477	3	GP9	EMD	1959	

Nos.	Qty.	Model	Builder	Date	Notes
1483, 1484	2	GP9	EMD	1959	
6359-6364, 6384, 6386, 6387	9	SD40-2	EMD	1974	2
6601, 6602, 6605, 6606, 6609-6618, 6620, 6621	16	SD9	EMD	1952-1954	3
Total	**48**				

Notes:
1. Ex-Milwaukee Road, same numbers; rebuilt to 1750 h.p. by Milwaukee Road 1974-76; classified SD10
2. Ex-Nickel Plate 533, 800, 807, 813, 814
3. Ex-Chicago & North Western, same numbers

DAKOTA, MISSOURI VALLEY & WESTERN RAILROAD

Reporting marks: DMVW   **Miles:** 364

Address: P. O. Box 299, Morton, MN

The Dakota, Missouri Valley & Western operates on two widely separated lines in North Dakota: from Oakes through Wishek and Bismarck to Washburn, and from Flaxton west to Whitetail, Montana. Enginehouses are at Wishek, Bismarck, and Crosby, N.D.

Nos.	Qty.	Model	Builder	Date	Notes
323, 324, 856, 862	4	GP35	GMD	1964-1965	1
Total	**4**				

Notes:
1. 856 and 862 are ex-Chicago & North Western 856 and 862

DAKOTA RAIL INC.

Reporting marks: DAKR   **Miles:** 44

Address: 25 Adams St. North, Hutchinson, MN 55350-2653

Dakota Rail extends from Hutchinson to Wayzata, Minnesota. Engine facilities are at Hutchinson and Spring Park, Minn.

Nos.	Qty.	Model	Builder	Date	Notes
81A, 81C	2	F3Au	EMD	1949	1
1206, 1208	2	SW1200	EMD	1955	2
Total	**4**				

Notes:
1. Ex-Milwaukee Road 81A, 81C; rebuilt to F7 specifications by Milwaukee Road
2. Ex-Milwaukee Road 1206, 1208

DAKOTA SOUTHERN RAILWAY

Reporting marks: DSRC   **Miles:** 187
Address: P. O. Box 436, Chamberlain, SD 57325
The Dakota Southern extends from Mitchell to Kadoka, South Dakota, on a former Milwaukee Road line. The enginehouse is at Chamberlain, S.D.

Nos.	Qty.	Model	Builder	Date	Notes
75	1	70-ton	GE	1947	1
76	1	70-ton	GE	1953	2
103	1	S3	Alco	1952	3
213	1	C420	Alco	1964	4
506	1	SD9	EMD	1958	5
512	1	SD7	EMD	1952	6
522	1	SD7	EMD	1953	7
6925	1	DD40X	EMD	1970	8
Total	**8**				

Notes:
1. Ex-Sisseton Southern, Iowa Terminal, Marianna & Blountstown
2. Ex-Sisseton Southern, Iowa Terminal, Tidewater Southern 743
3. Ex-Sisseton Southern, City of Prineville 103, Oregon & Northwestern 102, Brooks Scanlon
4. Ex-Long Island 213
5. Ex-Wisconsin & Southern, formerly Milwaukee Road 506
6. Ex-Wisconsin & Southern, formerly Milwaukee Road 512
7. Ex-Wisconsin & Southern, formerly Milwaukee Road 522
8. Ex-Union Pacific 6925 (on display at Chamberlain)

D&I RAILROAD

Reporting marks: DAIR   **Miles:** 138
Address: P. O. Box 829, Sioux Falls, SD 57117
The D&I runs from Dell Rapids, South Dakota, through Sioux Falls to Sioux City, Iowa. Engine facilities are at Dell Rapids.

Nos.	Qty.	Model	Builder	Date	Notes
1, 2	2	GP9	EMD	1954	Ex-Union Pacific 148, 262
3	1	GP20	EMD	1960	Ex-Burlington Northern 2019
4-9	6	GP9	EMD	1959	Ex-Milwaukee Road 293, 328, 303, 321, 323, BN 1807
10	1	GP9	EMD	1956	Ex-Burlington Northern 1893
11	1	GP9	EMD	1954	Ex-Burlington Northern 1963
12	1	GP20	EMD	1961	Ex-Burlington Northern 2068
Total	**12**				

DARDANELLE & RUSSELLVILLE RAILROAD

Reporting marks: DR   **Miles:** 5
Address: P. O. Box 150, Dardanelle, AR 72834

The Dardanelle & Russellville operates between Russellville and North Dardanelle, Arkansas. The enginehouse is at North Dardanelle.

Nos.	Qty.	Model	Builder	Date	Notes
15	1	SW1	EMD	1949	
16	1	SW1	EMD	1942	
17	1	SW7	EMD	1950	
18	1	S1	Alco	1948	
19	1	S3	Alco	1950	
Total	**5**				

DAVENPORT, ROCK ISLAND & NORTH WESTERN RAILWAY

Reporting marks: DRI   **Miles:** 48

Address: 102 S. Harrison St., Davenport, IA 52801

The "Dry Line" is a switching and terminal company serving Davenport and Bettendorf, Iowa, and Rock Island, Illinois.

Nos.	Qty.	Model	Builder	Date	Notes
108, 109	2	SW7	EMD	1954	Ex-Burlington Northern
121	1	NW2	EMD	1950	Ex-Burlington Northern
602	1	SW1200	EMD	1954	Ex-Milwaukee Road
Total	**4**				

DELAWARE COAST LINE RAILROAD

Reporting marks: DCLR   **Miles:** 23

Address: 603 New St., Milford, DE 19963

The Delaware Coast Line operates from Ellendale to Milton, Delaware, and from Georgetown to Lewes, Del. The enginehouse is at Georgetown.

Nos.	Qty.	Model	Builder	Date	Notes
2	1	RS36	Alco	1962	1
10	1	25-ton	GE	1963	
17	1	T6	Alco	1958	2
19	1	T6	Alco	1958	3
23	1	RS1	Alco	1954	4
200	1	C420	Alco	1963	5
	1	40-ton	Plymouth	1935	6
Total	**7**				

Notes:
1. Ex-Atlantic & Danville
2. Ex-Eastern Shore 17, Conrail 9844
3. Ex-Maryland & Delaware 19, Pennsylvania 9846
4. Ex-Maryland & Delaware 23, Soo Line 351
5. Ex-Long Island 200
6. Ex-duPont 6

DELRAY CONNECTING RAILROAD

Reporting marks: DC   **Miles:** 16

Address: P. O. Box 32538, Detroit, MI 48232

Delray Connecting is a switching road in Detroit's downriver district.

Nos.	Qty.	Model	Builder	Date	Notes
1, 2	2	SW9	GMD	1956	
202	1	SW1	EMD	1955	
Total	**3**				

DELTA SOUTHERN RAILROAD

Reporting marks: DSRR   **Miles:** 103

Address: P. O. Box 1709, Tallulah, LA 71284-1709

Delta Southern operates two former Missouri Pacific routes: McGehee, Arkansas, to Tallulah, Louisiana, and Huttig, Ark., to Monroe, La. Enginehouses are at Tallulah and Sterlington.

Nos.	Qty.	Model	Builder	Date	Notes
100, 102-105	5	CF7	ATSF	1972-1976	
Total	**5**				

DELTA VALLEY & SOUTHERN RAILWAY

Reporting marks: DVS   **Miles:** 2

Address: 1 Park St., Wilson, AR 72395

Delta Valley & Southern operates between Delpro and Elkins, Arkansas. The enginehouse is at Evadale.

Nos.	Qty.	Model	Builder	Date	Notes
50	1	45-ton	GE	1954	
Total	**1**				

Like its merger partner, Southern Pacific, the Rio Grande favors the SD40T-2 for heavy freight service on mountain routes with many tunnels.

Charles W. McDonald

DENVER & RIO GRANDE WESTERN RAILROAD

Reporting marks: DRGW   **Miles:** 2,392
Address: 650 Davis Road, Salt Lake City, UT 84119

Nos.	Qty.	Model	Builder	Date	Notes
130-139	10	SW1200	EMD	1965	
140-149	10	SW1000	EMD	1966, 1968	
3001-3004, 3006-3008, 3010, 3011, 3013-3024, 3026-3028	24	GP30	EMD	1962-1963	
3029-3050	22	GP35	EMD	1964-1965	
3051-3057, 3059-3062, 3064-3093	41	GP40	EMD	1966-1971	
3094-3107, 3109-3130	36	GP40-2	EMD	1972-1983	

Nos.	Qty.	Model	Builder	Date	Notes
3131-3153	23	GP40	EMD	1968	1
3154-3156	3	GP60	EMD	1990	
5315-5340	26	SD45	EMD	1967-1968	
5341-5401, 5403-5407, 5409-5413	71	SD40T-2	EMD	1974-1980	
5501-5517	17	SD50	EMD	1984	
5571	1	F9A	EMD	1955	
5762, 5763	2	F9B	EMD	1955	
5903, 5904, 5912, 5913, 5922, 5924, 5931, 5934, 5941, 5942, 5951, 5953, 5954	13	GP9	EMD	1955-1956	
Total	**299**				

Notes:
1. Ex-Conrail

DE QUEEN & EASTERN RAILROAD

Reporting marks: DQE   **Miles:** 46
Address: 412 E. Lockesburg Ave., De Queen, AR 71832
The De Queen & Eastern extends from West Line to De Queen and Perkins, Arkansas. It is operated in conjunction with the Texas, Oklahoma & Eastern. Engine facilities are at De Queen and Dierks, Ark.

Nos.	Qty.	Model	Builder	Date	Notes
D-5	1	SW1200	EMD	1960	
D-6	1	GP35	EMD	1964	
D-7	1	GP40	EMD	1966	
Total	3				

Charles W. McDonald

It is relatively unusual to find a high-horsepower, turbocharged unit such as a GP35 on a short line, and all the more so considering that De Queen & Eastern D-6 is not a hand-me-down but was purchased new.

DENVER RAILWAY

Reporting marks: DRWY   **Miles:** 11
Address: 107 Fifth St., Castle Rock, CO 80104
The Denver Railway is a switching road at Denver, Colorado.

Nos.	Qty.	Model	Builder	Date	Notes
966	1	NW2	EMD	1948	Ex-Burlington Northern
988	1	SW1	EMD	1945	Ex-Burlington Northern
999	1	NW5	EMD	1946	Ex-Burlington Northern
Total	3				

DEPEW, LANCASTER & WESTERN RAILROAD

Reporting marks: DLWR   **Miles:** 5
Address: 8364 Lewiston Road, Batavia, NY 14020
The Depew, Lancaster & Western operates between Depew and Lancaster, New York. The enginehouse is at Depew.

Nos.	Qty.	Model	Builder	Date	Notes
3600, 3603, 3604	3	RS11	Alco	1956	1
4085	1	RS3	Alco	1952	2
9085	1	S4	Alco	1956	3
Total	5				

Notes:
1. Ex-Central Vermont, same numbers
2. Ex-Delaware & Hudson
3. Ex-Baltimore & Ohio

DETROIT & MACKINAC RAILWAY

See Lake State Railway, page 92.

DEVCO RAILWAY

Reporting marks: DVR **Miles:** 88
Address: P. O. Box 2500, Sydney, NS B1P 6K9, Canada
The Devco Railway operates between Sydney and Glace Bay, Nova Scotia.

Nos.	Qty.	Model	Builder	Date	Notes
20	1	40	EMD	1940	1
61	1	S1	Alco	1940	2
203, 205	2	RS1	Alco	1944, 1945	3
216-228	13	GP38-2	GMD	1979-1983	
300	1	RS1	Alco	1951	4
Total	**18**				

Notes:
1. Ex-Electro-Motive demonstrator 1134
2. Ex-Chicago & North Western 1201, stored unserviceable
3. Ex-Chicago & North Western 201 and 205, stored unserviceable
4. Ex-Soo Line

Dennis A. Smolinski; collection of J. David Ingles

Out near the eastern tip of Cape Breton Island, Devco Railway 205 is a long way from its first job on the Minneapolis & St. Louis.

DODGE CITY, FORD & BUCKLIN RAILROAD

Reporting marks: FBDC **Miles:** 25
Address: P. O. Box 714, Dodge City, KS 67801
The Dodge City, Ford & Bucklin extends from Dodge City to Bucklin, Kansas. The enginehouse is at Dodge City.

Nos.	Qty.	Model	Builder	Date	Notes
1, 2	2	45-ton	GE	1941, 1945	
1001	1	CF7	ATSF	1977	
1309	1	GP7	EMD	1951	
6601	1	S1	Alco	1941	
Total	**5**				

DULUTH & NORTHEASTERN RAILROAD

Reporting marks: DNE **Miles:** 11
Address: 207 Ave. C, Cloquet, MN 55720
The Duluth & Northeastern operates between Cloquet and Saginaw, Minnesota. The enginehouse is at Cloquet.

Nos.	Qty.	Model	Builder	Date	Notes
31-33	3	SW1	EMD	1940-1941	
35	1	SW1000	EMD	1967	
Total	**4**				

DULUTH, WINNIPEG & PACIFIC RAILWAY

Reporting marks: DWP   **Miles:** 165
Address: P. O. Box 3008, Superior, WI 54880

The Duluth, Winnipeg & Pacific, a Canadian National subsidiary, extends from Itasca, Wisconsin, through Duluth, Minnesota, to a connection with CN at Ranier near International Falls. The road's shops are in Superior, Wisconsin.

Nos.	Qty.	Model	Builder	Date	Notes
5902-5912	11	SD40	EMD	1966	1
Total	**11**				

Notes:
1. Ex-Grand Trunk Western 5902-5912

DUNN-ERWIN RAILWAY

Reporting marks: DER   **Miles:** 6
Address: P. O. Box 917, Aberdeen, NC 28315

The Dunn-Erwin Railway operates from Dunn to Erwin, North Carolina. It is a division of the Aberdeen & Rockfish. The enginehouse is at Dunn.

Nos.	Qty.	Model	Builder	Date	Notes
5072	1	NW2	EMD	1949	Ex-Chesapeake & Ohio 5072
Total	**1**				

DULUTH, MISSABE & IRON RANGE RAILWAY

Reporting marks: DMIR   **Miles:** 401
Address: P. O. Box A, Proctor, MN 55810

The Missabe Road carries iron ore from Minnesota's Mesabi and Vermilion ranges to docks at Duluth and Two Harbors, Minn. The line's shops are at Proctor, a few miles from Duluth.

Nos.	Qty.	Model	Builder	Date	Notes
129, 130, 134, 138, 139, 142, 144, 149, 150, 152, 153, 155-158	15	SD9r	EMD	1957-1958	
159, 161, 164, 166, 168, 170, 171	7	SD9	EMD	1959	
175, 182, 185, 187, 189, 193	6	SD18	EMD	1960	
201-208	8	SD38AC	EMD	1971	
209, 211-213, 215	5	SD38-2	EMD	1975	
301-320	20	SD9r	EMD	1979-1988	
Total	**61**				

Robert C. Del Grosso

The bulk of Duluth, Missabe & Iron Range's road power was and still is SD9s like No. 161. DM&IR purchased 74 such locomotives.

EAST CAMDEN & HIGHLAND RAILROAD

Reporting marks: EACH   **Miles:** 18
Address: P. O. Box 3180, East Camden, AR 71701

The East Camden & Highland operates between Eagle Mills and East Camden, Arkansas. The enginehouse is at East Camden.

Nos.	Qty.	Model	Builder	Date	Notes
60-62	3	NW2	EMD	1941-1949	
Total	3				

EAST COOPER & BERKELEY RAILROAD

Reporting marks: ECBR   **Miles:** 18
Address: 540 E. Bay St., Charleston, SC 29402

The East Cooper & Berkeley operates between State Junction, near Cordesville, South Carolina, and Charity Church. The enginehouse is at Charity Church.

Nos.	Qty.	Model	Builder	Date	Notes
2000, 2001	2	SW1001	EMD	1977	
5105	1	S4	Alco	1953	1
6152, 6155	2	GP9	EMD	1954, 1956	2
Total	5				

Notes:
1. Ex-Chesapeake & Ohio 5105
2. Ex-Chesapeake & Ohio 6152, 6155

EAST ERIE COMMERCIAL RAILROAD

Reporting marks: EEC   **Track miles:** 12
Address: P. O. Box 7305, Erie, PA 16510

The East Erie Commercial Railroad is a switching road at Erie, Pennsylvania. It is owned by General Electric.

Nos.	Qty.	Model	Builder	Date	Notes
18	1	70-ton	GE	1950	
21, 22	2	85-ton	GE	1980	
Total	3				

EAST JERSEY RAILROAD & TERMINAL COMPANY

Reporting marks: EJR   **Miles:** 2
Address: E. 22nd Street, Bayonne, NJ 07002

East Jersey Railroad & Terminal is a terminal road at Bayonne, New Jersey.

Nos.	Qty.	Model	Builder	Date	Notes
18	1	65-ton	GE	1950	
19	1	80-ton	GE	1948	
Total	2				

EAST PORTLAND TRACTION COMPANY

Reporting marks: EPTC   **Miles:** 7
Address: 1952 S.E. Ochoco St., Milwaukie, OR 97222

East Portland Traction is a switching road between East Portland and Milwaukie, Oregon. The enginehouse is at Milwaukie.

Nos.	Qty.	Model	Builder	Date	Notes
25	1	25-ton	GE	1942	
100	1	SW1	EMD	1952	Ex-Portland Traction
5100	1	70-ton	GE	1948	Ex-Southern Pacific 5100
Total	3				

EAST TENNESSEE RAILWAY

Reporting marks: ETRY   **Miles:** 11
Address: P. O. Box 1479, Johnson City, TN 37601

The East Tennessee Railway extends from Johnson City to Elizabethtown, Tennessee. The enginehouse is at Johnson City.

Nos.	Qty.	Model	Builder	Date	Notes
211	1	RS32	Alco	1962	Ex-Southern Pacific
212	1	RS32	Alco	1963	Ex-Maryland & Delaware
Total	2				

EASTERN ALABAMA RAILWAY

Reporting marks: EARY   **Miles:** 41
Address: 6740 Antioch, Suite 265, Merriam, KS 66204

Eastern Alabama Railway has two separate routes: Wellington to Anniston, Alabama, and Gantt's Junction through Sylacauga to Talladega, Ala. The enginehouse is at Sylacauga.

Nos.	Qty.	Model	Builder	Date	Notes
1510	1	GP7	EMD	1951	1
1511, 1550, 1551	3	GP7	EMD	1953	1
Total	4				

Notes:
1. Ex-Kyle Railroad

EASTERN ILLINOIS RAILROAD

Reporting marks: EIRC   **Miles:** 52
Address: P. O. Box 1132, Charleston, IL 61920

Eastern Illinois Railroad extends from Neoga to Metcalf, Illinois, on a former Nickel Plate route. The engine facilities are at Charleston.

Nos.	Qty.	Model	Builder	Date	Notes
7001	1	CF7	ATSF	1974	
Total	1				

EASTERN SHORE RAILROAD INC.

Reporting marks: ESHR   **Miles:** 70
Address: P. O. Box 312, Cape Charles, VA 23310-0312

The Eastern Shore extends from Pocomoke City, Maryland, to Cape Charles, Virginia, with a carfloat connection to Norfolk. The enginehouses are at Cape Charles and Little Creek, Va.

Nos.	Qty.	Model	Builder	Date	Notes
1600, 1603	2	GP8	ICG	1974	
8066, 8096	2	GP10	ICG	1974, 1977	
Total	4				

EL DORADO & WESSON RAILWAY

Reporting marks: EDW   **Miles:** 6
Address: P. O. Box 46, El Dorado, AR 71731

The El Dorado & Wesson operates between El Dorado and Newell, Arkansas. Engine facilities are at El Dorado.

Nos.	Qty.	Model	Builder	Date	Notes
20-24	5	S2	Alco	1952-1957	Held for disposal
25, 26	2	SW7	EMD	1950, 1951	
Total	7				

ELGIN, JOLIET & EASTERN RAILWAY

Reporting marks: EJE   **Miles:** 200
Address: P. O. Box 899, Gary, IN 46401

The "J" describes an arc around Chicago from Gary, Indiana, to Waukegan, Illinois. The road's shops are at Joliet, Illinois.

Nos.	Qty.	Model	Builder	Date	Notes
300-307	8	SW1200	EMD	1960	
308	1	SW9	EMD	1953	
309	1	NW2	EMD	1947	
310-323	14	SW1200	EMD	1964, 1966	
444, 445	2	SW1001	EMD	1971	
446, 451-453, 455-457	7	NW2	EMD	1939-1952	
459	1	SW1000	EMD	1971	
607, 609, 610, 612, 613	5	SD9	EMD	1957-1959	1
650-655	6	SD38	EMD	1970	
656, 657, 659-669	13	SD38-2	EMD	1974-1975	
Total	58				

Notes:
1. Ex-Duluth, Missabe & Iron Range

Escanaba & Lake Superior 101 wears an orange-and-green color scheme copied from the Great Northern.

Scott Hartley

ESCANABA & LAKE SUPERIOR RAILROAD

Reporting marks: ELS   **Miles:** 342
Address: 1 Larkin Plaza, Wells, MI 49894

The Escanaba & Lake Superior extends from Wells, Michigan, northwest to Ontonagon, and from Green Bay, Wisconsin, north to Republic, Mich. The two routes cross at Channing, Mich. The enginehouse is at Wells.

Nos.	Qty.	Model	Builder	Date	Notes
101	1	DS44-660	BLW	1947	
201, 202	2	DRS44-1000	BLW	1948	1
210, 212, 213	3	RS12	BLW	1952-1953	2
300	1	RS12	BLW	1953	2
400-402	3	GP38	EMD	1969	3
1200, 1201	2	SW8	EMD	1952	4
1220-1224	5	SD9	EMD	1955	4
Total	**12**				

Notes:
1. Ex-Calumet & Hecla
2. Ex-Seaboard Air Line
3. Ex-Conrail
4. Ex-Reserve Mining

ELLIS & EASTERN COMPANY

Reporting marks: EW   **Miles:** 17
Address: 1201 W. Russell, Sioux Falls, SD 57118
The Ellis & Eastern operates from Ellis through Sioux Falls to Brandon, South Dakota, on ex-Chicago & North Western track. The enginehouse is at Sioux Falls.

Nos.	Qty.	Model	Builder	Date	Notes
1509	1	SW1200	EMD	1957	
6718	1	SW900	GMD	1955	Ex-Canadian Pacific
Total	2				

ESSEX TERMINAL RAILWAY

Reporting marks: ETL   **Track miles:** 74
Address: 1601 Lincoln Road, Windsor, ON N8Y 4R8, Canada
The Essex Terminal is a switching line at Windsor, Ontario.

Nos.	Qty.	Model	Builder	Date	Notes
102	1	GP9	GMD	1963	1
104	1	SW8	GMD	1954	
105	1	SW1200	GMD	1956	
107	1	SW1500	EMD	1971	2
108	1	GP9	GMD	1960	3
Total	5				

Notes:
1. Ex-Algoma Central 172, last GP9 built
2. Ex-Electro-Motive plant switcher 113
3. Ex-Cartier Railway 59

EUREKA SOUTHERN RAILROAD

Reporting marks: EUKA   **Miles:** 171
Address: 130 A St., Eureka, CA 95502
The Eureka Southern extends north from Willits, California, to Eureka, with branches to Carlotta, Fairhaven, and Korblex. Engine facilities are at Eureka.

Nos.	Qty.	Model	Builder	Date	Notes
30, 31	2	GP38	EMD	1967	1

Nos.	Qty.	Model	Builder	Date	Notes
32, 33	2	GP38	EMD	1969	2
70	1	GP7	EMD	1953	
Total	5				

Notes:
1. Ex-Conrail 7667, 7668; formerly Pennsylvania-Reading Seashore Lines 2007, 2008
2. Ex-Conrail 7820, 7821; formerly Penn Central 7820, 7821

EVERETT RAILROAD

Reporting marks: EV   **Miles:** 15
Address: P. O. Box 361, Claysburg, PA 16625
The Everett Railroad operates from Brookes Mills to Sproul, Pennsylvania, and from Roaring Spring to Martinsburg and Curryville, Pa. Engine facilities are at Claysburg.

Nos.	Qty.	Model	Builder	Date	Notes
4	1	80-ton	GE	1943	Ex-U.S. Navy
8933	1	SW9	EMD	1951	Ex-Conrail, Lehigh Valley 282
8990	1	SW9	EMD	1953	Ex-Conrail, New York Central
Total	3				

FARMRAIL CORPORATION

Reporting marks: FRMC   **Miles:** 85
Address: 136 E. Frisco Ave., Clinton, OK 73601
Farmrail operates from Elk City to Clinton and Weatherford, Oklahoma, and provides contract haulage to Erick. The line was once part of the Choctaw Route of the Rock Island. The enginehouse is at Clinton.

Nos.	Qty.	Model	Builder	Date	Notes
280, 286, 297, 316, 317, 331	6	GP9	EMD	1954-1959	1
1555	1	GP7	EMD	1953	2
Total	7				

Notes:
1. Ex- Milwaukee Road
2. Ex-Burlington Northern

FLORIDA CENTRAL RAILROAD

Reporting marks: FCEN   **Miles:** 60
Address: P. O. Box 967, Plymouth, FL 32768
The Florida Central extends from Orlando to Umatilla, Florida, with branches to Sorrento and Forest City. The enginehouse is at Plymouth, Fla. The railroad is a subsidiary of Pinsly Railroad Co.

Nos.	Qty.	Model	Builder	Date	Notes
47, 49, 53	3	CF7	ATSF	1974-1977	1
55, 57	2	GP7r	EMD	1952	2
207	1	GP35	EMD	1963	3
Total	6				

Notes:
1. Ex-Santa Fe 2474, 2494, 2503
2. Ex-Santa Fe 2155, 2157, rebuilt 1979
3. Ex-Norfolk & Western 207

FLORIDA MIDLAND RAILROAD

Reporting marks: FMID   **Miles:** 40
Address: P. O. Box 967, Plymouth, FL 32768
The Florida Midland has three separate lines: Wildwood to Leesburg, Florida, West Lake Wales to Frostproof, and Winter Haven to Gordonville. The engine facilities are at Wildwood, West Lake Wales, and Winter Haven. The railroad is a subsidiary of Pinsly Railroad Co.

Nos.	Qty.	Model	Builder	Date	Notes
64	1	CF7	ATSF	1973	Ex-Santa Fe 2643
103	1	S2	Alco	1949	Ex-Frankfort & Cincinnati
107	1	S4	Alco	1951	Stored serviceable
2637	1	CF7	ATSF	1973	Ex-Santa Fe 2637
Total	4				

FLORIDA EAST COAST RAILWAY

Reporting marks: FEC   **Miles:** 541
Address: 1 Malaga St., St. Augustine, FL 32085
The main line of the Florida East Coast runs from Jacksonville south to Miami. In addition there are a few short branches. The road's locomotive shops are at New Smyrna Beach.

Nos.	Qty.	Model	Builder	Date	Notes
221, 226	2	SW9	EMD	1952	
229, 233	2	SW1200	EMD	1954	
401, 403-410	9	GP40	EMD	1971	
411-422, 424-429, 433, 434	20	GP40-2	EMD	1972-1986	1
435, 436	2	GP40u	EMD	1967, 1968	2
437	1	GP40-2r	EMD	1979	3
438, 439	2	GP40u	EMD	1968, 1966	4
501-511	11	GP38-2	EMD	1977-1978	
651, 654-676	24	GP9	EMD	1954, 1957	5
Total	73				

Notes:
1. 434 was the last GP40-2 built
2. Ex-Burlington Northern 3003 and 3031, upgraded to Dash 2 specifications by NRE in 1987 and 1988
3. Rebuilt from wrecked unit 423 by Republic Locomotive Works
4. Ex-Burlington Northern 3036 and Milwaukee Road 2006, upgraded to Dash 2 specifications by NRE in 1988
5. Some stored, possibly for disposal

FLORIDA NORTHERN RAILROAD

Reporting marks: FNOR   **Miles:** 27
Address: P. O. Box 967, Plymouth, FL 32768

The Florida Northern extends from Candler to Lowell, Florida. The enginehouse is at Ocala. The railroad is a subsidiary of Pinsly Railroad Co.

Nos.	Qty.	Model	Builder	Date	Notes
50	1	CF7	ATSF	1974	Ex-Santa Fe 2508
Total	**1**				

FLORIDA WEST COAST RAILROAD

Reporting marks: FWCR   **Miles:** 44
Address: P. O. Box 1267, Trenton, FL 32693

The Florida West Coast extends from Newberry to Shamrock and Chiefland, Florida. The enginehouse is at Trenton.

Nos.	Qty.	Model	Builder	Date	Notes
669	1	GP9	EMD	1959	
1337, 1353	2	GP18	EMD	1960, 1961	1
Total	**3**				

Notes:
1. Ex-Rock Island 1337, 1353

FLOYDADA & PLAINVIEW RAILROAD

Reporting marks: FAPR   **Miles:** 27
Address: 211 S. 6th St., Brownfield, TX 79316

The Floydada & Plainview operates between the Texas towns of its name on a former Santa Fe branch. The enginehouse is at Plainview.

Nos.	Qty.	Model	Builder	Date	Notes
91	1	GP7	EMD	1950	Ex-Missouri-Kansas-Texas 91
Total	**1**				

Louis A. Marre

Green-and-white Fordyce & Princeton 1805 is a former Illinois Central unit. Only 16 GP28s were built for service in the U. S., 12 of those for Illinois Central.

FORDYCE & PRINCETON RAILROAD

Reporting marks: FP   **Miles:** 57
Address: P. O. Box 757, Crossett, AR 71642

The Fordyce & Princeton extends from Fordyce to Whitlow Junction, Arkansas, then to Crossett by trackage rights on the affiliated Ashley, Drew & Northern. The enginehouse is at Fordyce.

Nos.	Qty.	Model	Builder	Date	Notes
1503	1	SW1500	EMD	1970	1
1504	1	SW1500	EMD	1966	2
1805	1	GP28	EMD	1964	3
Total	**3**				

Notes:
1. Ex-Ashley, Drew & Northern 1509
2. Ex-Rock Island 941
3. Ex-Illinois Central Gulf 9436

FORE RIVER RAILWAY

Reporting marks: FRY   **Miles:** 3
Address: 145 E. Howard St., Quincy, MA 02169
The Fore River operates between East Braintree and Quincy, Massachusetts. Engine facilities are at Quincy.

Nos.	Qty.	Model	Builder	Date	Notes
17	1	S6	Alco	1957	Ex-Southern Pacific
18	1	S4	Alco	1950	Ex-Portland Terminal, stored unserviceable
Total	**2**				

FORT SMITH RAILROAD

Reporting marks: FSR   **Miles:** 47
Address: 101 N. 10th St., Fort Smith, AR 72901
Fort Smith Railroad operates a former Missouri Pacific branch from Paris, Arkansas, west 47 miles to Fort Smith, where the enginehouse is located.

Nos.	Qty.	Model	Builder	Date	Notes
1791	1	GP9	EMD	1957	Ex-Burlington Northern
1902	1	GP9	EMD	1956	Ex-Burlington Northern
7802	1	RS3m	EMD	1955	Ex-Lamoille Valley
Total	**3**				

FOX RIVER VALLEY RAILROAD

Reporting marks: FRVR   **Miles:** 208
Address: 200 Dousman St., Green Bay, WI 54306
The Fox River Valley extends from Green Bay, Wisconsin, south through Appleton to Granville, then by trackage rights on Chicago & North Western to Butler; from Green Bay to Cleveland, Wis.; from Manitowoc to Two Rivers; and from New London to Kaukauna. The lines are all former C&NW track. Fox River Valley's enginehouse is at Green Bay. Purchase of the Fox River Valley by the Wisconsin Central is in process.

Nos.	Qty.	Model	Builder	Date	Notes
814, 815, 820	3	GP30	EMD	1963	1
831, 832	2	GP35	EMD	1964-1965	1
2401, 2402	2	SD24	EMD	1959	2
2500	1	SD35	EMD	1965	3

Nos	Qty	Model	Builder	Date	Notes
4119, 4133, 4146, 4151, 4159, 4310, 4320, 4327, 4329, 4330	10	GP7r	EMD	1950-1951	1
4501-4506, 4508, 4509, 4514	9	GP9u	EMD	1953	1
4929	1	GP9	EMD	1957	4
6052	1	GP9	EMD	1955	5
Total	**29**				

Notes:
1. Ex-Chicago & North Western, same numbers
2. Ex-Maryland Midland; previously Burlington Northern
3. Ex-Southern 3024
4. Ex-Central Vermont, same number
5. Ex-Grand Trunk Western, previously Detroit, Toledo & Ironton 952

FORT WORTH & DALLAS RAILROAD

Reporting marks: FWD   **Miles:** 1
Address: P. O. Box 122269, Fort Worth, TX 76121
The Fort Worth & Dallas is a switching line at Fort Worth, Texas.

Nos.	Qty.	Model	Builder	Date	Notes
107	1	70-ton	GE	1948	
Total	1				

GALVESTON RAILWAY

Reporting marks: GVSR   **Track miles:** 38
Address: P. O. Box 5985, Dothan, AL 36302
Galveston Railway is a switching railroad at Galveston, Texas.

Nos.	Qty.	Model	Builder	Date	Notes
301-305	5	SW1001	EMD	1975-1980	
Total	5				

GARDEN CITY WESTERN RAILWAY

Reporting marks: GCW   **Miles:** 14
Address: P. O. Box 838, Garden City, KS 67846
Garden City Western operates from Garden City to Wolf, Kansas. The enginehouse is at Garden City. The company provides locomotives for the affiliated Garden City Northern, which extends from Garden City to Shallowater, 31 miles.

Nos.	Qty.	Model	Builder	Date	Notes
200	1	GP9	EMD	1957	Ex-Burlington Northern
201	1	SW1	EMD	1952	
202	1	SW1	EMD	1949	Ex-Chicago Short Line
Total	3				

GATEWAY WESTERN RAILWAY

Reporting marks: GWWR   **Miles:** 402
Address: 15 Executive Drive, Fairview Heights, IL 62208

Gateway Western extends from Springfield, Illinois, to Kansas City, Missouri, with branches to East St. Louis, Ill., Jacksonville, Ill., and Fulton, Mo. The lines are ex-Illinois Central Gulf (earlier Gulf, Mobile & Ohio, and before that Alton).

Nos.	Qty.	Model	Builder	Date	Notes
1500-1504, 1506	6	SW1500	EMD	1973	1
2020-2034, 2036-2047	27	GP38	EMD	1969	2
2048	1	GP38AC	EMD	1970	3
3008, 3019, 3020	3	GP40	EMD	1971	4
Total	37				

Notes:
1. Ex-Pittsburgh & Lake Erie
2. Ex-Conrail; 2020, 2025, 2030, 2033, 2038, 2042, 2045 held for disposal
3. Ex-Illinois Central Gulf
4. Ex-Union Pacific, earlier Missouri Pacific and Western Pacific

GENESEE & WYOMING RAILROAD

Reporting marks: GNWR   **Miles:** 47
Address: 3546 Retsof Road, Retsof, NY 14539
The Genesee & Wyoming operates from Groveland to Rochester, New York, and from G&W Junction to Silver Springs, N.Y. It is affiliated with the Rochester & Southern and the Buffalo & Pittsburgh. The enginehouse is at Retsof, N.Y.

Nos.	Qty.	Model	Builder	Date	Notes
45, 46	2	MP15DC	EMD	1980	
47	1	SW1500	EMD	1969	1
50, 51	2	GP38	EMD	1970	2
62, 63	2	C424M	Alco	1964	3
Total	7				

Notes:
1. Ex-Houston Belt & Terminal 50
2. Ex-Conrail 7844, 7862
3. Ex-Erie Lackawanna 2479, 2482

GEORGETOWN RAILROAD

Reporting marks: GRR   **Miles:** 8

Address: P. O. Box 529, Georgetown, TX 78627-0529

The Georgetown Railroad operates between Kerr and Georgetown, Texas. The enginehouse is at Feld.

Nos.	Qty.	Model	Builder	Date	Notes
1005, 1007	2	S12	BLW	1952	
1009	1	VO1000	BLW	1946	
1010	1	SW1500	EMD	1971	

Nos.	Qty.	Model	Builder	Date	Notes
1011, 1012	2	MP15DC	EMD	1978, 1980	
9014, 9015	2	GP9	EMD	1954	1
9050-9057	8	GP20	EMD	1960	
S1	1	Slug	CRIP	1980	2
Total	**17**				

Notes:
1. Ex-Burlington Northern
2. Built by Rock Island's Silvis shops

GEORGIA & ALABAMA

See Georgia & Alabama Division, South Carolina Central Railroad, page 144.

GEORGIA CENTRAL RAILWAY

Reporting marks: GC   **Miles:** 232

Address: P. O. Box 466, Vidalia, GA 30474

The Georgia Central extends from Savannah to Rhine, Georgia, with a branch from Vidalia to Macon. Engine facilities are at Vidalia.

Nos.	Qty.	Model	Builder	Date	Notes
1001-1016	16	U30B	GE	1966-1974	
Total	**16**				

GEORGIA GREAT SOUTHERN

See South Carolina Central Railroad, Georgia Great Southern Division, page 145.

GEORGIA NORTHEASTERN RAILROAD

Reporting marks: GNRR   **Miles:** 66

Address: P. O. Box 387, Pulaski, TN 38478

The Georgia Northeastern operates between Elizabeth and Ellijay, Georgia, on a former Louisville & Nashville line. The engine facilities are at Marietta.

Nos.	Qty.	Model	Builder	Date	Notes
6516, 6576, 6585	3	GP9	EMD	1957-1958	1
8704, 8705	2	GP18	EMD	1960	2
Total	**5**				

Notes:
1. Ex-Baltimore & Ohio
2. Ex-Nickel Plate

GEORGIA SOUTHWESTERN

See South Carolina Central Railroad, Georgia Southwestern Division, page 145.

GEORGIA WOODLANDS RAILROAD

Reporting marks: GWRC   **Miles:** 17
Address: P. O. Box 549, Washington, GA 30673-0549
The Georgia Woodlands Railroad extends from Barnett to Washington, Georgia. Enginehouses are located at Washington and Barnett.

Nos.	Qty.	Model	Builder	Date	Notes
45	1	SW8	EMD	1952	1
6584	1	GP9	EMD	1958	2
Total	2				

Notes:
1. Ex-Chicago, West Pullman & Southern
2. Ex-CSX

GETTYSBURG RAILROAD

Reporting marks: GETY   **Miles:** 23
Address: Box 4745, Gettysburg, PA 17325
The Gettysburg Railroad operates between Mount Holly Springs and Gettysburg, Pennsylvania. The enginehouse is at Gettysburg.

Nos.	Qty.	Model	Builder	Date	Notes
14	1	GP9	EMD	1957	
70	1	RS36	Alco	1962	
Total	2				

GLOSTER SOUTHERN RAILROAD

Reporting marks: GLSR   **Miles:** 35
Address: Box 757, Crossett, AR 71635
The Gloster Southern operates two lines: from Gloster, Mississippi, south to Slaughter, Louisiana, and from Silver Creek to Columbia, Miss. Enginehouses are at Gloster and Columbia. The Gloster Southern is owned by the Georgia Pacific Corporation, which also owns the Ashley, Drew & Northern and the Fordyce & Princeton.

Nos.	Qty.	Model	Builder	Date	Notes
903	1	SW900	EMD	1959	Ex-Rock Island
1501, 1502	2	CF7	ATSF	1974, 1972	
Total	3				

GOLDEN TRIANGLE RAILROAD

Reporting marks: GTRA   **Miles:** 9
Address: P. O. Box 2210, Columbus, MS 39701
The Golden Triangle Railroad extends from Columbus to Trinity, Mississippi. The enginehouse is at Trinity.

Nos.	Qty.	Model	Builder	Date	Notes
1	1	MP15AC	EMD	1980	
810, 819	2	GP38-2	EMD	1981	
Total	3				

The Golden Triangle Railroad was built in 1981 and 1982 to serve a Weyerhaeuser paper mill. The first locomotive on the roster was MP15AC G-1, now numbered 1.

David Hurt

GOVERNMENT OF ONTARIO TRANSIT

Address: 1120 Finch Ave. West, Toronto, ON M3J 3J8, Canada
GO Transit operates rail commuter service in and around Toronto on 264 miles of Canadian National and CP Rail routes.

Nos.	Qty.	Model	Builder	Date	Notes
LOCOMOTIVES					
520-561	42	F59PH	GMD	1988-1990	
700-710	11	GP40-2W	GMD	1973-1975	
720-726	7	GP40M	EMD	1967	1
AUXILIARY POWER UNITS					
800-802	3	F7Bm	EMD	1951	2
900-908	9	FP7m	GMD	1952-1953	3
910, 911	2	F7Am	EMD	1952	4
Total	80				

Notes:
1. Ex-Rock Island 3005, 3002, 3003, 3001, 3006, 3004
2. Ex-Burlington Northern 717, 733, 737; converted 1982
3. Ex-Ontario Northland
4. Ex-Milwaukee Road 63A, 104A

GRAFTON & UPTON RAILROAD

Reporting marks: GU   **Miles:** 15
Address: 5 Depot St., Hopedale, MA 01747
The Grafton & Upton extends from Milford to North Grafton, Massachusetts. Enginehouses are at North Grafton and Hopedale.

Nos.	Qty.	Model	Builder	Date	Notes
9	1	44-ton	GE	1946	
1001	1	S4	Alco	1952	
Total	2				

GRAINBELT CORPORATION

Reporting marks: GNBC   **Miles:** 186
Address: P. O. Box 1750, Clinton, OK 73601
Grainbelt operates from Enid to Davidson, Oklahoma, on a former Frisco route. The enginehouse is at Clinton.

Nos.	Qty.	Model	Builder	Date	Notes
3648, 3871, 6083	3	GP9	EMD	1955-1959	1
Total	3				

Notes:
1. Ex-Southern Pacific 3648 and 3871 and Chesapeake & Ohio 6083

GRAYSONIA, NASHVILLE & ASHDOWN RAILROAD

Reporting marks: GNA   **Miles:** 32
Address: P. O. Box 588, Nashville, AR 71852
The Graysonia, Nashville & Ashdown operates from Nashville, Arkansas, to Ashdown. The enginehouse is at Nashville, Ark.

Nos.	Qty.	Model	Builder	Date	Notes
80	1	MP15DC	EMD	1980	
9605	1	SW1500	EMD	1960	
Total	2				

GREAT RIVER RAILROAD

Reporting marks: GTR   **Miles:** 32
Address: P. O. Box 460, Rosedale, MS 38769
The Great River Railroad extends from Great River Junction, Mississippi, to Rosedale. The enginehouse is at Rosedale.

Nos.	Qty.	Model	Builder	Date	Notes
2	1	S2	Alco	1943	Ex-U. S. Army
8341	1	S1	Alco	1941	Ex-U. S. Army
Total	2				

GRAND TRUNK WESTERN RAILROAD

Reporting marks: GTW   **Miles:** 935
Address: 409 W. Jameson St., Battle Creek, MI 49016

Grand Trunk Western's main line extends from Port Huron, Michigan, to Chicago, with branches from Port Huron and Durand to Detroit. A secondary main line, the former Detroit, Toledo & Ironton, extends from Detroit to Cincinnati. GTW's principal shops are at Battle Creek, Mich. GTW is a subsidiary of Canadian National Railways.

Nos.	Qty.	Model	Builder	Date	Notes
1000-1003	4	CS9	GTW	1979-1980	1
1502	1	SW1200	EMD	1955	
1512-1519	8	SW1200	EMD	1960	
4135-4139, 4427, 4428, 4432-4435, 4437-4439, 4444, 4446, 4539, 4540, 4543, 4554	20	GP9	EMD	1954-1958	
4600-4618	19	GP9r	EMD	1954-1957	2
4700-4707	8	GP18	EMD	1960	
4901-4903, 4907, 4909, 4910, 4917-4922, 4930-4933	16	GP9	EMD	1954-1957	
5700-5727	28	GP38-2	EMD	1972	2, 3
5800-5811	12	GP38AC	EMD	1971	
5812-5836, 5844-5861	43	GP38-2	EMD	1977-1978	4
5900, 5901, 5913-5929	19	SD40	EMD	1969-1970	
5930-5937	8	SD40-2	EMD	1975	5
6046, 6049	2	GP7	EMD	1952, 1955	6
6200-6206	7	GP38	EMD	1966, 1969	7
6207-6220	14	GP38AC	EMD	1970-1971	7
6221-6228	8	GP38-2	EMD	1975	7
6250-6254	5	SD38DC	EMD	1969, 1971	7
6400-6405	6	GP40	EMD	1968	7
6406-6425	20	GP40-2	EMD	1972-1979	7
7010	1	SW9	EMD	1956	
7011, 7012	2	SW9	EMD	1952	

Charles W. McDonald

Grand Trunk Western has two groups of GP38ACs on its roster. Units 6207-6220 are former Detroit, Toledo & Ironton engines; 5800-5811 were purchased by GTW.

Nos.	Qty.	Model	Builder	Date	Notes
7017, 7019	2	SW1200	EMD	1955	
7262-7268	7	SW900	EMD	1958	
Total	**260**				

Notes:
1. Rebuilt from Alco S4, 900 h.p. Caterpillar engine; stored
2. Rebuilt by GTW
3. Ex-Missouri Pacific
4. 5844-5849 are ex Pittsburgh & Lake Erie; 5850-5861 are ex-Rock Island
5. Ex-Union Pacific
6. Ex-Detroit & Toledo Shore Line
7. Ex-Detroit, Toledo & Ironton

GREAT SMOKY MOUNTAINS RAILWAY

Reporting marks: GSMR   **Miles:** 67
Address: 1 Maple St., Sylva, NC 28779

The Great Smoky Mountains Railway operates freight and excursion service on a former Southern Railway branch from Dillsboro to Murphy, North Carolina. The enginehouse is at Dillsboro.

Nos.	Qty.	Model	Builder	Date	Notes
200	1	GP9	EMD	1954	
711, 777	2	GP7	EMD	1954	
993	1	CF7	ATSF	1974	
Total	**4**				

GREAT WALTON RAILROAD

Reporting marks: GRWR   **Miles:** 37
Address: Box 711, Monroe, GA 30655

The Great Walton Railroad operates from Social Circle to Monroe, Georgia, on former Georgia Railroad track, and from Machen to Covington, Ga., on former Central of Georgia track. Enginehouses are at Social Circle and Covington.

GREATER WINNIPEG WATER DISTRICT RAILWAY

Reporting marks: GWWD   **Miles:** 97
Address: 1500 Plessis Road, Winnipeg, MB R2C 2Z9, Canada

The Great Winnipeg Water District Railway extends from Winnipeg east to Waugh, Manitoba, on Shoal Lake just west of the Ontario border.

Nos.	Qty.	Model	Builder	Date	Notes
100	1	44-ton	GE	1946	1
101	1	44-ton	GE	1949	1

Nos.	Qty.	Model	Builder	Date	Notes
5093	1	SW9	EMD	1953	1
6243, 6400, 6525	3	GP9	EMD	1955-1957	2
Total	**4**				

Notes:
1. Ex-Chesapeake & Ohio 5093
2. Ex-Chesapeake & Ohio 6243 and Baltimore & Ohio 6400 and 6525

GREAT WESTERN RAILWAY

Reporting marks: GWR   **Miles:** 111
Address: P. O. Box 537, Loveland, CO 80539

The Great Western operates a network of lines between Loveland and Longmont, Colorado, and a former Southern Pacific line from Alturas, California, to Lakeview, Oregon. The enginehouse for the Colorado lines is at Loveland.

Nos.	Qty.	Model	Builder	Date	Notes
211, 296, 319	3	GP9	EMD	1959	Ex-Union Pacific
711-714	4	GP9	EMD	1957	
1500, 1017, 1021	3	GP7	EMD	1954	Ex Burlington Northern
Total	**10**				

Nos.	Qty.	Model	Builder	Date	Notes
102	1	44-ton	GE	1945	1, 2
103	1	44-ton	GE	1947	3
200, 202	2	RS23	MLW	1960	4
501, 503	2	S13	MLW	1959	5
Total	**8**				

Notes:
1. Stored serviceable
2. Ex-Swift Packing Co.
3. Ex-Canadian National 1
4. Ex-Devco 200, 202
5. Ex-BC Rail 501, 503

Green Bay & Western 323, an Alco C420, is a former Lehigh & Hudson River unit.

Stanley H. Mailer

GREEN BAY & WESTERN RAILROAD

Reporting marks: GBW   **Miles:** 254
Address: P. O. Box 2507, Green Bay, WI 54306
The Green Bay & Western operates from Kewaunee, Wisconsin, west through Green Bay to East Winona, Wis. The road's principal shops are at Green Bay. Purchase of the Green Bay & Western by Wisconsin Central is in process.

Nos.	Qty.	Model	Builder	Date	Notes
305-308	4	RS3u	Alco	1951, 1955	1
309	1	RS11	Alco	1956	
311-314	4	C424	Alco	1963-1965	
316, 318	2	RS27	Alco	1962	2
319-322	4	C424	Alco	1963	3
323	1	C420	Alco	1966	4
Total	16				

Notes:
1. Rebuilt by GB&W to 2000 h.p., classified RS20
2. Ex-Chicago & North Western
3. Ex Pennsylvania, Erie Lackawanna, Erie Lackawanna, Reading
4. Ex-Lehigh & Hudson River

GREEN MOUNTAIN RAILROAD

Reporting marks: GMRC   **Miles:** 50
Address: P. O. Box 498, Bellows Falls, VT 05101

Green Mountain extends from Bellows Falls, Vermont, northwest to Rutland. The enginehouse is at North Walpole, New Hampshire, across the Connecticut River from Bellows Falls.

Nos.	Qty.	Model	Builder	Date	Notes
400	1	RS1	Alco	1948	1
405	1	RS1	Alco	1951	2
1848	1	GP9	EMD	1954	3
1849	1	GP9	EMD	1955	4
1850	1	GP9	EMD	1956	5
4261	1	E8A	EMD	1952	6
Total	6				

Notes:
1. Ex-Illinois Central Gulf 1053, Gulf, Mobile & Ohio 1053, Illinois Terminal 753
2. Ex-Rutland 405
3. Ex-Bangor & Aroostook 76
4. Ex-Burlington Northern 1849, Northern Pacific 228
5. Ex-Chesapeake & Ohio 6181
6. Ex-Massachusetts Bay Transportation Authority 4261, Penn Central 4261, Pennsylvania 5761A

GREENVILLE & NORTHERN RAILWAY

Reporting marks: GRN   **Miles:** 10
Address: P. O. Box 2165, Greenville, SC 29602

The Greenville & Northern extends from Greenville to Travelers Rest, South Carolina. The enginehouse is at Greenville.

Nos.	Qty.	Model	Builder	Date	Notes
70, 75	2	70-ton	GE	1948, 1951	
Total	2				

GUILFORD TRANSPORTATION INDUSTRIES

Reporting marks: GTI   **Miles:** 2,316
Address: Iron Horse Park, North Billerica, MA 01862

Through its Springfield Terminal subsidiary, Guilford operates the Boston & Maine and Maine Central railroads.

Road	Nos.	Qty.	Model	Builder	Date	Notes
ST	1	1	44-ton	GE	1942	1
ST	10-20	11	GP7	EMD	1950-1953	2
ST	21	1	GP9	GMD	1963	3
ST	40, 42	2	GP18	EMD	1961	4
B&M	100	1	Slug	EMD	1983	5
MEC	251	1	GP38	EMD	1966	
B&M	252, 254, 256, 258	4	GP38	EMD	1966	
MEC	259	1	GP38	EMD	1967	
B&M	260+262	2	GP38	EMD	1967	
MEC	282+289	4	U23B	GE	1969	6
B&M	300-317	10	GP40 2	EMD	1977	
B&M	321+342	12	GP40	EMD	1968	7
MEC	400-409	10	U18B	GE	1975	
MEC	450-455	6	C424m	Alco	1963	8
B&M	470	1	GP9r	EMD	1950	9
ST	614	1	SD35	EMD	1965	10
ST	616+646	11	SD24u	EMD	1959	11
B&M	663	1	U30C	GE	1972	12
ST	676+689	9	SD45	EMD	1967-1968	13
B&M	690-692	3	SD39	EMD	1969	14
MEC	693	1	SD39	EMD	1969	15
PTM	1101	1	S1	Alco	1950	16
ST	1400, 1407	2	SW1	EMD	1953	17
ST	1411	1	SW8	EMD	1953	18
ST	1424	1	SW9	EMD	1953	19
Total		91				

Notes:
1. Ex-Sabine River & Northern 141; earlier, San Francisco & Napa Valley 40
2. Ex-MEC 569-593
3. Ex-Algoma Central
4. Ex-B&M 1751, 1753
5. Rebuilt by B&M from Union Pacific GP9 134
6. Ex-Delaware & Hudson
7. Ex-Conrail, Delaware & Hudson
8. Ex-Conrail
9. Rebuilt 1982
10. Ex-Western Maryland 7534
11. Ex-Santa Fe, classed SD26
12. Ex-Detroit Edison 12
13. Ex-Norfolk Southern
14. Ex-Norfolk & Western 2961-2963; Illinois Terminal 2301-2303
15. Ex-Norfolk & Western 2966, Illinois Terminal 2306
16. Stored
17. Ex-B&M 1118, 1129
18. Ex-B&M 802
19. Ex-B&M 1229

Charles W. McDonald

Maine Central U18B No. 406 rides on Blomberg trucks from an EMD trade-in.

HAMPTON & BRANCHVILLE RAILROAD

Reporting marks: HB **Miles:** 45
Address: P. O. Box 56, Hampton, SC 29924
The Hampton & Branchville operates between Hampton and Canadys, South Carolina. The enginehouse is at Miley.

Nos.	Qty.	Model	Builder	Date	Notes
70	1	70-ton	GE	1951	

Nos.	Qty.	Model	Builder	Date	Notes
120	1	SW1000	EMD	1967	
667, 686, 859, 906	4	GP9	EMD	1959	1
5943, 6025, 6249	3	GP9	EMD	1955-1957	2
Total	**9**				

Notes:
1. Ex-Norfolk & Western, same numbers
2. Ex-Chesapeake & Ohio, same numbers

HARTFORD & SLOCOMB RAILROAD

Reporting marks: HS   **Miles:** 22
Address: P. O. Box 2243, Dothan, AL 36302

The Hartford & Slocomb extends from Dothan to Hartford, Alabama. The enginehouse is at Hartford.

Nos.	Qty.	Model	Builder	Date	Notes
1	1	S1	Alco	1942	
257	1	RS1	Alco	1949	1
913	1	RS1	Alco	1950	2
Total	**3**				

Notes:
1. Ex-Chicago & Western Indiana 257
2. Ex-Atlanta & St. Andrews Bay 913

Robert C. Del Grosso

The Alco RS1, exemplified by Hartford & Slocomb 257, was the first road-switcher: a switcher with road trucks and a longer frame with room for a steam generator to heat passenger cars.

HARTWELL RAILROAD

Reporting marks: HRT   **Miles:** 10
Address: P. O. Box 429, Hartwell, GA 30643

The Hartwell Railroad operates from Hartwell to Bowersville, Georgia. Engine facilities are at Hartwell. The railroad is affiliated with the Great Walton Railroad.

Nos.	Qty.	Model	Builder	Date	Notes
7	1	SW1	EMD	1949	Ex-New York Central 599
Total	**1**				

HILLSDALE COUNTY RAILWAY

Reporting marks: HCRC   **Miles:** 95
Address: 50 Monroe St., Hillsdale, MI 49242

The Hillsdale County Railway operates from Coldwater, Michigan, through Steubenville, Indiana, to Montpelier, Ohio. The enginehouse is at Hillsdale, Michigan.

Nos.	Qty.	Model	Builder	Date	Notes
1601, 1602	2	GP7	EMD	1053, 1052	1
1603	1	GP9	EMD	1954	2
1766, 1770	2	GP9	EMD	1957	3
Total	**5**				

Notes:
1. Ex-Reading 614, 605
2. Ex-Cleveland Union Terminal 5903
3. Ex-Burlington Northern 1766, 1770

HOLLIS & EASTERN RAILROAD

Reporting marks: HE   **Miles:** 14
Address: Drawer C, Duke, OK 73532

The Hollis & Eastern operates from Altus to Duke, Oklahoma. The enginehouse is at Altus.

Nos.	Qty.	Model	Builder	Date	Notes
39	1	SW1	EMD	1951	Ex-Chicago, West Puliman & Southern
2520	1	CF7	ATSF	1974	Ex-Santa Fe 2520
Total	**2**				

HOUSATONIC RAILROAD

Reporting marks: HRRC **Miles:** 34
Address: P. O. Box 1146, Canaan, CT 06018
The Housatonic extends from New Milford through Canaan, Connecticut, to Pittsfield, Massachusetts. The enginehouse is at Canaan.

Nos.	Qty.	Model	Builder	Date	Notes
7324	1	GP9	EMD	1955	Ex-Conrail 7324
9935	1	RS3M	Alco	1953	Rebuilt by Conrail; ex-CR 9935
Total	**2**				

HOUSTON BELT & TERMINAL RAILWAY

Reporting marks: HBT **Miles:** 279
Address: 202 Union Station Building Houston, TX 77002
Houston Belt & Terminal is a switching road at Houston.

Nos.	Qty.	Model	Builder	Date	Notes
24, 29-31	4	SW9	EMD	1951	
33-37	5	SW1200	EMD	1966	
40-42	3	SW1000	EMD	1968	
50-54	5	SW1500	EMD	1968	
55	1	SW1500	EMD	1971	
60-64	5	MP15	EMD	1971	
82, 83	2	Slug	HBT	1982, 1983	1
Total	**25**				

Notes:
1. Built on the frames of EMD switchers

HUNTSVILLE & MADISON COUNTY RAILROAD AUTHORITY

Reporting marks: HMRC **Miles:** 13
Address: P. O. Box 308, Huntsville, AL 35804
The Huntsville & Madison County extends from Huntsville to Norton, Alabama, on a former Louisville & Nashville branch. The enginehouse is at Huntsville.

Nos.	Qty.	Model	Builder	Date	Notes
527	1	NW2	EMD	1946	Ex-Burlington Northern 527
Total	**1**				

HURON & EASTERN RAILWAY

Reporting marks: HESR **Miles:** 141
Address: 644 E. Huron Ave., Bad Axe, MI 48413
The Huron & Eastern operates a network of former Chesapeake & Ohio and New York Central lines in the "thumb" of Michigan's lower peninsula. Enginehouses are at Bad Axe and Vassar, Mich.

Nos.	Qty.	Model	Builder	Date	Notes
100-103	4	GP9	EMD	1956	
104	1	GP7	EMD	1952	
105	1	GP9	EMD	1957	
Total	**6**				

HUTCHINSON & NORTHERN RAILWAY

Reporting marks: HN **Miles:** 6
Address: 1800 Carey Building, Hutchinson, KS 67501
The Hutchinson & Northern is a switching road at Hutchinson, Kansas.

Nos.	Qty.	Model	Builder	Date	Notes
5	1	S1	Alco	1945	
6	1	SW900	EMD	1957	Ex-Fernwood, Columbia & Gulf
7	1	SW900	EMD	1955	Ex-Baltimore & Ohio
Total	**3**				

ILLINOIS CENTRAL RAILROAD

Reporting marks: IC   **Miles:** 2,767
Address: 233 N. Michigan Ave., Chicago, IL 60601

Nos.	Qty.	Model	Builder	Date	Notes
1400-1412, 1414-1419, 1423, 1427, 1429-1511	104	SW14	ICG	1978-1982	1
2000-2037	38	SD20	ICG	1980-1982	2
2038-2041	4	SD20	ICG	1982	3
2250	1	GP30	EMD	1962	4
2520	1	GP35	EMD	1964	4
2550	1	GP35	EMD	1965	5
3029	1	GP40	EMD	1966	
3100-3110, 3112-3136	39	GP40r	EMD	1966-1969	6
6000-6005	6	SD40	EMD	1968	
6006-6011, 6013-6018	12	SD40A	EMD	1969	7
6024	1	SD40A	EMD	1969	7, 8
6030-6034	5	SD40-2	EMD	1974-1975	9
6050-6054, 6056-6061, 6063-6068, 6070	18	SD40r	EMD	1966	4
6071	1	SD40X	EMD	1964	10
6100	1	SD40	EMD		11
7700, 7701, 7704-7707, 7711, 7717, 7721-7732, 7734, 7739, 7741-7746, 7800, 7850, 7900, 7901, 7903, 7910, 7911, 7913, 7914, 7917, 7918, 7952, 7954, 7962, 7965, 7970, 7988, 7994, 7996	47	GP8	ICG	1974-1978	
8006, 8010, 8015, 8016, 8021, 8037, 8042, 8049, 8050, 8057, 8067, 8086, 8089, 8092, 8097, 8098, 8109, 8116, 8117, 8216, 8129, 8130, 8136, 8153, 8156, 8173, 8174, 8178, 8186, 8202, 8203, 8214, 8217, 8219, 8220, 8241, 8265, 8274, 8288, 8290, 8292, 8294, 8295, 8302, 8308-8314, 8316, 8319, 8321, 8323, 8325, 8330, 8331, 8335, 8343, 8346, 8354, 8358, 8359, 8365, 8371, 8394-8396, 8442, 8446, 8447, 8460, 8461, 8463-8466	78	GP10	ICG	1974-1978	
8701-8753	53	GP11	ICG	1979-1981	
9450, 9451	2	SD28	EMD	1965	12
9510	1	GP38	EMD	1970	
9520-9523, 9525, 9527-9531, 9533-9537, 9539	16	GP38	EMD	1969-1971	4
9540-9544, 9546-9550, 9552	11	GP38AC	EMD	1971	4
9560-9574	15	GP38-2	EMD	1972	4
9600-9632, 9634-9635	35	GP38-2	EMD	1974	
Total	491				

Notes:
1. Rebuilt NW2s, SW7s, and SW9s
2. Rebuilt SD24s; 2000 h.p.
3. Rebuilt SD35s; 2000 h.p.
4. Ex-Gulf, Mobile & Ohio
5. Ex-GP40X IC 3075, EMD demonstrator 433A
6. Rebuilt 1987-1991 by VMV
7. Built on SD45 frames to permit larger fuel tank
8. Rebuilt from 6012
9. 6034 rebuilt from 6069 (ex-GM&O 920)
10. Ex-GM&O 950, EMD demonstrator 434
11. Ex-Burlington Northern
12. Ex-Columbus & Greenville 701, 702

THE GP8, GP10, AND GP11

Illinois Central dieselized its freight trains relatively late with Electro-Motive hood units — 48 GP7s and 348 GP9s — and by the mid-1960s the oldest of those units were approaching the time for a major overhaul. In 1968 IC initiated the country's first major diesel locomotive rebuilding program at its Paducah, Kentucky, shops. The rebuilding was extensive, including an increase in power output, low nose (after the first five units), four exhaust stacks, and, also after the first few units, paper air filters, which are housed in a boxy structure atop the hood or on each side of it.

The rebuilt GP7s were designated GP8s, and the rebuilt GP9s (and a few GP18s) became GP10s. IC was pleased with the results of the program. It eventually bought GP7s and GP9s from other roads to rebuild and undertook locomotive rebuilding projects for other railroads. Paducah shops turned out a total of 109 GP8s, 327 GP10s, and 54 GP11s for Illinois Central and its successor Illinois Central Gulf between 1967 and 1977, plus many such units for other railroads. The program continued with the GP11, which used Dash 2 electrical components.

The identity of a given locomotive is difficult to trace through Paducah's rebuilding program, because the components became fairly well scrambled. The official identity of the predecessor of a given rebuilt locomotive is a matter of accounting rather than of what parts came from which.

When ICG began spinning off shortline and regional railroads, it often included GP8s and GP10s along with the routes, accounting for their presence on the rosters of such roads as Chicago, Central & Pacific and MidSouth.

Other models on IC's roster that are products of the Paducah rebuilding program are GP26, SD20, and SW14.

INDIAN CREEK RAILROAD

Reporting marks: ICRK **Miles:** 5
Address: R.R. 1, Frankton, IN 46044
The Indian Creek Railroad extends from Anderson to Frankton, Indiana. The enginehouse is at Frankton.

Nos.	Qty.	Model	Builder	Date	Notes
6002	1	RS11	EMD	1959	Ex-Southern Pacific
Total	1				

INDIANA & OHIO RAIL CORPORATION

Reporting marks: INOH **Miles:** 292
Address: 8901 Blue Ash Road, Cincinnati, OH 45242
The railroads of the Indiana & Ohio Rail Corporation — Indiana & Ohio Railway, Indiana & Ohio Central Railroad, Indiana & Ohio Eastern Railroad, and Indiana & Ohio Railroad — operate freight service on a number of routes in southeastern Indiana and southwestern Ohio. Motive power is pooled among the four railroads.

Nos.	Qty.	Model	Builder	Date	Notes
51-56	6	GP7	EMD	1950-1953	
61-65	5	GP9	EMD	1956	
71	1	GP18	EMD	1960	
81, 82	2	GP30	EMD	1962	1
Total	14				

Notes:
1. Ex-Norfolk & Western

TRAINS: J. David Ingles

Indiana Harbor Belt's roster is all-switcher, although many of its freight trains travel considerable distances.

INDIANA HARBOR BELT RAILROAD

Reporting marks: IHB **Miles:** 114
Address: P. O. Box 389, Hammond, IN 46325
The Indiana Harbor Belt extends from the industrial area at the south end of Lake Michigan west and north through Blue Island and La Grange, Illinois, to Franklin Park, Ill. The road's shops are at Hammond, Indiana.

Nos.	Qty.	Model	Builder	Date	Notes
8712, 8714-8727, 8730, 8732, 8738-8740, 8774, 8780, 8782-8792, 8811, 8825-8831, 8833, 8834					
	36	NW2	EMD	1948-1949	1
8835, 8858-8866, 8978, 8979	12	SW7	EMD	1949-1950	1
9002-9008	7	SW9	EMD	1953	

Nos.	Qty.	Model	Builder	Date	Notes
9200-9222	23	SW1500	EMD	1966-1970	
Total	78				

Notes:
1. Many of Indiana Harbor Belt's NW2s and SW7s have been rebuilt.

INDIANA HI-RAIL CORPORATION

Reporting marks: IHRC **Miles:** 526
Address: R.R. 1, Connersville, IN 47331
Indiana Hi-Rail operates freight service under its own name on 10 separate routes in Indiana, Illinois, and Ohio. It also operates the Poseyville & Owensville Railroad, an 11-mile line between the Indiana towns in its name; and the Spencerville & Elgin Railroad, which extends from Glenmore to Lima, Ohio, 30 miles.

Nos.	Qty.	Model	Builder	Date	Notes
119	1	S4	Alco	1951	Ex-Santa Fe 1519
167	1	RS1	Alco	1950	Ex-Long Island 467
216	1	SW1200	EMD	1957	Ex-Baltimore & Ohio 9616
221	1	SW1200	EMD	1957	Ex-Baltimore & Ohio 9621
223	1	SW1200	EMD	1966	Ex-Terminal Railroad Association of St. Louis 1223
224	1	SW1200	EMD	1965	Ex-Terminal Railroad Association of St. Louis 1234
303	1	GP20	EMD	1959	Ex-Burlington Northern 2003
310, 311	2	C420	Alco	1964	
315	1	C430	Alco	1966	
316	1	GP20	EMD	1959	Ex-Burlington Northern 2016
327	1	C425	Alco	1966	
332, 334	2	C430	Alco	1967	
342-345	4	GP7u	EMD	1951-1952	
352	1	RS11	Alco	1960	Ex-Louisville & Nashville 952
365	1	C430	Alco	1965	
371-373	3	GP35	EMD	1964	
442, 443	2	RSD15	Alco	1960	
Total	25				

INDIANA RAIL ROAD

Reporting marks: INRD **Miles:** 252
Address: P. O. Box 2464, Indianapolis, IN 46206

The Indiana Rail Road operates from Indianapolis through Newton, Illinois, to Browns, Ill., and from Indianapolis to Tipton, Indiana. The enginehouse is in Indianapolis.

Nos.	Qty.	Model	Builder	Date	Notes
100	1	E9A	EMD	1961	1
200-207, 2551	9	CF7	ATSF	1973-1975	
300	1	GP9	EMD	1957	2
7307, 7309	2	SD18	EMD	1963	3
Total	**13**				

Notes:
1. Ex-Milwaukee Road
2. Ex-Conrail
3. Ex-Chesapeake & Ohio

IOWA INTERSTATE RAILROAD

Reporting marks: IAIS **Miles:** 552
Address: 800 Webster St., Iowa City, IA 52240

Iowa Interstate extends from Blue Island (Chicago), Illinois, to Council Bluffs, Iowa, on what used to be the main line of the Rock Island. Enginehouses are at Council Bluffs and Iowa City, Iowa.

Nos.	Qty.	Model	Builder	Date	Notes
250	1	SW1200	EMD	1966	Ex-Missouri Pacific 1299
300	1	GP9	EMD	1955	Ex-Western Pacific 725
303	1	GP9	EMD	1956	Ex-Chesapeake & Ohio 6075
306	1	GP9	EMD	1955	Ex-Western Pacific
309	1	GP7	EMD	1953	Ex-Norfolk & Western 3479
325	1	GP7	EMD	1953	Ex-Union Pacific 725
400	1	GP7u	EMD	1952	Ex-Rock Island 4424
401-404	4	GP10	ICG	1968-72	1
405	1	GP7u	EMD	1975	Ex-Burlington Northern 1415
406	1	GP10	ICG	1972	Ex-Illinois Central Gulf 8257

Nos.	Qty.	Model	Builder	Date	Notes
407, 408	2	GP7u	EMD	1951	Ex-BN 1404, 1405
413, 414, 430, 431, 436	5	GP10	ICG	1969	1
451, 457, 464, 466, 468, 469, 471	7	GP8	ICG	1968-69	1
473	1	GP10	ICG	1969	1
476, 479, 481	3	GP8	ICG	1968-69	1
483, 484	2	GP10	ICG	1969	1
600	1	GP38	EMD	1967	Ex-North Louisiana & Gulf 46, Conrail 7660
601, 602	2	GP38	EMD	1966	Ex-Missouri Pacific 2002, 2005
900	1	RS36	Alco	1963	Ex-Delaware & Hudson 5015
Total	**37**				

Notes:
1. Previous numbers of ex-Illinois Central Gulf units

IAIS	ICG	IAIS	ICG	IAIS	ICG
401	8001	431	8031	471	7871
402	8243	436	8036	473	8073
403	8326	451	7851	476	7976
404	8004	457	7857	479	7979
406	8257	464	7864	481	7981
413	8113	466	7866	483	8083
414	8114	468	7868	484	8084
430	8030	469	7869		

IOWA TRACTION RAILROAD

Reporting marks: IATR **Miles:** 9
Address: P. O. Box 309, Mason City, IA 50401

Iowa Traction is an electric railroad extending from Mason City to Clear Lake, Iowa. The enginehouse is at Emery.

Nos.	Qty.	Model	Builder	Date
50, 51	2	50-ton electric	Baldwin-Westinghouse	1920, 1921
53	1	50-ton electric	Texas Electric Railway	1928
54	1	50-ton electric	Baldwin-Westinghouse	1919
60	1	60-ton electric	Baldwin-Westinghouse	1917
Total	5			

JACKSON & SOUTHERN RAILROAD

Reporting marks: JSRC   **Miles:** 18
Address: P. O. Box 871, Cairo, IL 62914

Jackson & Southern extends from Delta to Jackson, Missouri. The enginehouse is at Jackson. It also operates and provides locomotives for the Golden Cat Railroad (Delta to Newman Spur, Mo., 11 miles).

Nos.	Qty.	Model	Builder	Date	Notes
103	1	GP9	EMD	1955	Ex-Illinois Central 9089
Total	1				

J&J RAILROAD

Reporting marks: JJRD   **Miles:** 8
Address: P. O. Box 131, Hardin, KY 42048

The J&J Railroad operates between Hardin and Murray, Kentucky. The enginehouse is at Hardin.

Nos.	Qty.	Model	Builder	Date	Notes
1	1	SW1	EMD	1940	
Total	1				

JAXPORT TERMINAL RAILWAY

Reporting marks: JXPT   **Miles:** 8
Address: P. O. Box 2624, Victoria, TX 77902

Jaxport Terminal is a switching line at Jacksonville, Florida.

Nos.	Qty.	Model	Builder	Date	Notes
252	1	SW7	EMD	1950	
1056	1	GP18	EMD	1960	Ex-CSX
Total	2				

JEFFERSON WARRIOR RAILROAD

Reporting marks: JEFW   **Miles:** 44
Address: P. O. Box 5346, Birmingham, AL 35207

The Jefferson Warrior Railroad is a switching road extending between Birmingham and Bessie Coal Mine, Alabama. The enginehouse is at Birmingham.

Nos.	Qty.	Model	Builder	Date	Notes
51-54	4	SW1500	EMD	1972	1
55, 56	2	SW9	EMD	1952	2
Total	6				

Notes:
1. Ex-Jim Walter Industries
2. Ex-Woodward Iron Co., Florida East Coast

JK LINES

Reporting marks: JKL   **Miles:** 16
Address: P. O. Box 68, Monterey, IN 46960

JK Lines operates between North Judson and Monterey, Indiana. The enginehouse is at Monterey.

Nos.	Qty.	Model	Builder	Date	Notes
7302, 7311	2	SD18	EMD	1963	Ex-Chesapeake & Ohio
Total	2				

JUNIATA TERMINAL COMPANY

Reporting marks: JTFS   **Miles:** 1
Address: B and Venango Streets, Philadelphia, PA 19134

Juniata Terminal is a switching railroad serving a passenger-car repair shop in Philadelphia.

Nos.	Qty.	Model	Builder	Date	Notes
9251	1	NW2	EMD	1948	Ex-CR
Total	1				

KALAMAZOO, LAKE SHORE & CHICAGO RAILWAY

Reporting marks: KLSC   **Miles:** 15
Address: P. O. Box 178, Paw Paw, MI 49079

Kalamazoo, Lake Shore & Chicago operates freight and excursion trains between Hartford and Paw Paw, Michigan. Engine facilities are at Paw Paw.

Nos.	Qty.	Model	Builder	Date	Notes
85	1	GP7	EMD	1953	Ex-Santa Fe
95	1	E8	EMD		Ex-Chicago & North Western
2066, 2067	2	GP7	EMD	1951	Ex-Santa Fe
Total	**3**				

KANKAKEE, BEAVERVILLE & SOUTHERN RAILROAD

Reporting marks: KBSR   **Miles:** 85
Address: P. O. Box 136, Beaverville, IL 60912

Kankakee, Beaverville & Southern extends from Kankakee to Danville, Illinois, and from Iroquois Junction, Ill., to Swanington, Indiana. The enginehouse is at Iroquois Junction.

Nos.	Qty.	Model	Builder	Date	Notes
301, 312, 321	3	RS11	Alco	1956	1
315	1	C420	Alco	1962	2
318	1	RS11	Alco	1959	3
Total	**5**				

Notes:
1. Ex-Norfolk & Western
2. Ex-Louisville & Nashville
3. Ex-Burlington Northern

KANSAS CITY SOUTHERN RAILWAY

Reporting marks: KCS   **Miles:** 882
Address: 114 W. 11th St., Kansas City, MO 64105-1804

Kansas City Southern's two principal routes are from Kansas City south to Port Arthur, Texas, and from Dallas through Shreveport to New Orleans. Trackage rights extend KCS service from Kansas City to Topeka, Kansas, and Omaha and Lincoln, Nebraska, and from Beaumont, Texas, to Houston and Galveston. KCS's shops are at Shreveport, Louisiana.

Nos.	Qty.	Model	Builder	Date	Notes
607, 611, 614, 618, 619, 621, 628-635	14	SD40	EMD	1966-1971	
637-692	56	SD40-2	EMD	1972-1980	
700-703	4	SD40X	EMD	1978	1
704-713	10	SD50	EMD	1981	
714-759	46	SD60	GMD	1989-1991	
777-795	19	GP40	EMD	1967	2
796, 797	2	GP40-2	EMD	1981	
798, 799	2	GP40-2	EMD	1979	
1300, 1305, 1309	3	SW7	EMD	1950	3
1500-1511, 1513-1541	41	SW1500	EMD	1966-1972	
1542	1	SW1500	EMD	1970	4
4000-4011	12	GP38-2	EMD	1974, 1978	
4012-4023	12	GP38-2	EMD	1973	5
4024-4028	5	GP38-2	EMD	1973	6
4056, 4057	2	Slug	KCS	1980, 1979	7
4076, 4077	2	Slug	KCS	1977, 1978	8
4078-4080	3	Slug	C&NW	1961-1964	9
4211, 4222, 4223	3	NW2	EMD	1947, 1949	
4250, 4252-4257	7	Slug	KCS	1969-1973	10
4363-4366	4	MP15DC	EMD	1975	
4748-4768, 4771-4773, 4775, 4776	26	GP40	EMD	1968	11
Total	**284**				

Notes:
1. Only SD40Xs built
2. Ex-Illinois Central Gulf; may be renumbered 4777-4795
3. Stored unserviceable
4. Ex-Howe Coal 2
5. Ex-Boston & Maine 200-211
6. Ex-Soo Line 4500-4505, ex-Milwaukee Road 350-355
7. Rebuilt from F7As, stored unserviceable
8. Rebuilt from F7Bs in 1977 and 1978
9. Rebuilt from TR2Bs; ex-Chicago & North Western BU5, BU4, and BU9
10. Rebuilt from NW2s and SW1s
11. Previously numbered 748-776; ex-Conrail 3170+3187

KANSAS CITY TERMINAL RAILWAY

Reporting marks: KCT   **Track miles:** 99
Address: 345 Broadway, Kansas City, MO 64141
Kansas City Terminal is a switching road at Kansas City, Missouri.

Nos.	Qty.	Model	Builder	Date	Notes
70, 72, 73, 75-79	8	SW1200	EMD	1964	
Total	8				

KANSAS SOUTHWESTERN RAILWAY

Reporting marks: KSW   **Miles:** 298
Address: 215 W. Dewey, Wichita, KS 67202
Kansas Southwestern operates several former Missouri Pacific branches centered on Wichita, Kansas. The enginehouse is at Wichita.

Nos.	Qty.	Model	Builder	Date	Notes
4436	1	GP9	EMD	1954	1
4557, 4912, 4916	3	GP9	EMD	1957	1
Total	4				

Notes:
1. Ex-Grand Trunk Western

KELLEY'S CREEK & NORTHWESTERN RAILROAD

Reporting marks: KCNW   **Miles:** 7
Address: 2971 E. Du Pont Ave., Shrewsbury, WV 25184
Kelley's Creek & Northwestern extends from Cedar Grove to Lewis, West Virginia. The enginehouse is at Ward.

Nos.	Qty.	Model	Builder	Date	Notes
1, 2	2	MP15DC	EMD	1976, 1980	
7	1	70-ton	GE	1951	
Total	3				

KENTUCKY & TENNESSEE RAILWAY

Reporting marks: KT   **Miles:** 11
Address: P. O. Box 368, Stearns, KY 42647
The Kentucky & Tennessee operates between Stearns and Oz, Kentucky. The enginehouse is at Hemlock.

Nos.	Qty.	Model	Builder	Date	Notes
102	1	S2	Alco	1944	Ex-Denver & Rio Grande Western 118
105	1	S2	Alco	1949	Ex-Delaware & Hudson 3028
Total	2				

KEOKUK JUNCTION RAILWAY

Reporting marks: KJRY   **Miles:** 38
Address: 117 S. Water St., Keokuk, IA 52632
Keokuk Junction Railway extends from Keokuk, Iowa, to La Harpe, Illinois, with a branch to Warsaw, Ill.

Nos.	Qty.	Model	Builder	Date	Notes
20	1	S2	Alco	1941	1
252, 253	2	GP7rm	EMD	1951, 1952	2
405	1	NW2	EMD	1949	3
Total	4				

Notes:
1. Ex-Manufacturers Railway
2. Ex-Chicago & North Western, rebuilt 1980 by C&NW with a Cummins 1500 h.p. engine; classified HE15 by C&NW
3. Ex-Peoria & Pekin Union 405

KLAMATH NORTHERN RAILWAY

Reporting marks: KNOR   **Miles:** 11
Address: 1 Sawmill Road, Gilchrist, OR 97737
The Klamath Northern extends from Gilchrist Junction to Gilchrist, Oregon. The enginehouse is at Gilchrist.

Nos.	Qty.	Model	Builder	Date	Notes
207	1	125-ton	GE	1982	
Total	**1**				

KIAMICHI RAILROAD

Reporting marks: KRR   **Miles:** 230
Address: P. O. Box 786, Hugo, OK 74743
Kiamichi Railroad operates from Hope, Arkansas, to Madill, Oklahoma, and from Antlers, Okla., to Paris, Texas. The engine facilties are at Hugo, Okla.

Nos.	Qty.	Model	Builder	Date	Notes
701, 702, 705-707	5	GP7	EMD	1951-1953	
901-908	8	GP9	EMD	1954-1959	
3801-3803, 3805	4	GP35m	EMD	1964-1965	
4050	1	Slug	EMD	1948	1
4055	1	Slug	EMD	1947	2
Total	**19**				

Notes:
1. Rebuilt from F7A by Kansas City Southern
2. Rebuilt from F3A by Kansas City Southern

KNOX & KANE RAILROAD

Reporting marks: KKRR   **Miles:** 78
Address: P. O. Box 4745, Gettysburg, PA 17325
The Knox & Kane operates between Knox and Mount Jewett, Pennsylvania, on a former Baltimore & Ohio line. Enginehouses are at Marienville and Kane, Pa.

Nos.	Qty.	Model	Builder	Date	Notes
1	1	50-ton	Porter	1946	
44	1	S6	Alco	1957	
337	1	S12	BLW	1948	
Total	**3**				

Louis A. Marre

Kiamichi 3803 was formerly Missouri Pacific 2607. It is a GP35 that MoPac deturbocharged and derated to 2000 h.p.

KWT RAILWAY, INC.

Reporting marks: KWT   **Miles:** 62
Address: 908 Depot St., Paris, TN 38242

The KWT Railway extends from Bruceton, Tennessee, to Murray, Kentucky, with a branch from Paris to Henry, Tenn. The enginehouse is at Paris.

Nos.	Qty.	Model	Builder	Date	Notes
300	1	GP7	EMD	1953	Ex-Louisville & Nashville
301	1	GP9	EMD	1957	Ex-Milwaukee Road
302	1	GP9	EMD	1955	Ex-Burlington Northern
303	1	GP18	EMD	1960	Ex-Burlington Northern
Total	4				

KYLE RAILROAD

Reporting marks: KYLE   **Miles:** 406
Address: Third and Railroad Ave., Phillipsburg, KS 67661

The Kyle Railroad operates between Limon, Colorado, and Clay Center and Manaska, Kansas. The enginehouse is at Phillipsburg, Kan.

Nos.	Qty.	Model	Builder	Date	Notes
101-103	3	GP9r	EMD	1951	1
1101-1103	3	SW8	EMD	1952	2
1825-1829	5	GP28	EMD	1964	3
2035-2040	6	GP20	EMD	1961-1962	4
2170, 2171	2	GP20	EMD	1960	5
5321, 5322, 5331, 5332, 5808, 5810, 5819-5821, 5918, 5925, 5928	12	U30C	GE	1972-1973	6
Total	31				

Notes:
1. Ex-Union Pacific 224, 498, 308; rebuilt by M-K 1980
2. Ex-Southern Pacific 1126, 1125, 1121
3. Ex-Illinois Central 9430, 9431, 9432, 9437
4. Ex-Southern Pacific 4100, 4109, 4111, 4116, 4143, 4147
5. Ex-Burlington Northern 2015, 2023
6. Ex-Burlington Northern, same numbers

LACKAWANNA VALLEY RAILROAD

Reporting marks: LVAL   **Miles:** 24
Address: 310 Breck St., Scranton, PA 18505

The Lackawanna Valley extends from Minooka Junction to Carbondale, Pennsylvania. The enginehouse is in Scranton.

Nos.	Qty.	Model	Builder	Date	Notes
22, 43	2	RS18	MLW	1959	
901	1	U30B	GE	1967	
Total	3				

LAKE ERIE, FRANKLIN & CLARION RAILROAD

Reporting marks: LEF   **Miles:** 15
Address: East Wood Street, Clarion, PA 16214

The Lake Erie, Franklin & Clarion operates from Summerville to Clarion, Pennsylvania. The enginehouse is at Clarion.

Nos.	Qty.	Model	Builder	Date	Notes
25-28	4	MP15DC	EMD	1976-1979	
Total	4				

LAKE STATE RAILWAY

Reporting marks: LSRC   **Miles:** 338
Address: P.O. Box 250, Tawas City, MI 48763

The Lake State Railway, successor in February 1992 to the Detroit & Mackinac extends from Bay City, Michigan, to Grayling and from Pinconning to LaRocque, near Millersburg. Enginehouses are at Tawas City, Alpena, and Bay City.

Nos.	Qty.	Model	Builder	Date	Notes
10	1	44-ton	GE	1942	
181, 281, 381	3	C425	Alco	1966	Ex-Conrail
469	1	RS2	Alco	1946	
646	1	S1	Alco	1946	
974, 975	2	TE56-4A	MK	1974, 1975	Rebuilt RS2s
976	1	C420	Alco	1964	Ex-Long Island
977	1	RS2	Alco	1949	
1077	1	RS3	Alco	1955	Ex-Boston & Maine
1280	1	C425	Alco	1965	Ex-Conrail
Total	**12**				

LAKE SUPERIOR & ISHPEMING RAILROAD

Reporting marks: LSI   **Miles:** 50
Address: 105 E. Washington St., Marquette, MI 49885

The Lake Superior & Ishpeming extends from Marquette to Republic Mine, Michigan, with a branch from Eagle Mills to Tilden Mine. The enginehouse is at Eagle Mills.

Nos.	Qty.	Model	Builder	Date	Notes
2300-2304	5	U23C	GE	1968-1970	
3000-3011	12	U30C	GE	1974-1975	Ex-Burlington Northern
3050-3053	4	U30C	GE	1974	Ex-Burlington Northern
Total	**21**				

The arrival of ex-Burlington Northern GEs like 3000 and 3003 in 1989 caused the retirement of Lake Superior & Ishpeming's Alcos and gave RSD12 No. 2404 the uncomfortable distinction of being the last operating Alco diesel on the road.

Eric Hirsimaki

LAKE TERMINAL RAILROAD

Reporting marks: LT   **Miles:** 24
Address: P. O. Box 471, Greenville, PA 16125
Lake Terminal is a switching road at Lorain and South Lorain, Ohio.

Nos.	Qty.	Model	Builder	Date	Notes
821-825	5	SW8	EMD	1951	1
1011, 1014-1020	8	NW2	EMD	1948-1949	
1021	1	SW1001	EMD	1968	
1022	1	SW1000	EMD	1967	2
1023	1	SW1001	EMD	1975	
1201, 1202	2	SW9	EMD	1952	
1203-1205	3	SW1200	EMD	1954	3
S1-S3	3	Slug		1980-1983	
Total	**24**				

Notes:
1. 825 is ex-Donora Southern
2. Ex-Newburgh & South Shore
3. Ex-Union Railroad

LAMOILLE VALLEY RAILROAD

Reporting marks: LVRC   **Miles:** 99
Address: RFD 1, Box 790, Morrisville, VT 05661
The Lamoille Valley extends from Fonda Junction to St. Johnsbury, Vermont. The enginehouse is at Morrisville. Lamoille Valley also operates and equips the Twin State Railroad — a former Maine Central line from St. Johnsbury east to Whitefield, New Hampshire.

Nos.	Qty.	Model	Builder	Date	Notes
117	1	RS3	Alco	1953	
144	1	RS3	Alco	1952	
7801	1	RS3	Alco	1952	Ex-Delaware & Hudson 4068
7805	1	RS3	Alco	1955	Ex-Delaware & Hudson 4094
7961	1	GP8	ICG	1968	Ex-Illinois Central Gulf 7961
Total	**5**				

LANCASTER & CHESTER RAILWAY

Reporting marks: LC   **Miles:** 29
Address: P. O. Box 1450, Lancaster, SC 29720
The Lancaster & Chester operates between the South Carolina towns of its name. The enginehouse is at Lancaster.

Nos.	Qty.	Model	Builder	Date	Notes
90, 91	2	SW900	EMD	1965	
92	1	SW900	EMD	1959	
Total	**3**				

LANDISVILLE RAILROAD

Reporting marks: AMHR   **Miles:** 3
Address: P. O. Box 338, Landisville, PA 17538
The Landisville Railroad operates between Landisville and Silver Spring, Pennsylvania. The enginehouse is at Amherst, southwest of Landisville.

Nos.	Qty.	Model	Builder	Date	Notes
8526	1	44-ton	GE	1944	Ex-Ontario Midland
8661	1	SW600M	EMD	1955	Ex-Delaware Coast Line, Conrail
Total	**2**				

LAURINBURG & SOUTHERN RAILROAD

Reporting marks: LRS   **Miles:** 28
Address: P. O. Box 1929, Laurinburg, NC 28352
The Laurinburg & Southern operates from Johns to Raeford, North Carolina. The enginehouse is at Laurinburg. The locomotives shown below often work on other railroads under the same management — Nash County, Red Springs & Northern, and Yadkin Valley.

Nos.	Qty.	Model	Builder	Date	Notes
101, 103, 109, 110	4	70-ton	GE	1946-1951	
111	1	S2	Alco	1950	
112, 113	2	S4	Alco	1952	

Nos.	Qty.	Model	Builder	Date	Notes
114, 116-118, 120, 121	6	SW1	EMD	1949-1950	
123-130	8	NW2	EMD	1948-1949	
131, 132	2	S2	Alco	1948, 1946	
133	1	SW1	EMD	1947	
134	1	NW2	EMD	1946	
150, 151	2	25-ton	GE	1943	
Total	27				

Mike Small

Laurinburg & Southern 113, an Alco S4, was the first Alco unit added to the road's roster.

LEADVILLE-CLIMAX SHORTLINE RAILWAY

Reporting marks: LCSR   **Miles:** 14
Address: 326 Seventh St., Leadville, CO 80461
The Leadville-Climax Shortline extends from Leadville to Climax, Colorado, 14 miles. For years the line was an orphan branch of the Colorado & Southern; it was previously part of the narrow gauge Denver, South Park & Pacific. The company also operates excursion trains. The enginehouse is at Leadville.

Nos.	Qty.	Model	Builder	Date	Notes
1714, 1918	2	GP9	EMD	1955, 1957	Ex-Burlington Northern
Total	2				

LEWIS & CLARK RAILWAY

Reporting marks: LINC   **Miles:** 30
Address: 1000 E. Main St., Battle Ground, WA 98604
The Lewis & Clark Railway operates freight and excursion service between Rye Junction and Chelatchie, Washington, 29 miles. The enginehouse is at Chelatchie.

Nos.	Qty.	Model	Builder	Date	Notes
80, 81	2	SW8	EMD	1954	
Total	2				

LEWISBURG & BUFFALO CREEK RAILROAD

Reporting marks: LBCR   **Miles:** 2
Address: RD 3, Box 154, Lewisburg, PA 17837
The Lewisburg & Buffalo Creek operates between Winfield and West Milton, Pennsylvania. It is affiliated with the West Shore Railroad.

Nos.	Qty.	Model	Builder	Date	Notes
1500	1	SW8	EMD	1953	1
9425	1	SW1	EMD	1950	2
Total	2				

Notes:
1. Ex-Conrail 8618, Penn Central 8618
2. Ex-Penn Central 8425, Pennsylvania 9425

LITTLE KANAWHA RIVER RAILROAD

Reporting marks: LKRR   **Miles:** 3
Address: P. O. Box 525, Marietta, OH 45750
The Little Kanawha River Railroad is a switching line at South Parkerburg, West Virginia.

Nos.	Qty.	Model	Builder	Date	Notes
1205	1	SW1200	EMD	1955	Ex-Illinois Terminal
Total	1				

LITTLE ROCK PORT RAILWAY

Reporting marks: LRPA   **Miles:** 10
Address: 7500 Lindsey Road, Little Rock, AR 72206
The Little Rock Port Railway is a switching line at Little Rock, Arkansas.

LITTLE ROCK & WESTERN RAILWAY

Reporting marks: LRWN   **Miles:** 73
Address: P. O. Box 386, Perry, AR 72125
The Little Rock & Western operates from Danville, Arkansas, to North Little Rock on part of the former Rock Island route from Memphis, Tennessee, to Amarillo, Texas. The enginehouse is at Perry, Ark.

Nos.	Qty.	Model	Builder	Date	Notes
101	1	C420	Alco	1964	Ex-Long Island 207
102	1	C420	Alco	1966	Ex-Louisville & Nashville 1307
103	1	GP9	EMD	1957	Ex-Burlington Northern 1786
Total	3				

The reflective rectangles along the frame and the placement of the lettering on Little Rock & Western 101 are similar to Green Bay & Western practice. GB&W rebuilt the unit for LR&W owner Green Bay Packaging.

Nos.	Qty.	Model	Builder	Date	Notes
1032	1	S2	Alco	1947	
Total	1				

LIVONIA, AVON & LAKEVILLE RAILROAD

Reporting marks: LAL   **Miles:** 10
Address: P. O. Box 190-B, Lakeville, NY 14480
The Livonia, Avon & Lakeville extends from Avon to Lakeville, New York. The enginehouse is at Lakeville.

Nos.	Qty.	Model	Builder	Date	Notes
20	1	RS1	Alco	1949	Ex-Lake Erie, Franklin & Clarion 20
72	1	S2	Alco	1941	Ex-South Buffalo 72
425	1	C425	Alco	1965	Ex-Conrail 5086, New Haven 2557
Total	3				

Charles W. McDonald

LOGANSPORT & EEL RIVER SHORT LINE

Reporting marks: LER   **Miles:** 2
Address: P. O. Box 1005, Logansport, IN 46947
The Logansport & Eel River is a switching road at Logansport, Indiana.

Nos.	Qty.	Model	Builder	Date	Notes
1	1	80-ton	GE	1944	
2	1	45-ton	Vulcan	1942	
Total	**2**				

LONG ISLAND RAIL ROAD

Reporting marks: LI   **Miles:** 322
Address: Jamaica Station, Jamaica, NY 11435
The Long Island Rail Road is primarily a passenger railroad, and most of its passengers ride in electric multiple-unit cars. Long Island's routes extend from Penn Station in New York east to Montauk and Greenport; a dense network of branches covers the west end of Long Island. The road's shops are at Morris Park, N. Y.

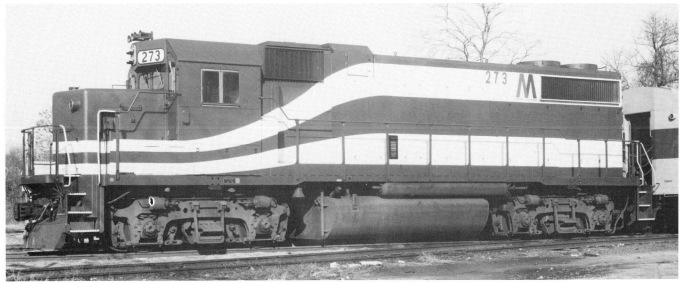

K. L. Douglas

The GP38-2, usually considered a freight locomotive, is a passenger hauler on the Long Island Rail Road.

Nos.	Qty.	Model	Builder	Date	Notes
LOCOMOTIVES					
100-107	8	SW1001	EMD	1977	
150-172	23	MP15AC	EMD	1977	
250-277	28	GP38-2	EMD	1976-1977	
AUXILIARY POWER UNITS					
605-609	5	FA2m	Alco	1956	
611, 612, 614-616	5	FA1m	Alco	1949	
619	1	F9Am	EMD	1954	
620-622	3	F7Am	EMD	1950	
Total	**73**				

LONGVIEW, PORTLAND & NORTHERN RAILWAY

Reporting marks: LPN   **Miles:** 4
Address: P. O. Box 22, Gardiner, OR 97441

The Longview, Portland & Northern operates between Gardiner Junction and Gardiner, Oregon. The enginehouse is at Gardiner.

Nos.	Qty.	Model	Builder	Date	Notes
111	1	S2	Alco	1949	
130	1	SW1500	EMD	1969	
Total	**2**				

LOS ANGELES JUNCTION RAILWAY

Reporting marks: LAJ   **Track miles:** 64
Address: 4433 Exchange Ave., Los Angeles, CA 90058

Los Angeles Junction is a switching road at Los Angeles. It is owned by the Santa Fe.

Nos.	Qty.	Model	Builder	Date	Notes
2563, 2568, 2571, 2619	4	CF7	ATSF	1972-1973	
Total	**4**				

LOUISIANA & DELTA RAILROAD

Reporting marks: LDRR   **Miles:** 100
Address: 402 W. Washington St., New Iberia, LA 70560

The Louisiana & Delta operates on four former Southern Pacific branches in Louisiana: New Iberia to Salt Mine, Baldwin to Weeks, Bayou Sale to Cabot, and Supreme to Thibodaux Junction. The enginehouse is at New Iberia.

Nos.	Qty.	Model	Builder	Date	Notes
1200	1	SW1200	EMD	1964	1
1500-1504	5	CF7	ATSF	1972, 1974	2
1751, 1752	2	GP9E	EMD	1956	3
Total	**8**				

Notes:
1. Ex-Southern Pacific 2250, leased to Conoco
2. Ex-Santa Fe 2618, 2622, 2470, 2471, 2489; 1500 is named Patoutville; 1501 is named City of New Iberia
3. Ex-Southern Pacific 3434 and 3790, rebuilt by SP in 1976, named Evangeline and North Bend

LOUISVILLE, NEW ALBANY & CORYDON RAILROAD

Reporting marks: LNAC   **Miles:** 8
Address: P. O. Box 10, Corydon, IN 47112

The Louisville, New Albany & Corydon extends from Corydon Junction to Corydon, Indiana. The enginehouse is at Corydon.

Nos.	Qty.	Model	Builder	Date	Notes
1	1	45-ton	GE	1951	Named Betty-Sue
101	1	45-ton	GE	1953	Ex-Ferdinand & Huntingburg
Total	**2**				

Long noted for its secondhand F units, Louisiana & North West has turned in recent years to hood units such as No. 50, a rebuilt GP7.

Charles W. McDonald

LOUISIANA & NORTH WEST RAILROAD

Reporting marks: LNW **Miles:** 62
Address: P. O. Box 89, Homer, LA 71040

Louisiana & North West operates freight service between McNeil, Arkansas, and Gibsland, Louisiana, 62 miles. Engine facilities are at McNeil, Gibsland, and Homer, La.

Nos.	Qty.	Model	Builder	Date	Notes
46, 48	1	F7A	EMD	1952, 1953	
50	1	GP7r	EMD	1953	Ex-Union Pacific 111
51	1	GP9r	EMD	1954	Ex-Union Pacific 286
52, 53	2	GP7r	EMD	1950, 1951	
54	1	GP35	EMD	1965	
Total	7				

LOWVILLE & BEAVER RIVER RAILROAD

Reporting marks: LBR **Miles:** 10
Address: 5515 Shady Ave., Lowville, NY 13367

The Lowville & Beaver River extends from Lowville to Croghan, New York, 10 miles. The enginehouse is at Lowville.

Nos.	Qty.	Model	Builder	Date	Notes
1947, 1950	2	44-ton	GE	1947, 1950	
Total	2				

MADISON RAILROAD

Reporting marks: CMPA   **Miles:** 31
Address: 511 State St., Madison, IN 47250

The Madison Railroad operates from North Vernon to Madison, Indiana. It is a division of the City of Madison Port Authority. The enginehouse is at Madison. The line has the steepest grade (5.89 percent) on a main route in the U.S.

Nos.	Qty.	Model	Builder	Date	Notes
3634	1	GP10	ICG	1971	Ex-Precision National
Total	1				

MAGMA ARIZONA RAILROAD

Reporting marks: MAA   **Miles:** 28
Address: P. O. Box 37, Superior, AZ 85273

Magma Arizona operates between Superior and Magma, Arizona, 28 miles. The enginehouse is at Superior.

Nos.	Qty.	Model	Builder	Date	Notes
8	1	S8	BLW	1952	Ex-Medford Corp.
0	1	S12	BLW	1953	Ex-McCloud River 31
10	1	DRS6-6-1500	BLW	1950	Ex-McCloud River 29
Total	3				

MAHONING VALLEY RAILWAY

Reporting marks: MVRY   **Miles:** 22
Address: P. O. Box 589, Campbell, OH 44405

The Mahoning Valley is a switching line at Youngstown, Struthers, and Campbell, Ohio.

Nos.	Qty.	Model	Builder	Date	Notes
466, 467	2	SW1	EMD	1949	Ex-Union Railroad
800, 801	2	SW8	EMD	1953	Ex-Aliquippa & Southern
1202	1	SW1200	EMD	1954	Ex-Aliquippa & Southern
1207	1	SW1200	EMD	1955	Ex-Aliquippa & Southern
Total	6				

MAINE COAST RAILROAD

Reporting marks: MC   **Miles:** 55
Address: RFD 1, Box 38C, Waldoboro, ME 04572

The Maine Coast Railroad operates former Maine Central routes from Brunswick to Rockland, Maine, 55 miles, and to Augusta, 33 miles. The enginehouse is at Rockland.

Nos.	Qty.	Model	Builder	Date	Notes
958	1	S1	Alco	1949	Ex-Maine Central 958
367	1	RS11	Alco	1956	Ex-Central Vermont
Total	2				

MANUFACTURERS RAILWAY

Reporting marks: MRS   **Track miles:** 42
Address: 2850 S. Broadway, St. Louis, MO 63118

The Manufacturers Railway is a switching line in St. Louis, Missouri. It is owned by Anheuser-Busch, Inc.

Nos.	Qty.	Model	Builder	Date	Notes
251, 252	2	MP15	EMD	1975	
250	1	CF7m	MK	1970	Rebuilt Alco S2
254	1	MP15DC	EMD	1983	
255	1	SW1500	EMD	1967	Ex-Electro-Motive 108
256	1	SW1500	EMD	1968	Ex-Indianapolis Union 26
257	1	SW1500	EMD	1970	
Total	7				

MANUFACTURERS' JUNCTION RAILWAY

Reporting marks: MJ   **Miles:** 2
Address: 2335 Cicero Ave., Chicago, IL 60650-2451

Manufacturers' Junction is a switching line at Cicero, Illinois.

Nos.	Qty.	Model	Builder	Date	Notes
6	1	SW1	EMD	1947	
7	1	SW1	EMD	1946	Ex-EMD demonstrator 700
Total	2				

MARINETTE, TOMAHAWK & WESTERN RAILROAD

Reporting marks: MTW   **Miles:** 13
Address: P. O. Box 310, Tomahawk, WI 54487
The Marinette, Tomahawk & Western extends from Tomahawk to Bradley and Wisconsin Dam, Wisconsin. The enginehouse is at Tomahawk.

Nos.	Qty.	Model	Builder	Date	Notes
80	1	SW7	EMD	1951	
83, 87	2	SW1500	EMD	1968	
Total	3				

MARYLAND & DELAWARE RAILROAD

Reporting marks: MDDE   **Miles:** 119
Address: 106 Railroad Ave., Federalsburg, MD 21632
The Maryland & Delaware operates three routes: Townsend, Delaware, to Chestertown and Centreville, Maryland; Seaford, Del., to Cambridge and Preston, Md.; and Frankford, Del., to Snow Hill, Md. The enginehouse is at Federalsburg, Md.

Nos.	Qty.	Model	Builder	Date	Notes
1201, 1202, 1203	3	RS3m	Alco		Ex-Conrail
2531, 2628	2	CF7	ATSF	1978	Ex-Santa Fe
Total	5				

MARYLAND & PENNSYLVANIA RAILROAD

Reporting marks: MPA   **Miles:** 25
Address: 96 S. George St., Suite 520, York, PA 17401
Maryland & Pennsylvania operates from York, Pennsylvania, to East York on its original line that years ago reached to Baltimore, and from York to Hanover on a former Pennsylvania Railroad line. The enginehouse is at York.

Nos.	Qty.	Model	Builder	Date	Notes
82, 84	2	SW9	EMD	1951, 1952	
86	1	GP7	EMD	1953	
1500, 1502, 1504	3	CF7	ATSF	1974, 1977	
1750, 1752, 1754, 1756	4	GP9	EMD	1956, 1957	
Total	10				

The chevron-shaped rain gutter over the cab window of Maryland & Pennsylvania GP7 No. 86 attests to its previous ownership by the Reading.

Charles W. McDonald

Sidney W. Earle Jr.

MARC 71, shown at Baltimore on CSX track (formerly Baltimore & Ohio) is a GP40 that has been rebuilt for commuter service by Morrison-Knudsen.

MARC
(Maryland State Railroad Administration)

Reporting marks: MARC   **Miles:** 150
Address: P. O. Box 8718, BWI Airport, MD 21240

MARC operates commuter service between Washington, D. C., and Perryville, Maryland, on Amtrak's Northeast Corridor line, and on two CSX routes: Washington-Baltimore and Washington-Martinsburg, West Virginia.

Nos.	Qty.	Model	Builder	Date	Notes
71-76	6	GP39H2	MK	1987	1
81-85	5	F9PH	EMD	1956	2
101, 102	2	RS3m	Alco	1951; 1953	3
4900-4904	5	AEM7	EMD	1986	4
Total	18				

Notes:
1. Rebuilt from ex-Baltimore & Ohio and Milwaukee Road GP40s by Morrison-Knudsen
2. Rebuilt from Baltimore & Ohio F7s in 1980 by Morrison-Knudsen
3. Ex-Conrail, rebuilt with 1200 h.p. EMD engines
4. Electric locomotives

MARYLAND MIDLAND RAILWAY

Reporting marks: MMID   **Miles:** 67
Address: P. O. Box A, Union Bridge, MD 21791

Maryland Midland extends from Emory Grove to Highfield, Maryland, and from Taneytown to Walkersville, Md. The enginehouse is at Union Bridge.

Nos.	Qty.	Model	Builder	Date	Notes
100, 101	2	F7A	EMD	1949	1
200, 201, 202	3	GP9	EMD	1957	2
Total	5				

Notes:
1. Ex-Chicago & North Western 4068A, 4083A
2. Ex-Norfolk & Western 703, 794, 812

MASSACHUSETTS CENTRAL RAILROAD

Reporting marks: MCER   **Miles:** 37
Address: 1 Wilbraham St., Palmer, MA 01069

The Massachusetts Central operates from Palmer to South Barre, Massachusetts, and from Forest Lake Junction to Bondsville. The enginehouse is at Palmer.

Nos.	Qty.	Model	Builder	Date	Notes
401	1	44-ton	Whitcomb		
2069	1	GP7r	EMD	1952	Ex-Santa Fe 2069
2100	1	NW5	EMD	1948	Ex-Southern 2100
2443	1	CF7	ATSF	1977	Ex-Santa Fe 2443
Total	4				

MASSACHUSETTS BAY TRANSPORTATION AUTHORITY

Reporting marks: MBTA   **Miles:** 263
Address: 10 Park Plaza, Boston, MA 02116

MBTA operates commuter service from Boston's North Station to Rockport, Ipswich, Haverhill, Lowell, and Fitchburg, Massachusetts, and from South Station to Framingham, Needham Heights, Forge Park, and Stoughton, Mass., and Providence, Rhode Island.

Nos.	Qty.	Model	Builder	Date	Notes
1000-1017	18	F40PH	EMD	1978, 1980	
1025-1033	9	F40PHM-2C	MK	1991	
1050-1075	25	F40PH-2C	EMD	1987-1988	
902, 904, 1921	3	GP9	EMD		
Total	55				

MASSENA TERMINAL RAILROAD

Reporting marks: MSTR   **Miles:** 9
Address: P. O. Box 347, Massena, NY 13662

The Massena Terminal is a switching road at Massena, New York.

Nos.	Qty.	Model	Builder	Date	Notes
12, 13	2	MP15	EMD	1974	
Total	2				

MCCLOUD RIVER RAILROAD

Reporting marks: MCR   **Miles:** 96
Address: P. O. Box 1500, McCloud, CA 96057

The McCloud River extends from Mount Shasta east to Burney, California, 78 miles, with a 19-mile branch to Hambone. The enginehouse is at McCloud.

Nos.	Qty.	Model	Builder	Date	Notes
36-38	3	SD38	EMD	1969	
39	1	SD38-2	EMD	1974	
Total	4				

MCCORMICK, ASHLAND CITY & NASHVILLE RAILROAD

Reporting marks: MACO   **Miles:** 23
Address: 2 Brentwood Commons, Suite 150, Brentwood, TN 37027

The McCormick, Ashland City & Nashville extends from Nashville to Chapmansboro, Tennessee, 19 miles. The line is part of the former Tennessee Central. The enginehouse is at Nashville.

Nos.	Qty.	Model	Builder	Date	Notes
100	1	GP18	EMD	1960	Ex-Rock Island 1333
Total	1				

MCLAUGHLIN LINE RAILROAD

Reporting marks: MCLR   **Track miles:** 5
Address: 710 N. Warren Ave., Apollo, PA 15615

The McLaughlin Line is a switching road at Apollo, Pennsylvania.

Nos.	Qty.	Model	Builder	Date	Notes
4	1	80-ton	GE	1958	
7	1	65-ton	Porter	1944	
12	1	80-ton	GE	1958	
Total	3				

MERIDIAN & BIGBEE RAILROAD

Reporting marks: MBRR   **Miles:** 51
Address: 199 22nd Ave. S., Meridian, MS 39301

The Meridian & Bigbee runs from Meridian, Mississippi, to Myrtlewood, Alabama. The enginehouse is at Meridian.

Nos.	Qty.	Model	Builder	Date	Notes
101	1	GP7	EMD	1952	
102	1	GP9	EMD	1957	
103	1	GP7r	EMD	1951	1
104	1	GP9	EMD	1957	2
105	1	GP7	EMD	1951	3

Nos.	Qty.	Model	Builder	Date	Notes
106	1	CF7	ATSF	1977	4
8244	1	GP7	EMD	1953	5
Total	7				

Notes:
1. Ex-Illinois Central 8951, acquired 1964; rebuilt by Illinois Central Gulf 1976
2. Ex-New York Central 6067
3. Ex-Tennessee, Alabama & Georgia 709, Southern 8236
4. Ex-Santa Fe 2442
5. Ex-Southern 8244, stored unserviceable for parts

MG RAIL

Reporting marks: MGRI **Miles:** 8
Address: 720 Olive St., Suite 2800, St. Louis, MO 63101
MG Rail is a terminal road at Jeffersonville, Indiana.

Nos.	Qty.	Model	Builder	Date	Notes
1210	1	SW9	EMD	1952	Ex-Terminal RR Association 1210
Total	1				

MICHIGAN SHORES RAILROAD

Reporting marks: MSR **Miles:** 8
Address: 434 East Grove St., Greenville, MI 48838
Michigan Shores Railroad is a switching line at Muskegon, Michigan.

Nos.	Qty.	Model	Builder	Date	Notes
1077	1	RS3	Alco	1955	Ex-Detroit & Mackinac 1077
Total	1				

MICHIGAN SOUTHERN RAILROAD

Reporting marks: MSO **Miles:** 24
Address: P. O. Box 239, White Pigeon, MI 49099
Michigan Southern extends from Sturgis to Coldwater, Michigan, on a former Conrail line that was part of the original Michigan Southern Railroad (chartered 1837). Engine facilities are at Sturgis.

Nos.	Qty.	Model	Builder	Date	Notes
66	1	S2	Alco	1949	Ex-South Buffalo
466	1	RS2	Alco	1946	Ex-Detroit & Mackinac 466;
Total	2				first RS2 built

METRA
(Chicago Commuter Rail Service Board)

Miles: 396

Address: 547 W. Jackson Blvd., Chicago, IL 60606
Metra operates diesel-powered commuter service from Chicago to Orland Park, Joliet, Aurora, Geneva, Big Timber, Harvard, McHenry, and Fox Lake, Illinois, and Kenosha, Wisconsin. Metra operates electric commuter service from Chicago to South Chicago, University Park, and Blue Island, Ill.

Nos.	Qty.	Model	Builder	Date	Notes
100-149	50	F40PH	EMD	1977-1980	1
150-173	24	F40PH-2	EMD	1981-1983	2
174-184	11	F40PH-2	EMD	1989	2
185-214	30	F40PH-M	EMD	1991-1992	3
600-614	15	F40C	EMD	1974	4
9900-9908, 9910-9925	25	E9Ar	EMD	1949-1956	5
Total	155				

Notes:
1. 3200 h.p.
2. 3000 h.p.
3. Slant-front cab
4. Ex-Milwaukee Road
5. Ex-Burlington Northern E8s and E9s, rebuilt to E9 specifications by Morrison-Knudsen; also appear on BN roster

Metro-North, an agency of the state of New York, teams up with NJ Transit, an arm of the New Jersey Department of Transportation, to operate commuter service between Port Jervis, N.Y., and Hoboken, N.J. GP40FH-2 No. 4185 is one of 6 Metro-North locomotives assigned to that service.

METRO-NORTH COMMUTER RAILROAD

Reporting marks: MNCR   **Miles:** 660
Address: 347 Madison Ave., New York, NY 10017

Metro-North operates commuter service from New York's Grand Central Terminal to Poughkeepsie and Dover Plains, New York, and, in conjunction with the Connecticut Department of Transportation, to New Haven, Connecticut. Metro-North also operates commuter service between Suffern and Port Jervis, N.Y., in conjunction with NJ Transit.

Nos.	Qty.	Model	Builder	Date	Notes
401-403	3	E10B	GE	1952	1
543	1	GP7u	EMD	1953	2
605	1	RS3m	Alco	1952	3
750	1	GP9	EMD	1955	4
801-807	7	B23-7	GE	1978	5
2001+2033	31	FL9	EMD	1957-1960	

Nos.	Qty.	Model	Builder	Date	Notes
2040, 2041	2	FL9RHAC	RLW	1991	6
4184-4189	6	GP40FH-2	M-K	1988-1990	7
5022, 5023, 5042, 5047	4	FL9	EMD	1957-1960	8
Total	**48**				

Notes:
1. Electric; ex-Conrail 4750, 4751, 4753, Niagara Junction 14, 16, 17
2. Ex-Conrail 5432, New York Central 5770
3. Ex-Conrail 9905, Lackawanna 912; rebuilt with 1200 h.p. EMD engine
4. Ex-Conrail 7508, New York Central 5935
5. Ex-Conrail 1903-1909
6. Rebuilt by Republic Locomotive Works with a 3200 h.p., 12-cylinder EMD 710 engine and AC traction motors. 2040 is ex-2009, previously Penn Central 5034, New Haven 2034
7. Rebuilt GP40s for Suffern-Port Jervis service
8. Rebuilt 1991 by Republic Locomotive Works; painted New Haven colors

MID-ATLANTIC RAILROAD

Reporting marks: MRR   **Miles:** 89
Address: P. O. Box 368, Chadbourn, NC 28436
Mid-Atlantic's lines run between Mullins, South Carolina, and Whiteville, North Carolina, 37 miles, and from Conway, S. C., to Chadbourn, N. C., 39 miles. The enginehouse is at Chadbourn.

Nos.	Qty.	Model	Builder	Date	Notes
950, 951, 958	3	GP18	EMD	1961	Ex-Norfolk & Western, same numbers
Total	3				

MIDDLETOWN & HUMMELSTOWN RAILROAD

Reporting marks: MIDH   **Miles:** 9
Address: P. O. Box G, Hummelstown, PA 17036
The Middletown & Hummelstown connects the two Pennsylvania towns of its name. The enginehouse is at Middletown.

Nos.	Qty.	Model	Builder	Date	Notes
1	1	65-ton	GE	1941	Ex-U. S. Army
Total	1				

MIDDLETOWN & NEW JERSEY RAILWAY

Reporting marks: MNJ   **Miles:** 15
Address: 140 East Main St., Middletown, NY 10940
Middletown & New Jersey extends about 15 miles southwest from Middletown, New York. However, it rarely operates more than a few miles out of Middletown.

Nos.	Qty.	Model	Builder	Date	Notes
1, 2	2	44-ton	GE	1946, 1947	
Total	2				

MIDLAND TERMINAL COMPANY

Reporting marks: MDLR   **Miles:** 11
Address: 1200 Midland Ave., Midland, PA 15059
The Midland Terminal is a switching railroad at Midland, Pennsylvania.

Nos.	Qty.	Model	Builder	Date	Notes
431	1	NW2	EMD	1941	
801	1	SW8	EMD	1951	
Total	2				

MIDLOUISIANA RAIL

Reporting marks: MDR   **Miles:** 64
Address: P. O. Box 2028, Hodge, LA 71247
MidLouisiana Rail operates freight service from Gibsland, Louisiana, south through Hodge, location of the enginehouse, to Winnfield, 64 miles. The line north of Hodge is the former North Louisiana & Gulf Railroad; from Hodge south it is the former Central Louisiana & Gulf (previously a branch of the Rock Island). MidLouisiana is a subsidiary of MidSouth Rail Corporation.

Nos.	Qty.	Model	Builder	Date	Notes
42-45	4	MP15	EMD	1975	Ex-North Louisiana & Gulf
Total	4				

MID-MICHIGAN RAILROAD

Reporting marks: MMRR   **Miles:** 67
Address: 432 E. Grove St., Greenville, MI 48838
Mid-Michigan operates two former CSX lines (ex-Chesapeake & Ohio, previously Pere Marquette): from Elmdale to Greenville, Michigan, 32 miles, and from Saginaw through Alma to Elwell, Mich., 35 miles. Enginehouses are at Greenville and St. Louis, Mich.

Nos.	Qty.	Model	Builder	Date	Notes
24	1	GP9	EMD	1958	Ex-Nickel Plate 517
5967	1	GP9	EMD	1955	Ex-Chesapeake & Ohio 5967
Total	2				

MID-MICHIGAN RAILROAD, TEXAS NORTHEASTERN DIVISION

Reporting marks: TNER   **Miles:** 184
Address: P. O. Box 1296, Sherman, TX 75091-9702

Mid-Michigan's Texas Northeastern Division extends from Texarkana to Sherman, Texas, with a branch from Bells to Trenton. The enginehouse is at Sherman, Texas.

Nos.	Qty.	Model	Builder	Date	Notes
115	1	GP7	EMD	1951	1
173	1	GP9	EMD	1956	
1229, 1237	2	SSB1200	ATSF		2
2153, 2166, 2219	3	GP7r	EMD	1951-1953	2
Total	7				

Notes:
1. Ex-Missouri-Kansas-Texas
2. Ex-Santa Fe

MIDWEST COAL HANDLING COMPANY

Reporting marks: MWCL   **Miles:** 7
Address: 1901 Lantaff Blvd., Madisonville, KY 42431

Midwest Coal Handling operates between Drakesboro and Central City, Kentucky, 7 miles, on a former Louisville & Nashville line. The enginehouse is at Drakesboro.

Nos.	Qty.	Model	Builder	Date	Notes
48	1	RSD1	Alco	1942	1
2005	1	GP7	EMD	1951	2
2492, 2495, 2508, 2511	4	CF7	ATSF	1974	2
2627	1	CF7	ATSF	1972	2
Total	7				

Notes:
1. Ex-U. S. Army
2. Ex-Santa Fe, same numbers

Louis Saillard

The angled cab roof identifies MidSouth 7001 as a fairly late CF7. The paint scheme is dark gray with a green stripe, barely visible in black-and-white reproduction.

MIDSOUTH RAIL CORPORATION

Reporting marks: MSRC   **Miles:** 1,085
P. O. Box 1232, Jackson, MS 39215-1232

MidSouth operates from Shreveport, Louisiana, to Meridian, Mississippi; from Hattiesburg to Gulfport, Miss.; and from Redwood to LeTourneau, Miss. MidSouth's subsidiary SouthRail operates from Waynesboro to Corinth, Miss.; from Bay Springs to Aberdeen, Miss.; from Woodland to Middleton and Corinth, Miss.; and from Artesia, Miss., to Brookwood, Ala. MidSouth's shops are at Jackson, Miss.

Nos.	Qty.	Model	Builder	Date	Notes
1001-1083	83	GP10	ICG	1968-1979	1
1801-1804	4	GP18	EMD	1960, 1963	1
7001-7015	15	CF7	ATSF	1972-1977	2
9000-9005	5	GP9	EMD	1956-1959	
Total	107				

Notes:
1. Ex-Illinois Central Gulf
2. Ex-Santa Fe

MINERAL WELLS & EASTERN RAILWAY

Reporting marks: MWRY   **Miles:** 23
Address: P. O. Box 2624, Victoria, TX 77902
Mineral Wells & Eastern extends from Weatherford to Mineral Wells, Texas, 22 miles. The enginehouse is at Mineral Wells.

Nos.	Qty.	Model	Builder	Date	Notes
701	1	GP30	EMD	1963	Ex-Union Pacific 701
Total	1				

MINNESOTA COMMERCIAL RAILWAY

Reporting marks: MNNR   **Track miles:** 121
Address: 508 Cleveland Ave. N., St. Paul, MN 55114
Minnesota Commercial Railway (formerly Minnesota Transfer Railway) is a switching road in Minneapolis and St. Paul.

Nos.	Qty.	Model	Builder	Date	Notes
100	1	GP7	EMD	1952	1
302-304, 306	4	SW1500	EMD	1968-1970	2
400	1	NW2	EMD	1949	3
484	1	CF7	ATSF	1975	
Total	7				

Notes:
1. Ex-Bangor & Aroostook 74
2. Ex-Minnesota Transfer 302-304, 306
3. Ex-Elgin, Joliet & Eastern 441

MINNESOTA, DAKOTA & WESTERN RAILWAY

Reporting marks: MDW   **Miles:** 4
Address: P. O. Box 19, International Falls, MN 56649
The Minnesota, Dakota & Western is a 4-mile switching road at International Falls, Minnesota.

Nos.	Qty.	Model	Builder	Date	Notes
16-20	5	S2	Alco	1947-1951	
Total	5				

MISSISSIPPI & SKUNA VALLEY RAILROAD

Reporting marks: MSV   **Miles:** 22
Address: P. O. Box 265, Bruce, MS 38915
The Mississippi & Skuna Valley operates from Bruce Junction to Bruce, Mississippi. The enginehouse is at Bruce.

Nos.	Qty.	Model	Builder	Date	Notes
D-4	1	SW900	EMD	1952	
D-5	1	CF7	ATSF	1978	
Total	2				

MISSISSIPPIAN RAILWAY

Reporting marks: MISS   **Miles:** 24
Address: 935 S. Main St., Amory, MS 38821
The Mississippian Railway extends from Amory to Fulton, Mississippi. The enginehouse is at Amory.

Nos.	Qty.	Model	Builder	Date	Notes
1	1	SW1	EMD	1952	
314	1	S1	Alco	1947	Stored unserviceable
321	1	S1	Alco	1950	
Total	3				

MISSISSIPPI DELTA RAILROAD

Reporting marks: MSDR   **Miles:** 58
Address: P. O. Box 1446, Clarksdale, MS 38614
Mississippi Delta operates from Swan Lake to Lula and Jonestown, Mississippi. The enginehouse is at Clarksdale, Miss.

Nos.	Qty.	Model	Builder	Date	Notes
7738	1	GP8	ICG	1975	Ex-Illinois Central Gulf 7738
8068	1	GP10	ICG	1970	Ex-Illinois Central Gulf 8068
Total	2				

MISSISSIPPI EXPORT RAILROAD

Reporting marks: MSE **Miles:** 42
Address: P. O. Box 743, Moss Point, MS 39563
Mississippi Export runs freight service between Evanston and Pascagoula, Mississippi. The enginehouse is at Moss Point, Miss.

Nos.	Qty.	Model	Builder	Date	Notes
60	1	GP9	EMD	1959	
64	1	SW1500	EMD	1973	
65, 66	2	GP38-2	EMD	1975, 1979	
Total	4				

MNVA RAILROAD

Reporting marks: MNVA **Miles:** 142
Address: 262 First St., Morton, MN 56270
MNVA Railroad operates from Hanley Falls, Minnesota, through Norwood to Minneapolis on a former Minneapolis & St. Louis route. The enginehouse is at Morton, Minn.

Nos.	Qty.	Model	Builder	Date	Notes
315	1	GP9	EMD	1959	Ex-Milwaukee Road
321	1	GP9u	EMD	1963	Ex-Milwaukee Road

MODESTO & EMPIRE TRACTION COMPANY

Reporting marks: MET **Miles:** 5
Address: P. O. Box 3106, Modesto, CA 95353
The Modesto & Empire Traction Company extends from Modesto, California, east to a connection with the Santa Fe at Empire. The enginehouse is at Modesto.

Nos.	Qty.	Model	Builder	Date	Notes
600	1	70-ton	GE	1947	
601	1	70-ton	GE	1949	

Nos.	Qty.	Model	Builder	Date	Notes
322	1	SD35	EMD	1965	Ex-Louisville & Nashville 1216
428	1	NW2	EMD	1949	
862	1	GP35	EMD	1963	Ex-Chicago & North Western 862
904	1	SD24	EMD	1963	Ex-Kennecott Copper 904
Total	6				

MOHAWK, ADIRONDACK & NORTHERN RAILROAD

Reporting marks: MHWA **Miles:** 108
Address: 8364 Lewiston Road, Batavia, NY 14020
Mohawk, Adirondack & Northern operates former Conrail (ex-New York Central routes) from Carthage, New York, to Lowville, 17 miles; from Carthage to Newton Falls, 46 miles; and from Utica to Lyons Falls, 45 miles.

Nos.	Qty.	Model	Builder	Date	Notes
804-806	3	C425	Alco	1964	1
Total	3				

Notes:
1. Ex-BC Rail, previously Erie Lackawanna

Nos.	Qty.	Model	Builder	Date	Notes
602	1	70-ton	GE	1952	
603	1	70-ton	GE	1955	Ex-Cherry River Boom & Lumber
604	1	70-ton	GE	1950	Ex-North Louisiana & Gulf
605	1	70-ton	GE	1947	Ex-Southwest Portland Cement
606	1	70-ton	GE	1950	Ex-Southern Pacific
607	1	70-ton	GE	1952	Ex-Oregon, Pacific & Eastern
608	1	70-ton	GE	1955	Ex-Oregon, Pacific & Eastern
609	1	70-ton	GE	1955	Ex-Klamath Northern
Total	10				

MONONGAHELA CONNECTING RAILROAD

Reporting marks: MCRR   **Track miles:** 43
Address: 3540 Second Ave., Pittsburgh, PA 15219
Monongahela Connecting is a switching line at Pittsburgh.

Nos.	Qty.	Model	Builder	Date	Notes
400	1	T6	Alco	1968	
422	1	SW1001	EMD	1974	
423	1	SW1001	EMD	1975	
701	1	C415	Alco	1968	
Total	**4**				

MONONGAHELA RAILWAY

Reporting marks: MGA   **Miles:** 162
Address: 53 Market St., Brownsville, PA 15417
Monongahela Railway is a subsidiary of Conrail. It extends from Brownsville Junction, Pennsylvania, to Blacksville, Pa., and Keyport, West Virginia.

Nos.	Qty.	Model	Builder	Date	Notes
2300-2310	11	S23-7B	GE	1990	1
6600, 6609	2	C30-7A	GE	1984	2
Total	**13**				

Notes:
1. U23Bs remanufactured by GE
2. Leased from Conrail; also on Conrail roster

MONTANA RAIL LINK

Reporting marks: MRL   **Miles:** 940
Address: P. O. Box 8779, Missoula, MT 59807
Montana Rail Link extends from Huntley, Montana, to Sandpoint, Idaho, and by trackage rights on Burlington Northern to Spokane, Washington. It has six branches in Montana, all former Northern Pacific routes, as is the main line. The road's shops are at Missoula and Laurel, Mont.

Nos.	Qty.	Model	Builder	Date	Notes
11	1	NW2u	EMD	1939	1
12-15	4	SW1200	EMD	1956-1957	2
16	1	SW9	EMD	1952	3
17, 18	2	SW1200	EMD	1957	4
104-117, 119-127	23	GP9	EMD	1954-1958	5
151	1	GP9u	EMD	1957	6
200, 201, 204-206, 208-211, 213-216, 218	14	SD40	EMD	1967-1971	7
250	1	SD40-2	EMD	1974	8
251, 252	2	SD40r	EMD	1966	9
253-258	6	SD40r	EMD	1967-1971	10
290	1	SD40r	EMD	1966	11
301-308	8	SD45-2	EMD	1974	12
351-358	8	SD45r	EMD	1967-1971	13
401, 402	2	GP35	EMD	1964	14
600-602	3	SD9	EMD	1956, 1957	15
603	1	SD9	EMD	1956	16
604	1	SD9	EMD	1956	17
605-607	3	SD9	EMD	1953, 1954	18
608-610	3	SD9	EMD	1957	19
700	1	E9A	EMD	1961	20
701, 702	2	SD35	EMD	1965	21
703	1	SD35	EMD	1964	22
704, 705	2	SD35	EMD	1965	23
Total	**91**				

Notes:
1. Ex-Burlington Northern 19; rebuilt to 1200 h.p. by BN
2. Ex-Burlington Northern 208, 220, 203, 217
3. Ex-Burlington Northern 269
4. Ex-Burlington Northern 215, 216

5. Ex-Burlington Northern
6. Ex-Burlington Northern 1744; rebuilt 1991 by MRL to 1900 h.p.
7. Ex-Burlington Northern 6301-6345 series
8. Ex-Burlington Northern 6377
9. Ex-Union Pacific 3004, 3024, rebuilt 1990, classified SD40-2XR
10. Ex-Burlington Northern, rebuilt 1990-1991, classified SD40-2XR
11. Ex-Burlington Northern 6395, rebuilt 1991, classified SD40-2XR
12. Ex-CSX 8975-8982
13. Ex-Burlington Northern rebuilt by MRL; classified SD45-2XR
14. Ex-Grand Trunk Western 6353, 6355; previously Detroit, Toledo & Ironton 353, 355
15. Ex-Elgin, Joliet & Eastern 604, 606, 603
16. Ex-Southern Pacific 4361
17. Ex-Elgin, Joliet & Eastern 605
18. Ex-Southern Pacific 1541, 4316, 1515
19. Ex-Elgin, Joliet & Eastern 611, 609, 610
20. Ex-Milwaukee Road 37C
21. Ex-Norfolk & Western 1566, 1546
22. Ex-Chesapeake & Ohio 4591
23. Ex-Norfolk & Western 1543, 1553

Kyle Brehm

Montana Rail Link 401, an ex-Grand Trunk Western GP35, leads a multi-domed excursion train along a branch near Hamilton, Mont.

MONTANA WESTERN RAILWAY

Reporting marks: MWRR   **Miles:** 52

Address: 700½ Railroad St., Butte, MT 59701

Montana Western extends from Garrison to Butte, Montana. The enginehouse is at Butte.

Nos.	Qty.	Model	Builder	Date	Notes
103	1	GP7	EMD	1954	1
104-107	4	GP9	EMD	1957	1
201-203	3	GP9	EMD	1957	2
Total	**8**				

Notes:
1. Ex-Butte, Anaconda & Pacific; leased from Rarus Railway
2. Ex-CSX

MOSCOW, CAMDEN & SAN AUGUSTINE RAILROAD

Reporting marks: MCSA   **Miles:** 7

Address: P. O. Box 128, Camden, TX 75934

The Moscow, Camden & San Augustine operates freight service between Moscow and Camden, Texas, 7 miles. The enginehouse is at Camden.

Nos.	Qty.	Model	Builder	Date	Notes
1	1	SW1200	EMD	1954	
3	1	SW900	EMD	1957	
Total	**2**				

The locomotive — No. 1303, the first to be painted for CTCUM — and the cars are ex-Canadian Pacific, and the track they are on at Dorion, Quebec, is stil CP Rail property.

Alex Mayes

COMMISSION DE TRANSPORT DE LA COMMUNAUTÉ URBAINE DE MONTREAL
(Montreal Urban Community Transit Commission)

Miles: 58

Address: 159 St. Antoine St., Montreal, PQ H2N 1M3 Canada

MUCTC operates commuter service on two routes out of Montreal: Windsor Station to Rigaud on CP Rail (diesel-powered) and Central Station to Deux Montagnes on Canadian National (electrified).

Nos.	Qty.	Model	Builder	Date	Notes
DIESEL LOCOMOTIVES					
1300-1305	6	FP7	GMD	1952	1
1306	1	FP7	GMD	1951	2
1310-1313	4	GP9u	GMD	1959	3

Nos.	Qty.	Model	Builder	Date	Notes
ELECTRIC LOCOMOTIVES					
6710-6715	6		GE	1914-1917	4, 6

Nos.	Qty.	Model	Builder	Date	Notes
6716, 6717, 6722, 6723	4		EE	1924	4, 6
6725-6727	3		GE	1950	5, 6
Total	**24**				

Notes:
1. Ex-Canadian Pacific 4070-4075
2. Ex-Canadian Pacific 4040
3. Ex-Canadian National 4346, 4307, 4299, 4309
4. Boxcab B+B units
5. Center-cab B-B units
6. Electric units for commuter service through the Mount Royal Tunnel are leased from Canadian National Railways. They also appear on the Canadian National roster.

MOUNT HOOD RAILROAD

Reporting marks: MH   **Miles:** 21
Address: 110 Railroad Ave., Hood River, OR 97031
The Mount Hood Railroad operates freight and excursion service between Hood River and Parkdale, Oregon. The enginehouse is at Hood River.

Nos.	Qty.	Model	Builder	Date	Notes
88, 89	2	GP9	EMD	1955, 1959	
Total	**2**				

MOUNT VERNON TERMINAL RAILWAY

Reporting marks: MVT   **Miles:** 3
Address: P. O. Box 216, Clear Lake, WA 98235
Mount Vernon Terminal is a switching railroad at Mount Vernon, Washington.

Nos.	Qty.	Model	Builder	Date	Notes
505	1	VO1000	BLW	1944	1
Total	**1**				

Notes:
1. Ex-U. S. Navy 11 (from Bremerton, Washington)

MUNCIE & WESTERN RAILROAD

Reporting marks: MWR   **Miles:** 4
Address: 1509 S. Macedonia, Muncie, IN 47307
The Muncie & Western is a switching road at Muncie, Indiana.

Nos.	Qty.	Model	Builder	Date	Notes
7	1	65-ton	GE	1944	
8	1	70-ton	GE	1946	
Total	**2**				

MUNICIPALITY OF EAST TROY, WISCONSIN, RAILROAD

Reporting marks: METW   **Miles:** 7
Address: 222 N. Charles St., Waukesha, WI 53186
The Municipality of East Troy Railroad extends from East Troy, Wisconsin, to a connection with Wisconsin Central at Mukwonago. The line, a remnant of the Milwaukee Electric, is still electrified, and the Wisconsin Trolley Museum operates excursion trains on the line. The enginehouse is at East Troy.

Nos.	Qty.	Model	Builder	Date	Notes
4	1	80-ton	GE	1944	1
Total	**1**				

Notes:
1. Ex-Minnesota, Dakota & Western 11

NAPA VALLEY RAILROAD

Reporting marks: NVRR   **Miles:** 22
Address: 1275 McKinstry St., Napa, CA 94559
Napa Valley Railroad runs a dinner train, the Napa Valley Wine Train, between Napa and St. Helena, California, and also operates freight service. The enginehouse is at Napa.

Nos.	Qty.	Model	Builder	Date	Notes
50	1	44-ton	GE	1941	1
70-73	4	FPA4	MLW	1958-1959	2
Total	**5**				

Notes:
1. Ex-Camino, Placerville, & Lake Tahoe 102
2. Ex-VIA Rail Canada 6760, 6775, 6787, 6790; 70 and 72 are equipped with HEP

NASH COUNTY RAILROAD

Reporting marks: NCYR   **Miles:** 19

Address: P. O. Box 1929, Laurinburg, NC 28352

Nash Country Railroad extends from Rocky Mount to Spring Hope, North Carolina. The enginehouse is at Rocky Mount.

Nos.	Qty.	Model	Builder	Date	Notes
115, 119	2	SW1	EMD	1949	1
122	1	NW2	EMD	1948	2
Total	**3**				

Notes:

1. Ex-Union Railroad 456, 464
2. Ex-Chicago, Madison & Northern 201, Baltimore & Ohio 9547

NASHVILLE & EASTERN RAILROAD

Reporting marks: NERR   **Miles:** 131

Address: 206 South Maple St. Lebanon, TN 37087

The Nashville & Eastern operates a portion of the old Tennessee Central from Vine Hill, Tennessee, to Monterey, with branches to Old Hickory and Carthage. The enginehouse is at Vine Hill.

Nos.	Qty.	Model	Builder	Date	Notes
1801-1803, 1850-1853	7	RSD12	Alco	1956-1959	1
2525	1	CF7	ATSF	1974	2
5323, 5328, 5338-5340, 5343-5345	8	U30B	GE	1966-1972	3
Total	**16**				

Notes:

1. Ex-Lake Superior & Ishpeming, same numbers
2. Ex-Santa Fe, same number
3. Ex Burlington Northern, same numbers

NATCHEZ TRACE RAILROAD

Reporting marks: NTR   **Miles:** 57

Address: P. O. Box 477, Holly Springs, MS 38635

Natchez Trace Railroad operates from Grand Junction, Tennessee, to Oxford, Mississippi. The enginehouse is at Holly Springs, Miss.

Nos.	Qty.	Model	Builder	Date	Notes
8274, 8281	2	GP7	EMD	1951, 1952	1
Total	**2**				

Notes:

1. Ex-Southern 8274, 8281

NATIONAL RAILWAYS OF MEXICO

Reporting marks: NDM   **Miles:** 12,712

Address: 1766 Avenida Jesús García 140, Mexico DF 06358, Mexico

In 1986 National Railways of Mexico merged the four regional railroads of that country: Chihuahua Pacific, Pacific Railroad, Sonora-Baja California, and United South Eastern Railways. For some years prior to 1986, new power had been arriving painted NdeM colors and numbered in NdeM's numbering scheme, but the process of consolidating rosters is a slow one (and information trickles north across the border even more slowly). The consolidated roster appears below; rosters for the regional railroads appear on the update pages at the back of the book.

Nos.	Qty.	Model	Builder	Date	Notes
1/01-1/07	7	B23-7	GE	1981	
17, 19	2	PA1u	Alco	1948, 1947	1, 2
201, 204, 217, 220	4	C420	Alco	1964	3
601-618	18	C628	Alco	1964	1
802	1	GA8	EMD	1964	4
803	1	GA18	EMD	1971	4
932	1	U30C	GE	1968	5
2203	1	FTA	EMD	1945	11, 23
3204, 3205, 3215, 3219, 3234	5	C424	MLW	1966	6
5301-5306	6	S6	Alco	1955-1960	7
5310-5313	4	S6	Alco		2
5400, 5401, 5403, 5409,					

Nos.	Qty.	Model	Builder	Date	Notes
5410, 5412, 5416	7	GA8	EMD	1964, 1967	8
5418-5421, 5427	5	GA8	EMD	1964, 1967	7, 8
5602, 5634, 5639, 5654, 5656	5	RS1	Alco	1950-1959	
5700, 5701	2	C420	Alco	1965	7
5800, 5802-5806, 5808-5814, 5816-5818, 5821, 5823, 5826-5828, 5830, 5832-5834, 5837, 5842-5844, 5846-5848, 5851-5858, 5861, 5863, 5864, 5866-5889	67	G12	EMD	1956-1964	
5902	1	RSD35	Alco	1963	
6208A	1	F2A	EMD	1946	9
6300-6307	8	GP28	EMD	1964-1965	10

6700-6703, 6705-6715, 6717-6736, 6738-6749, 6751-6758, 6760-

Nos.	Qty.	Model	Builder	Date	Notes
6762, 6764-6768, 6780-6799	93	C30-7	GE	1979-1980	
6809, 6814	2	AS616	BLW	1954	12
7000-7028	29	GP40-2	EMD	1975-1982	10
7029-7043	15	GP40-2	EMD	1972-1982	11
7100-7103, 7105-7107	7	GP9	EMD	1956-1958	
7200-7205, 7207, 7209-7211, 7213-7215, 7219, 7220, 7222-7224, 7226, 7229-7231, 7234-7241, 7243, 7244, 7248-7250, 7252, 7254-7256, 7258-7261, 7263-7265, 7268-7283, 7285-7288, 7290, 7292	68	RS11	Alco	1958-1964	
7295	1	RS11	Alco	1956	7

7402-7407, 7410-7412, 7414, 7416-7418, 7420-7425, 7427, 7429,

National Railways of Mexico occasionally went shopping north of two borders — in Canada — mostly for passenger cars but sometimes for locomotives. Number 9500 was the first of a batch of 72 M424Ws built in 1980 and 1981 for Mexican railroads: 53 for NdeM, 16 for Pacific Railway, and 3 for United South Eastern Railways.

Pierre Alain Patenaude

Nos.	Qty.	Model	Builder	Date	Notes
7430, 7432, 7433, 7435-7455, 7457, 7459, 7460, 7462-7472	59	RSD12	Alco	1961-1964	
7474-7485, 7487	13	RSD12	Alco	1959-1961	2
7504, 7507, 7511, 7516, 7518, 7521, 7531, 7534	8	GP18	EMD	1961-1963	
7537-7539	3	GP18	EMD	1961	11
7600-7610	11	M420TR	MLW	1975	2
7700, 7702, 7703, 7705-7712	11	RSD5ru	Alco		2
7800-7810	11	RSD5ru	Alco		2
7902-7904, 7906	4	U30C	GE	1969, 1971	2
8000-8002, 8004-8008, 8010-8013	12	M636	MLW	1972-1973	2
8100, 8104, 8106-8112, 8114-8116, 8122, 8124, 8126, 8128, 8130-8132, 8135-8140, 8143, 8144	27	C424	Alco	1964	
8200-8204, 8206-8213, 8216-8244, 8246, 8248, 8250-8254	49	GP35	EMD	1964-1965	13
8255, 8256	2	GP38	EMD	1971	7
8257, 8258	2	GP35	EMD	1965	11
8300-8303, 8305, 8309-8311, 8313-8315, 8317-8325, 8327-8331	25	C628	Alco	1967-1968	
8332-8337	6	C628	Alco	1966, 1968	2
8400-8505, 8407-8409	9	GP40	EMD	1967	
8410-8417	8	GP40	EMD	1971	10
8500-8512, 8514-8521	21	SD40	EMD	1968	
8522-8535	14	SDP40	EMD	1968, 1970	
8536-8544, 8546-8557, 8559-8575	38	SD40	EMD	1971-1972	
8576-8585	10	SD40	GMD	1972	
8600, 8601, 8603-8611, 8613, 8614, 8616-8618	16	C630	MLW	1972	

Nos.	Qty.	Model	Builder	Date	Notes
8700-8722, 8724-8747, 8749-8761, 8763-8796, 8798	95	SD40-2	EMD	1972-1980	
8800-8859	60	SW1504	EMD	1973	
8900-8905, 8907-8911, 8913-8917, 8922, 8926-8929, 8933, 8934, 8937	24	U36C	GE	1973	
8938-8943, 8945, 8946, 8948, 8950-8952, 8954-8957	16	U36CG	GE	1973	
8958-8969, 8972, 8973, 8975-8978, 8980-8982, 8984-8986	24	U36C	GE	1974	
8987	1	U33Cu	GE	1971	14
8989, 8991-8996	7	U36C	GE	1975	2
9000-9005, 9007-9010, 9012, 9014, 9015, 9017, 9019, 9020, 9022, 9024-9034, 9036, 9037, 9039, 9040, 9043, 9044	34	U18B	GE	1974	
9100-9102, 9104-9108, 9110-9115, 9117, 9119-9129	26	U23B	GE	1975	
9130-9133, 9135-9160, 9162, 9164, 9167, 9168, 9070-9080	45	B23-7	GE	1980-1981	
9181	1	B23-7	GE	1979	7
9182-9191	10	U23B	GE	1975	2
9200, 9202, 9204-9219	18	GP38-2P	EMD	1975	
9220-9225, 9227, 9228, 9230-9241, 9243-9257, 9259-9273, 9275-9279, 9281, 9282, 9284-9291, 9293, 9296-9299	70	GP38-2	EMD	1975-1976	
9300-9316	17	U36C	GE	1975	
9317-9341	25	C36-7	GE	1979-1980	
9400-9410, 9412-9414	14	GP38-2	EMD	1975-1979	
9415-9427	13	GP38-2	EMD	1972-1982	7

Nos.	Qty.	Model	Builder	Date	Notes
9428-9439	12	GP38-2	EMD	1979-1980	10
9500-9552	53	M424W	BBD	1980-1981	
9553-9555	3	M424W	MLW	1981	7
9556-9567	12	M424W	MLW	1980-1981	2
9600-9605, 9607-9616, 9618-9652	51	C30-7	GE	1980-1981	
9653-9660, 9662-9667, 9669-9680, 9682-9687	32	C30-7	GE	1979, 1981	2
9801-9825	25	MP15DC	EMD	1982	
9901-9909	9	GP38-2	EMD	1982	
10001-10030, 10032-10044, 10046-10052	50	B23-7	GE	1981-1982	15
11001-11064, 11066-11166	165	C30-7	GE	1982-1985	17
11149-11152	4	C30-7	GE	1986	16
12001-12003, 12005-12011	101	B23-7	GE	1982	15
13001-13004	4	SD40-2	EMD	1986	2
13005-13008	4	SD40-2	EMD	1972	7
13009-13048	40	SD40-2	EMD	1972	18
13049-13076	28	SD40-2			19
14000-14046	47	C30-S7	GE	1990-1991	20
DH 801, DH 802	2	SD45	EMD	1966	21
EA001-EA039	39	E60C	GE	1982-1983	22
Total	1,901				

Notes:
1. Ex-Delaware & Hudson
2. Ex-Pacific Railroad (Ferrocarril del Pacifico)
3. Ex-Long Island, same numbers
4. Ex-Coahuila y Zacatecas, originally 36"gauge
5. Ex-Chicago & North Western
6. Ex-Canadian National
7. Ex-United South Eastern Railways (Unidos del Sureste)
8. 36" gauge
9. Stored unserviceable
10. Ex-Chihuahua Pacific
11. Ex-Sonora-Baja California
12. Repowered with EMD engines
13. Many rebuilt to GP38 specifications
14. Ex-Southern Pacific 8718, rebuilt to 3600 h.p.
15. Assembled in Mexico
16. Ex-Pacific Railroad (Ferrocarril del Pacifico; to be renumbered
17. Assembled in Mexico; 11132-11135 are ex-Pacific Railroad
18. Ex-Milwaukee Road, rebuilt by MK and VMV
19. SD40s and SD45s rebuilt to SD40-2 specifications by VMV
20. 14000-14022 assembled in Mexico
21. Ex-EMD 4354, 4352
22. Electric locomotives
23. Ex-Northern Pacific 6010D, restored to operating condition

NDC RAILROAD

Reporting marks: NDCR   **Miles:** 0.5
Address: P. O. Box 7, Route 329, Northampton, PA 18067
NDC Railroad operates a half mile of track for switching at Northampton, Pennsylvania.

Nos.	Qty.	Model	Builder	Date	Notes
51	1	S6	Alco	1956	
99	1	RS3m	Alco	1953	
101	1	S6	Alco	1956	
Total	3				

NEWBURGH & SOUTH SHORE RAILROAD

Reporting marks: NSR   **Track miles:** 9
Address: 2728 E. 104th St., Chicago, IL 60617-5766
Newburgh & South Shore is a switching railroad at Cleveland, Ohio.

Nos.	Qty.	Model	Builder	Date	Notes
1019, 1021	2	SW1001	EMD	1971, 1975	
Total	2				

117

NEW ENGLAND SOUTHERN RAILROAD

Reporting marks: NEGS   **Miles:** 45
Address: P. O. Box 2106, Concord, NH 03301
New England Southern extends from Manchester to Meredith, New Hampshire, with a branch from Concord to Penacook. The enginehouse is at Concord.

Nos.	Qty.	Model	Builder	Date	Notes
302	1	GP7	EMD	1950	Ex-Rock Island 438
303	1	S4	Alco	1950	Ex-Green Mountain 303
503	1	GP18	EMD	1960	Ex-Rock Island 1341
Total	3				

NEW HAMPSHIRE & VERMONT RAILROAD

Reporting marks: NHVT   **Miles:** 81
Address: RFD 1, Box 790, Morrisville, VT 05661
The New Hampshire & Vermont operates former Boston & Maine track from Woodsville to Berlin, New Hampshire, 60 miles, with a 20-mile branch to Groveton, New Hampshire. The enginehouse is at Whitefield, N. H.

Nos.	Qty.	Model	Builder	Date	Notes
405	1	RS11	Alco	1961	
Total	1				

NEW HAMPSHIRE NORTHCOAST CORPORATION

Reporting marks: NHN   **Miles:** 30
Address: P. O. Box 429, Ossipee, NH 03864
New Hampshire Northcoast operates from Ossipee, New Hampshire, south to Rochester, 30 miles, and 12 miles beyond to Dover by trackage rights on Boston & Maine. The enginehouse is at Ossipee.

Nos.	Qty.	Model	Builder	Date	Notes
1755-1759	5	GP9	EMD	1957	
Total	5				

NEW JERSEY TRANSIT RAIL OPERATIONS

Reporting marks: NJT   **Miles:** 1,163 [check]
Address: 1160 Raymond Blvd., Newark, NJ 07101
NJ Transit operates commuter service on several routes: New York (Penn Station) to Trenton; New York and Hoboken to Bay Head; Newark to High Bridge; Hoboken to Gladstone, Dover, Netcong, Suffern, N. Y., Port Jervis, N. Y. (in conjunction with Metro-North), and Spring Valley, N. Y.; and Lindenwold to Atlantic City.

Nos.	Qty.	Model	Builder	Date	Notes
436, 438	2	SW9	EMD	1952	1
503	1	SW1500	EMD	1971	
958-963, 967, 973	8	E60CP	GE	1974	2
4100-4112	13	GP40P	EMD	1968	
4113-4129	17	F40PH-2	EMD	1981	
4130-4144	15	GP40FH-2	M-K	1987-1989	3

Alex Mayes

The E60 was something of a stopgap on Amtrak's roster, soon replaced by smaller, lighter AEM7s. Several displaced Amtrak E60s found a second job hauling commuters for NJ Transit.

Nos.	Qty.	Model	Builder	Date	Notes
4151-4182	32	U34CH	GE	1971-1973	
4400-4414	15	ALP44	ABB	1990	
5681, 5902	2	GP7	EMD	1952	4
7000	1	GP9	EMD	1955	4
Total	**106**				

Notes:
1. Ex-Conrail
2. Electric locomotives; ex-Amtrak
3. Rebuilt GP40s
4. Owned by United Railroads Historical Society.

NEW ORLEANS LOWER COAST RAILROAD

Reporting marks: NOLR   **Miles:** 24
Address: 1502 Belle Chase Highway South, Belle Chase, LA 70037
New Orleans Lower Coast extends from Gretna to Myrtle Grove, Louisiana. The enginehouse is at Belle Chase.

Nos.	Qty.	Model	Builder	Date	Notes
8375, 8377	2	GP10	ICG	1974, 1975	
Total	**2**				

NEW ORLEANS PUBLIC BELT RAILROAD

Reporting marks: NOPB   **Track miles:** 123
Address: P. O. Box 51658, New Orleans, LA 70151
New Orleans Public Belt is a switching road at New Orleans, Louisiana.

Nos.	Qty.	Model	Builder	Date	Notes
103, 105, 106	3	SW1001	EMD	1971	
151-153	3	SW1500	EMD	1971	
Total	**6**				

NEW ORLEANS UNION PASSENGER TERMINAL

Reporting marks: NOUT   **Track miles:** 23
Address: 1001 Loyola Ave., New Orleans, LA 70113
New Orleans Union Passenger Terminal owns the rail passenger station in New Orleans. Its rail facilities are operated by Amtrak.

Nos.	Qty.	Model	Builder	Date	Notes
1	1	SW8	EMD	1953	
Total	**1**				

NEW YORK & LAKE ERIE RAILROAD

Reporting marks: NYLE   **Miles:** 50
Address: 50 Commercial St., Gowanda, NY 14070
The New York & Lake Erie extends from Gowanda to Salamanca, New York, with a branch from Dayton to Waterboro. The enginehouse is at Gowanda.

Nos.	Qty.	Model	Builder	Date	Notes
85	1	S2	Alco	1950	1
1013	1	C425	Alco	1965	2
1700	1	132-ton	GE	1940	3
6101	1	C425	Alco	1966	4
Total	**4**				

Notes:
1. Ex-South Buffalo
2. Ex-Norfolk & Western 1013
3. Ex-Wellsville, Addison & Galeton, formerly Ford Motor Co.; stored unserviceable
4. Ex-Conrail, Pennsylvania

NEW YORK, SUSQUEHANNA & WESTERN RAILWAY

Reporting marks: NYSW   **Miles:** 261
Address: Ridgefield Park, NJ 07660

The Susquehanna extends from Croxton, New Jersey, to Warwick, New York, and from Binghamton, N. Y., to Utica and Jamesville. It has operating rights on Conrail between Warwick and Binghamton. Engine facilities are at Little Ferry, N. J., and Binghamton, N. Y.

Nos.	Qty.	Model	Builder	Date	Notes
116	1	NW2	EMD	1948	1
120	1	SW9	EMD	1953	2
1800, 1802, 1804	3	GP18	EMD	1962	
2010	1	C420	Alco	1967	3
2012	1	GP38	EMD	1968	4
3000, 3006	2	C430	Alco	1967	5
3612, 3614, 3618, 3620, 3622, 3624, 3626, 3630, 3632, 3634	10	SD45	EMD	1970-1971	6
3636, 3638	2	F45	EMD	1970	7
6366, 6515	2	SD45	EMD	1971	8
Total	**23**				

Notes:
1. Ex-Conrail 9264, Penn Central 8684, New York Central 9501, New York, Ontario & Western 116
2. Ex-Chesapeake & Ohio 5091
3. Ex-Virginia & Maryland 202, Long Island 221
4. Ex-Baltimore & Ohio 3800
5. Ex-New York Central 2050, 2056
6. Ex-Burlington Northern 6480, 6486, 6500, 6503, 6509, 6513, 6514, 6521, 6525, 6542
7. Ex-Burlington Northern 6640, 6644
8. Ex-Burlington Northern, same numbers.

NICOLET BADGER NORTHERN RAILROAD

Reporting marks: NBNR   **Miles:** 38
Address: P. O. Box 5, Laona, WI 54541

Nicolet Badger Northern operates freight and excursion service between Wabeno and Tipler, Wisconsin. The enginehouse is at Laona.

Nos.	Qty.	Model	Builder	Date	Notes
62	1	NW2	EMD	1948	1
103	1	65-ton	Whitcomb	1944	2
Total	**2**				

Notes:
1. Ex-Marinette, Tomahawk & Western
2. Ex-U. S. Army, Laona & Northern

NIMISHILLEN & TUSCARAWAS RAILWAY

Reporting marks: NTRY   **Miles:** 61
Address: P. O. Box 24700, Canton, OH 44701-4700

The Nimishillen & Tuscarawas is a switching road serving Massillon and Canton, Ohio.

Nos.	Qty.	Model	Builder	Date	Notes
1203	1	SW1200	EMD	1954	
1208, 1210	2	SW1200	EMD	1956	
1211, 1283, 1285	3	SW1200	EMD	1957	
Total	**6**				

NITTANY & BALD EAGLE RAILROAD

Reporting marks: NBER   **Miles:** 46
Address: 356 Priestley Ave., Northumberland, PA 16823

The Nittany & Bald Eagle extends from Tyrone to Mill Hall, Pennsylvania, with a branch from Milesburg to Lemont. The enginehouse is at Bellefonte.

Nos.	Qty.	Model	Builder	Date	Notes
2427, 2429	2	CF7	ATSF	1977	Ex-Santa Fe 2427, 2429
Total	**2**				

NORFOLK SOUTHERN RAILWAY

Reporting marks: NS   **Miles:** 17,505
Address: Three Commercial Place, Norfolk, VA 23510-2191

Nos.	Qty.	Model	Builder	Date	Notes
50-53, 55-57	7	SD9M	EMD	1957	1
67, 68, 70-83	16	SW1500	EMD	1966-1974	2
100-104	5	TC10	N&W	1983-1986	3
115, 116	2	F40PH	EMD	1977	4
198, 199	2	SD9	EMD	1955	2
673, 696	2	GP9	EMD	1959	1
1002, 1004, 1006	3	SW1	EMD	1947, 1950	2
1007	1	SW1	EMD	1947	5
1012	1	SW1	EMD	1947	2
1209	1	SW1200	EMD	1955	1
1329-1342, 1344-1355, 1357-1388	58	GP40	EMD	1966-1967	1
1580-1624	45	SD40	EMD	1966, 1971	1
1625-1652	27	SD40-2	EMD	1973-1974	1
1733	1	SW1500	EMD	1968	2
1765-1770, 1784, 1800	8	SD45	EMD	1969-1970	1
2000-2003	4	GP9r	EMD	1959	1
2008, 2009	2	GP9u	EMD	1959	1
2105	1	SW1	EMD	1950	1
2290-2293, 2296	5	SW1500	EMD	1970	1
2300-2310, 2312-2321, 2323-2333, 2335-2347	45	SW1500	EMD	1968-1970	2
2348-2375, 2377-2435	87	MP15DC	EMD	1977-1982	2
2645, 2650-2657, 2659-2662, 2664, 2666, 2671-2673, 2675, 2676, 2679, 2680, 2682-2684, 2686, 2687, 2691, 2695, 2699, 2701-2703, 2705-2715	44	GP35	EMD	1965-1966	2
2717-2798, 2800-2822	105	GP38	EMD	1969-1970	2
2823-2878	56	GP38AC	EMD	1971	2
2879-2886	8	GP38	EMD	1966, 1968	2

EMD

The 36 GP59s that Norfolk Southern purchased in 1989 are notable on the NS roster for having low short hoods. Two subtle design elements worth noting are the rounded corners of the cab and the angle in the front wall of the cab, the latter a reminder of the GP30.

Nos.	Qty.	Model	Builder	Date	Notes
2952	1	SD9	EMD	1957	1
3161	1	SD45	EMD	1970	2
3170-3180, 3182-3190, 3192-3197, 3199, 3200	28	SD40	EMD	1971-1972	2
3201-3216, 3218-3243, 3255, 3256, 3258-3328	115	SD40-2	EMD	1972-1979	2
3500-3521	22	B30-7A	GE	1982	2
3522-3566	45	Dash 8-32B	GE	1989	
3815-3820	6	B36-7	GE	1981	2
3900-3939, 3941-3966, 3968, 3969	68	U23B	GE	1972-1977	2
3970-3993, 3995-4023	53	B23-7	GE	1978-1981	2

Nos.	Qty.	Model	Builder	Date	Notes
4100-4109, 4111-4159	59	GP38AC	EMD	1971	1
4160-4163	4	GP38-2	EMD	1977	6
4600-4605	6	GP49	EMD	1980	2
4606-4635, 4637-4641	35	GP59	EMD	1989	
5000-5020, 5024-					
5162, 5165-5256	252	GP38-2	EMD	1972-1979	2
6073-6133, 6135-6206	133	SD40-2	EMD	1975-1980	1
6500-6505	6	SD50	EMD	1980	1
6506-6525	20	SD50	EMD	1984	2
6550-6700	151	SD60	EMD	1984-1991	
7000-7002	3	GP40X	EMD	1978	2
7003-7092	90	GP50	EMD	1980-1981	2
7101-7150	50	GP60	EMD	1991	
8003-8074, 8076-8082	79	C30-7	GE	1978-1979	1
8500-8542	43	C36-7	GE	1981-1984	
8550-8588, 8590-					
8640, 8643-8688	136	C39-8	GE	1984-1987	
8689-8713	25	Dash 8-40C	GE	1990-1991	
9710-9741	32	Slug	N&W	1982-1988	
9818-9820, 9826,					
9827, 9830, 9831, 9833-					
9837, 9839-9841	15	Slug	SOU	1968-1977	
9842-9851	10	Slug	N&W	1982-1985	
9852-9855	4	Slug	GE	1990	
9900, 9902-9923	23	Slug	N&W	1981-1982	
9950, 9951	2	Slug	N&W	1976, 1977	7
Total	**2,053**				

Notes:
1. Ex-Norfolk & Western
2. Ex-Southern Railway (including subsidiaries)
3. GP9 rebuilt with 1000 h.p. Caterpillar engines for yard service
4. Leased from Metra; listed on Metra roster
5. Ex-Atlantic & East Carolina; 1007 is the oldest unit on NS roster
6. Ex-Illinois Terminal
7. Rebuilt from FM Train Masters

NORTH CAROLINA & VIRGINIA RAILROAD

Reporting marks: NCVA   **Miles:** 52
Address: 214 N. Railroad St., Ahoskie, NC 27910
The North Carolina & Virginia operates from Boykins, Virginia, south to Kelford, North Carolina, 30 miles, then northeast 22 miles to Tunis, N. C. The enginehouse is at Ahoskie, N. C.

Nos.	Qty.	Model	Builder	Date	Notes
23	1	GP7	EMD	1950	
6244	1	GP9	EMD	1959	Ex-Chesapeake & Ohio 6244
6515	1	GP9	EMD	1959	Ex-Baltimore & Ohio 6515
Total	**3**				

NORTH CAROLINA & VIRGINIA RAILROAD, VIRGINIA SOUTHERN DIVISION

Reporting marks: NCVB   **Miles:** 78
Address: P. O. Box 12, Keysville, VA 23947
The Virginia Southern Division of the North Carolina & Virginia extends from O&H Junction, near Oxford, North Carolina, to Burkeville, Virginia, 78 miles. The enginehouse is at Keysville, Va.

Nos.	Qty.	Model	Builder	Date	Notes
618	1	GP9	EMD	1956	Ex-Baltimore & Ohio 6618
Total	**1**				

NORTH CAROLINA PORTS RAILWAY COMMISSION

Reporting marks: NCPR   **Miles:** 6
Address: 1717 Woodbine St., Wilmington, NC 28401
North Carolina Ports Railway Commission performs switching at Wilmington, N. C.

Nos.	Qty.	Model	Builder	Date	Notes
1201, 1202	2	SW1200	EMD	1966	1
1203-1205	3	SW1200	EMD	1966	2
Total	5				

Notes:

1. Ex-Missouri Pacific 1201, 1202
2. Ex-Missouri Pacific 1203-1205; leased to Wilmington Terminal Railroad and also shown on that roster

NORTH SHORE RAILROAD

Reporting marks: NSHR   **Miles:** 2

Address: 356 Priestly Ave., Northumberland, PA 17857

The North Shore operates between Northumberland and Hicks Ferry, Pennsylvania. The enginehouse is at Northumberland.

Nos.	Qty.	Model	Builder	Date	Notes
365	1	SW8	EMD	1952	1
446	1	SW9	EMD	1952	2
Total	2				

Notes:

1. Ex-Conrail 8669, Lehigh Valley 263
2. Ex-Conrail 8983, Penn Central 8983

NORTHEAST KANSAS & MISSOURI RAILWAY

Reporting marks: NEKM   **Miles:** 108

Address: P. O. Box 476, Hiawatha, KS 66434

The Northeast Kansas & Missouri extends 108 miles west from St. Joseph, Missouri, to Upland, Kansas, just east of Marysville. The enginehouse is at Hiawatha, Kansas.

Nos.	Qty.	Model	Builder	Date	Notes
2167, 2210	2	GP7r	EMD	1952	Ex-Santa Fe
Total	2				

NORTHWESTERN OKLAHOMA RAILROAD

Reporting marks: NOKL   **Miles:** 9

Address: P. O. Box 1131, Woodward, OK 73801

The Northwestern Oklahoma operates from Warner to Fisk, Oklahoma. The enginehouse is at Woodward.

Nos.	Qty.	Model	Builder	Date	Notes
1	1	44-ton	GE	1946	
Total	1				

OAKLAND TERMINAL RAILWAY

Reporting marks: OTR   **Track miles:** 29

Address: P. O. Box 24352, Oakland, CA 94623

The Oakland Terminal is a switching railroad at Oakland, California. It is jointly owned by Union Pacific and Santa Fe.

Nos.	Qty.	Model	Builder	Date	Notes
2197	1	GP7	EMD	1952	1
Total	1				

Notes:

1. Ex-Santa Fe, rebuilt 1980; also on Alameda Belt Line roster

OCTORARO RAILWAY

Reporting marks: OCTR   **Miles:** 56

Address: P.O. Box 146, Kennett Square, PA 19348

The Octoraro Railway operates a former Reading branch from Wilmington, Delaware, north to South Modena, Pennsylvania, 30 miles, and part of a former Pennsylvania Railroad branch from Chadds Ford Junction southwest to Avondale and Nottingham, 26 miles. The enginehouse is at Kennett Square, Pa.

Nos.	Qty.	Model	Builder	Date	Notes
2, 5	2	RS2	Alco	1949	1
55	1	SW1	EMD	1950	
134	1	RS3	Alco	1951	
341, 346	2	GP7r	EMD	1951, 1953	
4103, 4118	2	RS3	Alco	1952	2
Total	8				

Notes:
1. Ex-Toledo, Peoria & Western 202, 205, stored unserviceable
2. Ex-Delaware & Hudson 4103, 4118

OGEECHEE RAILWAY

Reporting marks: OGEE   **Miles:** 50
Address: Route 5, P. O. Box 172-B, Sylvania, GA 30467
The Ogeechee Railway has four separate routes in Georgia: Dover to Metter, 29 miles; Ardmore to Sylvania, 21 miles; Cochran to Hartford, 10 miles; and Fort Valley to Perry, 13 miles.

Nos.	Qty.	Model	Builder	Date	Notes
903	1	SW8	EMD	1952	1
1001, 1002	2	SW8	GMD	1951	2
1210	1	NW2	EMD	1949	
Total	4				

Notes:
1. Ex-Rock Island 825
2. Ex-Canadian National 6707, 6505

OHI-RAIL CORPORATION

Reporting marks: OHIC   **Miles:** 39
Address: 6200 Salineville Road N.E., Mechanicstown, OH 44651
Ohi-Rail operates from Minerva to Hopedale, Ohio. The enginehouse is at Mechanicstown.

Nos.	Qty.	Model	Builder	Date	Notes
101, 102	2	S2	Alco	1945-1946	1
Total	2				

Notes:
1. Ex-Fairport, Painesville & Eastern

OHIO CENTRAL RAILROAD INC.

Reporting marks: OHCR   **Miles:** 152
Address: P. O. Box 564, Sugar Creek, OH 44681
The Ohio Central extends from Zanesville to Brewster, Ohio. The enginehouse is at Coshocton.

Nos.	Qty.	Model	Builder	Date	Notes
12	1	S1	Alco	1950	Ex-Timken 5912
52	1	SW9	EMD	1953	Ex-Southern Pacific 2302
91	1	GP9	EMD	1957	Ex-Burlington Northern 1761
94	1	GP9	EMD	1959	Ex-Baltimore & Ohio 6594
95	1	GP9	EMD	1956	Ex-Baltimore & Ohio 6499
1501	1	GP7	EMD	1953	Ex-Pittsburgh & Lake Erie 1501
Total	6				

OHIO SOUTHERN RAILROAD

Reporting marks: OSRR   **Miles:** 66
Address: P. O. Box 564, Sugar Creek, OH 44681
The Ohio Southern operates from New Lexington to Zanesville, Ohio.

Nos.	Qty.	Model	Builder	Date	Notes
14	1	S2	Alco	1941	1
19	1	S4	Alco	1946	2
Total	2				

Notes:
1. Ex-Newburgh & South Shore 1014
2. Ex-Norfolk & Western 2069

OIL CREEK & TITUSVILLE LINES

Reporting marks: OCTL   **Miles:** 16
Address: 50 Commercial St., Gowanda, NY 14070

The Oil Creek & Titusville operates from Rouseville to Titusville, Pennsylvania. The enginehouse is at Titusville.

Nos.	Qty.	Model	Builder	Date	Notes
75	1	S2	Alco	1947	1
Total					

1. Ex-South Buffalo; previously New York & Lake Erie

OLD AUGUSTA RAILROAD

Reporting marks: OAR   **Miles:** 3
Address: P. O. Box 329, New Augusta, MS 39462

The Old Augusta Railroad extends from New Augusta to Augusta, Mississippi. The enginehouse is at Augusta.

Nos.	Qty.	Model	Builder	Date	Notes
100	1	NW2	EMD	1946	Ex-Conrail 9175
200	1	MP15	EMD	1974	Ex-Conrail 9630, Reading 2780
Total	2				

OLYMPIC RAILROAD

Reporting marks: OLYR   **Miles:** 4
Address: 6429 129th Ave. S.E., Bellevue, WA 98006

The Olympic Railroad is a terminal road at Port Townsend, Washington. It has no rail connection; it connects with Union Pacific and Burlington Northern at Seattle by carfloat.

Nos.	Qty.	Model	Builder	Date	Notes
52	1	SW1	EMD	1940	Ex-Seattle & North Coast 51
Total	1				

OMAHA, LINCOLN & BEATRICE RAILWAY

Reporting marks: OLB   **Miles:** 3
Address: 1815 Y St., Lincoln, NE 68501

The Omaha, Lincoln & Beatrice is a switching railroad at Lincoln, Nebraska.

Nos.	Qty.	Model	Builder	Date	Notes
101	1	44-ton	GE	1950	
102	1	70-ton	GE	1957	Ex-Sioux City Terminal
103	1	SW12	RLW	1990	Rebuilt Ex-Rock Island switcher
Total	3				

ONTARIO CENTRAL RAILROAD

Reporting marks: ONCT   **Miles:** 13
Address: 48 Belden Ave., Sodus, NY 14551

Ontario Central operates a 13-mile portion of the former Lehigh Valley main line from East Shortsville to West Victor, New York. The enginehouse is at Victor.

Nos.	Qty.	Model	Builder	Date	Notes
86	1	RS36	Alco	1962	Ex-Nickel Plate 865
Total	1				

ONTARIO MIDLAND RAILROAD

Reporting marks: OMID   **Miles:** 54
Address: 48 Belden Ave., Sodus, NY 14551

Ontario Midland extends from Newark, New York, north to Sodus Point, 15 miles, and from West Webster to Red Creek, N.Y., 42 miles. The two routes cross at Wallington. The enginehouse is at Victor.

Nos.	Qty.	Model	Builder	Date	Notes
36	1	RS11	Alco	1957	Ex-Norfolk & Western 361
40	1	RS36	Alco	1962	Ex-Norfolk & Western 408
Total	2				

ONTARIO NORTHLAND RAILWAY

Reporting marks: ONT **Miles:** 754
Address: 555 Oak St. E., North Bay, ON P1B 8L3, Canada

The Ontario Northland operates freight and passenger service from North Bay, Ontario, north through Cochrane to tidewater at Moosonee. There are branches to Elk Lake, Timmins, and Iroquois Falls, Ont., and Noranda, Quebec. The road's principal shops are at North Bay.

Nos.	Qty.	Model	Builder	Date	Notes
1508, 1509, 1517, 1520, 1521	5	FP7	GMD	1952, 1953	
1600-1605	6	GP9	GMD	1956-1957	
1730-1737	8	SD40-2	GMD	1973-1974	
1800-1809	10	GP38-2	GMD	1974-1984	
1984-1987	4	FP7m	GMD	1951-1953	1
Total	33				

Notes:
1. Ex-1519, 1518, 1501, 1510; rebuilt for use with TEE trainsets; now haul trains of rebuilt ex-GO Transit cars

OREGON, PACIFIC & EASTERN RAILWAY

Reporting marks: OPE **Miles:** 18
Address: 101 S. 10th St., Cottage Grove, OR 97424

The Oregon, Pacific & Eastern operates from Cottage Grove to Culp Creek, Oregon. The enginehouse is at Cottage Grove.

Nos.	Qty.	Model	Builder	Date	Notes
21	1	S2	Alco	1943	Ex-San Francisco Belt
602	1	SW8	EMD	1952	Ex-Yreka Western
Total	2				

OSAGE RAILROAD

Reporting marks: ORR **Miles:** 35
Address: 1604 S. Spruce St., Coffeyville, KS 67337

The Osage Railroad operates a segment of the former Midland Valley between Tulsa and Barnsdall, Oklahoma.

Nos.	Qty.	Model	Builder	Date	Notes
1000	1	CF7	ATSF	1971	
Total	1				

OTTER TAIL VALLEY RAILROAD

Reporting marks: OTVR **Miles:** 54
Address: 200 N. Mill St., Fergus Falls, MN 56537

Otter Tail Valley operates a former Great Northern route from Fergus Falls to Moorhead, Minnesota.

Nos.	Qty.	Model	Builder	Date	Notes
181, 189, 192, 194	4	GP18	EMD	1963	1
Total	4				

Notes:
1. Ex-Southern Railway, same numbers; previously Norfolk Southern

OUACHITA RAILROAD

Reporting marks: OUCH **Miles:** 26
Address: P. O. Box 150, Dardanelle, AR 72834-0150

The Ouachita Railroad operates a former Rock Island branch between El Dorado Arkansas, and Lille, Louisiana, 26 miles. The enginehouse is at El Dorado.

Nos.	Qty.	Model	Builder	Date	Notes
63	1	SW9	EMD	1952	Ex-Kirby Lumber 1000
64	1	NW2	EMD	1939	Ex-Great Northern 109
Total	2				

PADUCAH & LOUISVILLE RAILWAY

Reporting marks: PAL　　　　　**Miles:** 356
Address: 1500 Kentucky Ave., Paducah, KY 42001

The Paducah & Louisville extends between the Kentucky cities of its name, with branches to Central City, Blackford, Kevil, and Clayburn. The shops are at Paducah.

Nos.	Qty.	Model	Builder	Date	Notes
1236	1	SW9	EMD	1951	1
1300B, 1302-1309	9	SW13	ICG	1972-1973	2, 3
1425	1	SW14	ICG	1980	3
1500, 2003	2	GP7	EMD	1951	4
2257, 2259, 2267, 2276	4	GP30	EMD	1963	5
2529	1	GP35	EMD	1965	5
7733	1	GP10	ICG	1975	
7955	1	GP8	ICG	1970	
8008, 8124, 8132, 8154	4	GP10	ICG	1974-1975	
8183	1	GP10	ICG	1972	
8200, 8201	2	GP7	EMD	1950	3
8237, 8300-8303, 8305, 8306	7	GP10	ICG	1972-1974	
8307	1	GP8	ICG	1975	
8308, 8310-8312	4	GP10	ICG	1975	
8313	1	GP8	ICG	1974	
8314	1	GP10	ICG	1973	
8315, 8316	2	GP8	ICG	1973, 1971	
8317, 8318	2	GP10	ICG	1975, 1972	
8320, 8321	2	GP8	ICG	1973	
8322, 8324	2	GP10	ICG	1974, 1972	
8325	1	GP8	ICG	1973	
8326, 8327	2	GP10	ICG	1975, 1972	
8328	1	GP8	ICG	1974	
8329-8332	4	GP10	ICG	1972-1975	
8333	1	GP8	ICG	1971	
8335-8337	3	GP10	ICG	1972-1973	
8338	1	GP8	ICG	1978	
8339-8343, 8345, 8351, 8361, 8370	9	GP10	ICG	1970-1975	
Total	71				

Notes:
1. Ex-Louisiana Midland 10, rebuilt 1967 by IC
2. 1300B is a cabless booster unit
3. Ex-Illinois Central Gulf; Illinois Central
4. Ex-Santa Fe
5. Ex-Illinois Central Gulf; Gulf, Mobile & Ohio

PARR TERMINAL RAILROAD

Reporting marks: PRT　　　　　**Miles:** 2
Address: 402 Wright Ave., Richmond, CA 94804

Parr Terminal is a switching line at Richmond, California.

Nos.	Qty.	Model	Builder	Date	Notes
1	1	30-ton	Plymouth	1929	
2	1	S2	Alco	1948	
3	1	30-ton	Plymouth	1943	Ex-U. S. Army
2285	1	SW1200	EMD	1965	Ex-Southern Pacific
Total	4				

PATAPSCO & BACK RIVERS RAILROAD

Reporting marks: PBR　　　　　**Miles:** 10
Address: P. O. Box 9166, Baltimore, MD 21222

Patapsco & Back Rivers is a switching line in the Sparrows Point section of Baltimore. It is owned by Bethlehem Steel Company.

Nos.	Qty.	Model	Builder	Date	Notes
12-19	8	Slug	P&BR		1
112	1	SW7	EMD	1951	
113, 115	2	SW9	EMD	1952	
116, 122	2	SW7	EMD	1950	

Nos.	Qty.	Model	Builder	Date	Notes
123, 124	2	SW9	EMD	1952, 1951	
128, 132	2	SW1200	EMD	1957	
135	1	SW9	EMD	1951	
140, 141	2	VO1000	BLW	1940	
144	1	DS44-1000	BLW	1949	
147	1	VO1000	BLW	1942	
201, 202	2	SW1200	EMD	1956	
205, 206	2	SW9	EMD	1951-1952	
218	1	SW1500	EMD	1969	
904	1	SW9	EMD	1952	
936	1	SW7	EMD	1951	
Total	29				

Notes:
1. Rebuilt from Baldwin switchers

PEARL RIVER VALLEY RAILROAD

Reporting marks: PRV   **Miles:** 5
Address: P. O. Box 190, Picayune, MS 39466
The Pearl River Valley operates from Nicholson to Goodyear, Mississippi. The enginehouse is at Picayune.

Nos.	Qty.	Model	Builder	Date	Notes
101	1	65-ton	GE	1949	
Total	1				

PECOS VALLEY SOUTHERN RAILWAY

Reporting marks: PVS   **Miles:** 34
Address: P. O. Box 349, Pecos, TX 79772
Pecos Valley Southern extends from Pecos, Texas, to Saragosa, 29 miles. The enginehouse is at Pecos.

Nos.	Qty.	Model	Builder	Date	Notes
7	1	70-ton	GE	1949	
9	1	SW900m	EMD	1957	
Total	2				

PEE DEE RIVER RAILWAY

Reporting marks: PDRR   **Miles:** 16
Address: P. O. Box 917, Aberdeen, NC 28315
The Pee Dee River Railway operates from McColl to Marlboro Mills, South Carolina. The enginehouse is at Tatum. The road is affiliated with the Aberdeen & Rockfish.

Nos.	Qty.	Model	Builder	Date	Notes
5081	1	SW9	EMD	1952	Ex-Chesapeake & Ohio 5081
Total	1				

PEND OREILLE VALLEY RAILROAD

Reporting marks: POVA   **Miles:** 62
Address: P. O. Box 565, Newport, WA 99156
The Pend Oreille Valley extends from Newport to Metaline Falls, Washington, on a former Milwaukee Road branch. The enginehouse is at Metaline Falls.

Nos.	Qty.	Model	Builder	Date	Notes
101	1	GP9	EMD	1955	Ex-Burlington Northern 1846
102	1	GP9	EMD	1957	Ex-Burlington Northern 1735
Total	2				

PENINSULA TERMINAL

Reporting marks: PT   **Miles:** 2
Address: 2416 N. Marine Drive, Portland, OR 97217
Peninsula Terminal is a switching railroad at North Portland, Oregon.

Nos.	Qty.	Model	Builder	Date	Notes
3	1	70-ton	GE	1956	
901	1	45-ton	GE	1943	
Total	2				

PEORIA & PEKIN UNION RAILWAY

Reporting marks: PPU   **Miles:** 12
Address: 101 Wesley Road, Creve Coeur, IL 61611
The Peoria & Pekin Union is a switching line at Peoria, Illinois. The road's shop is at Creve Coeur (East Peoria).

Nos.	Qty.	Model	Builder	Date	Notes
600, 601	2	NW2	EMD	1947	Ex-400, 401
602	1	SW7	EMD	1950	Ex-410
603	1	NW2	EMD	1949	Ex-404
605	1	SW9	EMD	1951	Ex-411
607	1	NW2	EMD	1948	Ex-402
609	1	SW1200	EMD	1965	Ex-500
Total	**7**				

PHILADELPHIA, BETHLEHEM & NEW ENGLAND RAILROAD

Reporting marks: PBNE   **Miles:** 8
Address: 1744 E. Fourth St., Bethlehem, PA 18015
The Philadelphia, Bethlehem & New England is a switching road at Bethlehem, Pennsylvania. It is owned by Bethlehem Steel Company.

Nos.	Qty.	Model	Builder	Date	Notes
10-14	5	Slug	Alco		
21-25, 27, 28	7	NW2	EMD	1941-1949	
31-34	4	SW7	EMD	1950	
35-38	4	SW9	EMD	1951-1952	
39-43	5	SW1200	EMD	1956-1957	
50	1	SW900m	EMD	1956	
51, 52	2	SW900	EMD	1955	
Total	**28**				

PICKENS RAILROAD

Reporting marks: PICK   **Miles:** 7
Address: 402 Cedar Rock St., Pickens, SC 29671
The Pickens Railroad runs between Easley and Pickens, South Carolina. The enginehouse is at Pickens.

Nos.	Qty.	Model	Builder	Date	Notes
2	1	VO660	BLW	1946	1
5	1	DS44-750	BLW	1951	2
Total	**2**				

Notes:
1. Built new for Pickens
2. Ex-Youngstown Sheet & Tube 1

PIGEON RIVER RAILROAD

Reporting marks: PGRV   **Miles:** 14
Address: P. O. Box 123, South Milford, IN 46786
The Pigeon River Railroad operates from Wolcottville, Indiana, to Montpelier, Ohio, 14 miles on its own line and approximately 25 miles by trackage rights on the Hillsdale County Railway.

Nos.	Qty.	Model	Builder	Date	Notes
47	1	GP7	EMD	1952	Ex-Detroit & Toledo Shore Line 47
Total	**1**				

PIONEER & FAYETTE RAILROAD

Reporting marks: PF   **Miles:** 0.5
Address: 414 E. Main St., Fayette, OH 43521
Pioneer & Fayette operates a half mile of track at Franklin Junction, near Alvordton, Ohio.

Nos.	Qty.	Model	Builder	Date	Notes
102	1	25-ton	Plymouth	1936	
Total	**1**				

PIONEER VALLEY RAILROAD

Reporting marks: PVRR   **Miles:** 24

Address: P. O. Box 995, Westfield, MA 01086

The Pioneer Valley extends from Westfield, Massachusetts, to Holyoke and Easthampton. The enginehouse is at Westfield.

Nos.	Qty.	Model	Builder	Date	Notes
106	1	S2	Alco	1949	1
2558, 2565, 2597, 2647	4	CF7	ATSF	1974-1978	2
Total	5				

Notes:
1. Ex-Frankfort & Cincinnati, Chesapeake & Ohio
2. Ex-Santa Fe, same numbers

PITTSBURGH, ALLEGHENY & MCKEES ROCKS RAILROAD

Reporting marks: PAM   **Miles:** 5

Address: 180 Nichol Ave., McKees Rocks, PA 15136

The Pittsburgh, Allegheny & McKees Rocks is a switching line at McKees Rocks, Pennsylvania.

Nos.	Qty.	Model	Builder	Date	Notes
17, 20	2	70-ton	GE	1950, 1951	
Total	2				

PITTSBURGH & LAKE ERIE RAILROAD

Reporting marks: PLE   **Miles:** 404

Address: Commerce Court, 4 Station Square, Pittsburgh, PA 15219

Pittsburgh & Lake Erie's main line extends from Youngstown, Ohio, through Pittsburgh to Connellsville, Pennsylvania. It has a branch from McKeesport, Pa., to Brownsville, and it reaches to Ashtabula, Ohio, and Buffalo, New York, by trackage rights on Conrail and Norfolk Southern. P&LE's shops are at McKees Rocks, Pa.

Nos.	Qty.	Model	Builder	Date	Notes
1536+1573	8	SW1500	EMD	1972-1973	

Nos.	Qty.	Model	Builder	Date	Notes
1585, 1586, 1589-1595, 1597	10	MP15DC	EMD	1975	
2025-2041	17	GP38	EMD	1969	Ex-Conrail
2057-2060	4	GP38-2	EMD	1976	Ex-Rock Island
Total	39				

PITTSBURGH & OHIO VALLEY RAILWAY

Reporting marks: POV   **Miles:** 7

Address: Neville Island, Pittsburgh, PA 15225

The Pittsburgh & Ohio Valley is a switching line on Neville Island which is in the Ohio River northwest of downtown Pittsburgh.

Nos.	Qty.	Model	Builder	Date	Notes
3	1	SW9	EMD	1951	
4	1	SW8	EMD	1952	
5	1	SW1	EMD	1951	
6	1	SW900	EMD	1956	
Total	4				

PITTSBURG & SHAWMUT RAILROAD

Reporting marks: PS   **Miles:** 96

Address: Box 45, RD 8, Kittanning, PA 16201

The Pittsburg & Shawmut extends from Freeport Junction to Brockway Pennsyvlania. The enginehouse is at Brookville. Pittsburg & Shawmut operates and provides locomotives for the affiliated Red Bank Railroad which runs between Lawsonham and Sligo, Pa.

Nos.	Qty.	Model	Builder	Date	Notes
10, 11	2	GP7	EMD	1953, 1951	1
1774-1776, 1816, 1851, 1865, 1866, 1891, 1949	9	SW9	EMD	1953	
Total	11				

Notes:
1. Ex-Conrail 5672, 5818

PITTSBURGH, CHARTIERS & YOUGHIOGHENY RAILWAY

Reporting marks: PCY   **Miles:** 13
Address: 208 Island Ave., McKees Rocks, PA 15236
The Pittsburgh, Chartiers & Youghiogheny operates from Neville Island to Carnegie and McKees Rocks, Pennsylvania, 12 miles. The enginehouse is at McKees Rocks.

Nos.	Qty.	Model	Builder	Date	Notes
5	1	SW9	EMD	1951	
Total	1				

POCONO NORTHEAST RAILWAY

Reporting marks: PNER   **Track miles:** 93
Address: 1004 Exeter Ave., Exeter, PA 18643
Pocono Northeast operates freight service on several routes in the area of Wilkes-Barre and Scranton, Pennsylvania. Engine facilities are at Scranton.

Nos.	Qty.	Model	Builder	Date	Notes
87	1	NW2	EMD	1948	
183	1	SW7	EMD	1951	
601	1	SW1	EMD	1942	
1751	1	GP9	EMD	1959	
Total	4				

POINT COMFORT & NORTHERN RAILWAY

Reporting marks: PCN   **Miles:** 14
Address: P. O. Box 238, Lolita, TX 77971
The Point Comfort & Northern extends from Lolita to Point Comfort, Texas. The enginehouse is at Lolita.

Nos.	Qty.	Model	Builder	Date	Notes
12	1	SW1500	EMD	1974	
13-15	3	MP15	EMD	1980	
3726	1	GP38	EMD	1969	Ex-GATX
3731	1	GP38	EMD	1969	Ex-GATX
Total	6				

PORT BIENVILLE RAILROAD

Reporting marks: PBVR   **Miles:** 9
Address: 1002 Royal Gardens Drive, Madison, AL 35758
The Port Bienville Railroad operates between Ansley and Port Bienville, Mississippi. The enginehouse is at Port Bienville.

Nos.	Qty.	Model	Builder	Date	Notes
522, 524	2	SW1	EMD	1939, 1946	1
Total	2				

Notes:
1. Ex-Columbus & Greenville 522, 524; Illinois Central 601, 611

PORT JERSEY RAILROAD

Reporting marks: PJR   **Miles:** 2
Address: 203 Port Jersey Blvd., Jersey City, NJ 07305
The Port Jersey Railroad is a switching railroad at Bayonne, New Jersey.

Nos.	Qty.	Model	Builder	Date	Notes
1197	1	SW1200	EMD	1963	Ex-Missouri Pacific 1197
Total	1				

PORT OF PALM BEACH DISTRICT RAILWAY

Reporting marks: PPBD   **Miles:** 4
Address: P. O. Box 9935, Riviera Beach, FL 33419

Port of Palm Beach District Railway is a switching railroad in the Riviera Beach district of Palm Beach, Florida.

Nos.	Qty.	Model	Builder	Date	Notes
238	1	SW1	EMD	1941	Ex-Elgin, Joliet & Eastern 238
Total	1				

PORT OF TILLAMOOK BAY RAILROAD

Reporting marks: POTB   **Miles:** 96
Address: 4000 Blimp Blvd., Tillamook, OR 97141

The Port of Tillamook Bay Railroad operates between Hillsboro and Tillamook, Oregon, on a former Southern Pacific line. Enginehouses are at Tillamook and Banks.

Nos.	Qty.	Model	Builder	Date	Notes
4368, 4381, 4414	3	SD9	EMD	1955-1956	1
Total	3				

Notes:
1. Ex-Southern Pacific, same numbers

PORT ROYAL RAILROAD

Reporting marks: PRYL   **Miles:** 26
Address: P. O. Box 279, Charleston, SC 29402

The Port Royal Railroad operates from Yemassee to Port Royal, South Carolina. Engine facilities are at Port Royal.

Nos.	Qty.	Model	Builder	Date	Notes
6513	1	GP9	EMD	1957	Ex-Baltimore & Ohio 6513
6554	1	GP9	EMD	1959	Ex-Baltimore & Ohio 6554
Total	2				

PORT TERMINAL RAILROAD OF SOUTH CAROLINA

Reporting marks: PTSC   **Miles:** 5
Address: 540 E. Bay St., Charleston, SC 29402

The Port Terminal Railroad of South Carolina is a switching railroad at North Charleston, South Carolina.

Nos.	Qty.	Model	Builder	Date	Notes
1002	1	S4	Alco	1953	
2002	1	SW1001	EMD	1977	
Total	2				

PORT TERMINAL RAILROAD ASSOCIATION

Reporting marks: PTRA   **Track miles:** 177
Address: 501 Crawford St., Suite 423, Houston, TX 77002

The Port Terminal Railroad Association is a terminal road serving the north and south sides of the Houston Ship Channel. Locomotives are supplied by its owners

Road	Nos.	Qty.	Model	Builder	Date
ATSF	2117, 2183, 2204, 2208, 2238	5	GP7r	EMD	1951-1952
BN	33, 45	2	SW1500	EMD	1968
UP	1268, 1269, 1272, 1273	4	SW10	UP	1984
UP	1592	1	GP15-1	EMD	1977
UP	1683, 1692	2	GP15-1	EMD	1982
UP	2217	1	GP38-2	EMD	1972
SP	3794	1	GP9r	EMD	1956
SP	4116, 4119, 4121, 4143, 4151	5	GP20r	EMD	1962
Total		21			

PORT UTILITIES COMMISSION OF CHARLESTON, S. C.

Reporting marks: PUCC   **Miles:** 6
Address: 540 E. Bay St., Charleston, SC 29402
The Port Utilities Commission of Charleston, S. C., operates switching service at Charleston, South Carolina.

Nos.	Qty.	Model	Builder	Date	Notes
1001	1	SW1001	EMD	1975	
1003	1	S2	Alco	1946	
Total	2				

PRESCOTT & NORTHWESTERN RAILROAD

Reporting marks: PNW   **Miles:** 31
Address: P. O. Box 747, Prescott, AR 71857
The Prescott & Northwestern extends from Prescott to Highland, Arkansas. The enginehouse is at Prescott.

Nos.	Qty.	Model	Builder	Date	Notes
23-25	3	70-ton	GE	1954-1956	
Total	3				

PROVIDENCE & WORCESTER RAILROAD

Reporting marks: PW   **Miles:** 365
Address: 382 Southbridge St., Worcester, MA 01601
The Providence & Worcester has routes that radiate from Worcester, Massachusetts to Providence, Rhode Island; Groton, Connecticut; and Gardner, Mass. It operates freight service on Amtrak's Northeast Corridor route between Providence and New Haven, Conn. Enginehouses are at Worcester, Mass., and Valley Falls, R. I.

Nos.	Qty.	Model	Builder	Date	Notes
1702	1	GP9	EMD	1956	1
1801	1	U18B	GE	1976	
2001-2005	5	M420W	MLW	1974-1975	2
2006-2009	4	GP38-2	EMD	1980-1981	
2010, 2011	2	GP38	EMD	1969	
2201	1	B23-7	GE	1978	
Total	14				

Notes:
1. Ex-Conrail
2. The only MLW units built for a U. S. railroad

Jason Pelletier

Providence & Worcester's five M420Ws are the only Montreal-built units built for a U. S. railroad. They foreshadowed by a decade the "safety cab" now appearing on U. S. road units.

QUINCY RAILROAD

Reporting marks: QRR   **Miles:** 4

Address: P. O. Box 420, Quincy, CA 95971

The Quincy Railroad extends from Quincy, California, to a junction with Union Pacific (formerly Western Pacific) at Quincy Junction. The enginehouse is at Quincy.

Nos.	Qty.	Model	Builder	Date	Notes
3	1	44-ton	GE	1945	Ex-U. S. Army
4	1	S1	Alco	1942	
Total	**2**				

QUEBEC NORTH SHORE & LABRADOR RAILWAY

Reporting marks: QNSL   **Miles:** 357

Address: P. O. Box 1000, Sept-Isles, PQ P4R 4L5, Canada

The Quebec North Shore & Labrador extends north from Sept-Isles, Quebec, to Labrador City, Labrador, and Schefferville, Que. The road's shops are at Sept-Iles.

Nos.	Qty.	Model	Builder	Date	Notes
149, 151, 155, 160, 165, 168, 174-176	9	GP9	GMD	1956-1960	1
200-203	4	SD40	GMD	1969	2
219, 220	2	SD40	GMD	1971	2
221-264	44	SD40-2	GMD	1972-1975	
Total	**59**				

Notes:

1. All stored unserviceable except 155
2. Stored

RAHWAY VALLEY RAILROAD

Reporting marks: RV   **Miles:** 5

Address: P. O. Box 156, Kenilworth, NJ 07033

The Rahway Valley extends from Cranford to Springfield, New Jersey. The enginehouse is at Kenilworth.

Nos.	Qty.	Model	Builder	Date	Notes
16, 17	2	70-ton	GE	1951, 1954	
Total	**2**				

RAILROAD SWITCHING SERVICE OF MISSOURI

Reporting marks: RSM   **Miles:** 2

Address: 720 Olive St., St. Louis, MO 63101

Railroad Switching Service of Missouri is a switching line in St. Louis.

Nos.	Qty.	Model	Builder	Date	Notes
1209	1	SW1200	EMD	1952	Ex-Terminal Railroad Association of St. Louis 1209
Total	**1**				

RARUS RAILWAY

Reporting marks: RARW   **Track miles:** 75

Address: P. O. Box 1070, Anaconda, MT 59711

Rarus Railway operates between Butte and Anaconda, Montana, on the line of the former Butte, Anaconda & Pacific. The enginehouse is at Anaconda.

Nos.	Qty.	Model	Builder	Date	Notes
102	1	GP7	EMD	1952	1
103	1	GP7	EMD	1953	1
104-107	4	GP9	EMD	1957	1
301	1	GP7	EMD	1953	2
Total	**7**				

Notes:

1. Ex-Butte, Anaconda & Pacific, same numbers
2. Ex-Burlington Northern 1634, Northern Pacific 560

READING, BLUE MOUNTAIN & NORTHERN RAILROAD

See Blue Mountain & Reading, page 31.

RED SPRINGS & NORTHERN RAILROAD

Reporting marks: RSNR   **Miles:** 13
Address: P. O. Box 1929, Laurinburg, NC 28352
The Red Springs & Northern operates from Parkton to Red Springs, North Carolina. The enginehouse is at Red Springs.

Nos.	Qty.	Model	Builder	Date	Notes
104	1	70-ton	GE	1950	Ex-Lancaster & Chester 62
Total	1				

RED RIVER VALLEY & WESTERN RAILROAD

Reporting marks: RRVW   **Miles:** 667
Address: 5th and Minnesota Avenues, Breckenridge, MN 56520
Red River Valley & Western operates a network of former Burlington Northern lines (mostly ex-Northern Pacific) in North Dakota: Breckenridge, Minnesota-Davenport, N. D.-Casselton-Marion; Wahpeton-Oakes-La Moure-Jamestown-Carrington-Minnewaukan; Horace-Davenport-La Moure-Edgeley; Pingree-Regan; Carrington-Turtle Lake; and Oberon-Esmond. Engine facilities are at Breckenridge and Carrington.

Nos.	Qty.	Model	Builder	Date	Notes
300-309	10	CF7	ATSF	1970-1977	
Total	10				

RICHMOND, FREDERICKSBURG & POTOMAC RAILROAD

Reporting marks: RFP   **Miles:** 114
Address: 2134 W. Laburnum Ave., Richmond, VA 23227
The Richmond, Fredericksburg & Potomac, now part of CSX, extends from Washington, D. C., to Richmond, Virginia. The road's shops are at Richmond.

Nos.	Qty.	Model	Builder	Date	Notes
4-7	4	SW1500	EMD	1967	
81-85	5	SW1200	EMD	1965	1
91	1	SW1500	EMD	1966	
121-127	7	GP40	EMD	1966-1967	
131, 133, 134, 136, 138	5	GP35	EMD	1965	
141-147	7	GP40-2	EMD	1972	
S1, S2	2	Slug	Chrome	1985	2
Total	31				

Notes:
1. Stored serviceable
2. Built from RF&P Alco S2s

Charles W. McDonald

The days are numbered for the blue-and-gray paint scheme of the Richmond, Fredericksburg & Potomac; the road's operations are quickly being absorbed by CSX.

RIVER TERMINAL RAILWAY

Reporting marks: RT   **Track miles:** 30
Address: P. O. Box 6073, Cleveland, OH 44101
River Terminal is a switching road at Cleveland.

Nos.	Qty.	Model	Builder	Date	Notes
61	1	SW7	EMD	1951	
63, 64	2	SW1200	EMD	1959	
76	1	SW7	EMD	1951	
92, 98, 100	3	SW900	EMD	1955, 1959	
101-109	9	SW1001	EMD	1968-1978	
Total	**16**				

R. J. CORMAN RAILROAD

Reporting marks: RJCR   **Miles:** 82
Address: 603 Third St., Bardstown, KY 40004
The R. J. Corman Railroad operates from Bardstown Junction to Wickland, Kentucky. The enginehouse is at Bardstown.

Nos.	Qty.	Model	Builder	Date	Notes
3501	1	GP30	EMD	1963	
9002, 9005-9010	7	GP9	EMD	1956-1958	1
Total	**8**				

Notes:
1. Ex-Chesapeake & Ohio 6066, Baltimore & Ohio 6509, 6521, 6533, 6586, 6523, 6592

ROBERVAL & SAGUENAY RAILWAY

Reporting marks: RS   **Miles:** 58
Address: C.P. 1277, Jonquiere, PQ G7S 4K8, Canada
The Roberval & Saguenay extends from Port Alfred to Kenogami,

Pierre Alain Patenaude

Montreal built only two M420TRs, both for Roberval & Saguenay. They were delivered with Zero Weight Transfer trucks that were ill-suited for switching duties and were replaced by conventional Type B trucks. The M420TR resembles an MP15AC; the Type 2 M420TR, 15 of which were built for Mexico's Ferrocarril del Pacifico, is 7 feet longer and bulkier in appearance.

Quebec, and from Saguenay Power Junction to Alma, Quebec. The shops are at Arvida.

Nos.	Qty.	Model	Builder	Date	Notes
24, 25	2	RS18	MLW	1960, 1965	
26, 27	2	M420TR	MLW	1972	
32	1	RS11	Alco	1959	1
33-41	9	C420	Alco	1964	2
Total	**14**				

Notes:
1. Ex-Southern Pacific 2931
2. Ex-Long Island 209, 214-216, 212, 218, 219, 206, 211

R. J. CORMAN RAILROAD, MEMPHIS DIVISION

Reporting marks: RJCM   **Miles:** 92
Address: 147 E. First St., Guthrie, KY 42234
R. J. Corman's Memphis Division extends from Zinc, Tennessee, through Guthrie, Kentucky, to Bowling Green, Ky. The enginehouse is at Guthrie.

Nos.	Qty.	Model	Builder	Date	Notes
1940	1	FP7	EMD	1950	Ex-Southern 6141
1941	1	FP7	EMD	1950	Ex-Southern 6138
9001	1	GP9	EMD	1955	Ex-Chesapeake & Ohio 5956
Total	**3**				

R. J. CORMAN RAILROAD, OHIO DIVISION

Reporting marks: RJCO   **Miles:** 50
Address: Route 5, P. O. Box 498, Dover, OH 44662
R. J. Corman's Ohio Division extends from Warwick, Ohio, south through Massillon to Uhrichsville. The enginehouse is at Dover.

Nos.	Qty.	Model	Builder	Date	Notes
9003	1	GP9	EMD	1955	Ex-Chesapeake & Ohio 5999
9004	1	GP9	EMD	1956	Ex-Baltimore & Ohio 6469
Total	**2**				

ROCHESTER & SOUTHERN RAILROAD

Reporting marks: RSR   **Miles:** 107
Address: 1372 Brooks Ave., Rochester, NY 14624
Rochester & Southern operates from Rochester to East Salamanca, New York. It is affiliated with the Buffalo & Pittsburgh and the Genesee & Wyoming. The enginehouse is at Rochester.

Nos.	Qty.	Model	Builder	Date	Notes
101-106	6	GP40	EMD	1968	
107, 108	2	SW1200	EMD	1964	
7803, 7822	2	GP38	EMD	1969	Ex-Conrail
Total	**10**				

R. P. Campbell

The Rockdale, Sandow & Southern shares a red, black, and white paint scheme with two other roads owned by Alcoa, the Point Comfort & Northern and the Bauxite & Northern.

ROCKDALE, SANDOW & SOUTHERN RAILROAD

Reporting marks: RSS   **Miles:** 34
Address: P. O. Box 387, Rockdale, TX 76567
The Rockdale, Sandow & Southern extends from Marjorie to Sandow, Texas. The enginehouse is at Rockdale.

Nos.	Qty.	Model	Builder	Date	Notes
13-15	3	MP15	EMD	1974	
Total	**3**				

SABINE RIVER & NORTHERN RAILROAD

Reporting marks: SRN **Miles:** 40
Address: P. O. Box 5000, Orange, TX 77630
The Sabine River & Northern extends from Bessmay to Orange, Texas, with a branch to Evadale. The enginehouse is at Julford.

Nos.	Qty.	Model	Builder	Date	Notes
1208	1	SW1200	EMD	1965	1
1505, 1506	2	GP7	EMD	1953	2
1751	1	GP9	EMD	1954	3
1757	1	GP9	EMD	1955	4
1759	1	GP9	EMD	1954	3
Total	**6**				

Notes:
1. Ex-Missouri Pacific 1208
2. Ex Reading 613, 609
3. Ex-Inland Container, Burlington Northern
4. Ex-Detroit, Toledo & Ironton 988

ST. LAWRENCE & ATLANTIC RAILROAD

Reporting marks: SLR **Miles:** 165
Address: Exchange Street, Berlin, NH 03570
St. Lawrence & Atlantic operates a former Canadian National (ex-Grand Trunk) line from Portland, Maine, to Norton, Vermont, on the Canadian border. Engine facilities are at Lewiston Junction, Maine.

Nos.	Qty.	Model	Builder	Date	Notes
1758	1	GP9	EMD	1954	1
4442, 4445, 4447, 4450	4	GP9	EMD	1956	2
9625-9627	3	MP15	EMD	1974	3
Total	**8**				

Notes:
1. Ex-Grand Trunk Western 4441
2. Ex-Central Vermont, same numbers
3. Ex-Conrail 9625-9627, Reading 2775-2777

ST. LAWRENCE & RAQUETTE RIVER RAILROAD

Reporting marks: SLRR **Miles:** 31
Address: 50 Commercial St., Gowanda, NY 14070
The St. Lawrence & Raquette River operates a line owned by the Ogdensburg Bridge & Port Authority from Norfolk through Norwood to Ogdensburg, New York.

Nos.	Qty.	Model	Builder	Date	Notes
10	1	70-ton	GE	1956	Ex-Norwood & St. Lawrence
12	1	SW7	EMD	1951	Ex-CSX
14	1	SW8	EMD	1959	Ex-Rock Island
Total	**3**				

ST. MARIES RIVER RAILROAD

Reporting marks: STMA **Miles:** 71
Address: P. O. Box 619, St. Maries, ID 83861-0619
The St. Maries River Railroad operates from Bovil to Plummer, Idaho. The enginehouse is at St. Maries.

Nos.	Qty.	Model	Builder	Date	Notes
101-103	3	GP9	EMD	1959	1
501, 502	2	SW1200	EMD	1954	2
Total	**5**				

Notes:
1. Ex-Milwaukee Road 292, 301, 322
2. Ex-Milwaukee Road 612, 618

ST. MARYS RAILROAD

Reporting marks: SM **Miles:** 11
Address: P. O. Box 520, St. Marys, GA 31558
The St. Marys Railroad extends from St. Marys to Kingsland, Georgia. The enginehouse is at St. Marys.

Nos.	Qty.	Model	Builder	Date	Notes
503	1	SW1500	EMD	1971	
504, 505	2	MP15DC	EMD	1974, 1976	
Total	3				

SALT LAKE, GARFIELD & WESTERN RAILWAY

Reporting marks: SLGW   **Miles:** 17
Address: 1200 W. North Temple, Salt Lake City, UT 84116
The Salt Lake, Garfield & Western extends from Salt Lake City west to Saltair, Utah. The engine facilities are in Salt Lake City.

Nos.	Qty.	Model	Builder	Date	Notes
D1	1	44-ton	GE	1941	
D4	1	S6	Alco	1955	
D5	1	65-ton	GE	1943	Ex-U. S. Army
Total	3				

SANDERSVILLE RAILROAD

Reporting marks: SAN   **Miles:** 9
Address: P. O. Box 269, Sandersville, GA 31082
The Sandersville Railroad extends from Tennille through Sandersville to Kaolin, Georgia. Engine facilities are at Sandersville.

Nos.	Qty.	Model	Builder	Date	Notes
90	1	Slug	Alco	1980	
100	1	SW1500	EMD	1967	
200	1	SW1200	EMD	1964	
300, 400, 500	3	SW1500	EMD	1969-1970	
Total	6				

SAN DIEGO & IMPERIAL VALLEY RAILROAD

Reporting marks: SDIV   **Miles:** 145
Address: 743 Imperial Ave., San Diego, CA 92101

The San Diego & Imperial Valley extends from San Diego to El Cajon and to Plaster City. The enginehouse is at San Ysidro, Calif.

Nos.	Qty.	Model	Builder	Date	Notes
1229, 1237	2	SSB1200	ATSF	1974	
1438	1	GP20	EMD	1961	1
2151, 3162, 4168	3	GP7r	EMD	1951-1952	2
5911	1	GP9	EMD	1955	3
Total	7				

Notes:
1. Ex-Union Pacific 483
2. Ex-Santa Fe, rebuilt 1978-1979
3. Ex-Denver & Rio Grande Western 5911

SAND SPRINGS RAILWAY

Reporting marks: SS   **Miles:** 5
Address: 216 N. McKinley, Sand Springs, OK 74063
The Sand Springs extends from Tulsa to Sand Springs, Oklahoma. The enginehouse is at Sand Springs.

Nos.	Qty.	Model	Builder	Date	Notes
100-102	3	SW900	EMD	1956, 1957	
Total	3				

SAN FRANCISCO BELT RAILROAD

Reporting marks: SFBR   **Track miles:** 20
Address: 221 World Trade Center, San Francisco, CA 94111
The San Francisco Belt is a switching line operating along San Francisco's waterfront.

Nos.	Qty.	Model	Builder	Date	Notes
23-25	3	S2	Alco	1944-1945	
Total	3				

SAN LUIS CENTRAL RAILROAD

Reporting marks: SLC   **Miles:** 30
Address: P. O. Box 108, Monte Vista, CO 81144

The San Luis Central extends from Monte Vista to Center, Colorado. The enginehouse is at Monte Vista.

Nos.	Qty.	Model	Builder	Date	Notes
70	1	SW8	EMD	1952	Ex-Cincinnati Union Terminal 37
71	1	70-ton	GE	1955	
Total	2				

SAN MANUEL ARIZONA RAILROAD

Reporting marks: SMA   **Miles:** 29
Address: P. O. Box M, San Manuel, AZ 85631

The San Manuel Arizona Railroad operates from Hayden to San Manuel, Arizona. The enginehouse is at San Manuel.

Nos.	Qty.	Model	Builder	Date	Notes
8-10, 12	4	RS3	Alco	1951	
16, 17	2	GP38-2	EMD	1974	
Total	6				

SANTA CRUZ, BIG TREES & PACIFIC RAILWAY

Reporting marks: SCBG   **Miles:** 11
Address: P. O. Box G-1, Felton, CA 95018

The Santa Cruz, Big Trees & Pacific operates excursion and freight service between Olympia and Santa Cruz, California. The enginehouse is at Felton.

Nos.	Qty.	Model	Builder	Date	Notes
2600, 2641	2	CF7	ATSF	1971-1972	1
Total	2				

Notes:
1. Ex-Santa Fe 2600, 2641

SANTA MARIA VALLEY RAILROAD

Reporting marks: SMV   **Miles:** 18
Address: P. O. Box 340, Santa Maria, CA 93456

The Santa Maria Valley extends from Guadalupe to Gates, California. The enginehouse is at Santa Maria.

Nos.	Qty.	Model	Builder	Date	Notes
10, 20, 30, 40, 50	5	70-ton	GE	1948-1952	
70	1	70-ton	GE	1950	1
80	1	70-ton	GE	1953	2
Total		7			

Notes:
1. Ex-Arkansas & Ozarks
2. Ex-Fort Dodge, Des Moines & Southern

SAVANNAH STATE DOCKS RAILROAD

Reporting marks: SSDK   **Miles:** 24
Address: P. O. Box 2406, Savannah, GA 31402

Savannah State Docks Railroad is a switching line at Savannah and Port Wentworth, Georgia. The enginehouse is at Savannah.

Nos.	Qty.	Model	Builder	Date	Notes
1001, 1002	2	SW1001	EMD	1978	
Total	2				

SEAGRAVES, WHITEFACE & LUBBOCK RAILROAD

Reporting marks: SWGR   **Miles:** 11
Address: 211 S. 6th St., Brownfield, TX 79316

The Seagraves, Whiteface & Lubbock Railroad extends from Lubbock, Texas, southwest to Seagraves, with a branch from Doud to Whiteface and Coble. The enginehouses are at Brownfield and Levaland.

Nos.	Qty.	Model	Builder	Date	Notes
105, 113, 118	3	GP7	EMD	1951-1952	1
Total	3				

Notes:
1. Ex-Missouri-Kansas-Texas, same numbers

SEMINOLE GULF RAILROAD

Reporting marks: SGLR   **Miles:** 118
Address: 2830 Winkler Ave., Fort Myers, FL 33916
Seminole Gulf operates two lines in southwest Florida: from Arcadia through Fort Myers to Vanderbilt Beach, and from Oneco through Bradenton to Venice. Engine facilities are at Fort Myers and Oneco.

Nos.	Qty.	Model	Builder	Date	Notes
571, 575-578	5	GP9	EMD	1956-1958	
752-754	3	GP9	EMD	1955-1957	
Total	8				

SEQUATCHIE VALLEY RAILROAD

Reporting marks: SQVR   **Miles:** 40
Address: P. O. Box 1317, Shelbyville, TN 37160
The Sequatchie Valley Railroad extends from Bridgeport, Alabama, to Brush Creek, Tennessee. The enginehouse is at South Pittsburg, Tenn.

Nos.	Qty.	Model	Builder	Date	Notes
60A	1	F7A	EMD	1950	Ex-Milwaukee Road 60A
1186	1	SW9	EMD	1953	
1201	1	SW1200	EMD	1955	Ex-Illinois Terminal 1201
1210	1	SW1200	EMD	1955	Ex-Illinois Terminal 1210
1507	1	F7B	EMD	1953	Ex-Alaska Railroad 1507
1534	1	SD35	EMD	1965	Ex-Norfolk & Western 1534
1686	1	NW2	EMD	1948	
Total	7				

SIERRA RAILROAD

Reporting marks: SERA   **Miles:** 49
Address: 13645 Tuolumne Road, Sonora, CA 95370
The Sierra railroad extends from Oakdale, California, east to Standard. The enginehouse is at Oakdale.

Nos.	Qty.	Model	Builder	Date	Notes
40, 42	2	S12	BLW	1955	
44	1	S12	BLW	1951	Ex-Sharon Steel 2 (earlier 10)
Total	3				

Ronald N. Johnson

Two of Sierra's three green-and-white Baldwin S12s working in multiple lead a freight across a short bridge east of Oakdale.

141

SHAMOKIN VALLEY RAILROAD

Reporting marks: SVRR   **Miles:** 33
Address: 356 Priestley Ave., Northumberland, PA 17857
The Shamokin Valley Railroad operates from Mount Carmel Junction to Sunbury, Pennsylvania. The enginehouse is at Northumberland. The railroad is affiliated with the North Shore Railroad.

Nos.	Qty.	Model	Builder	Date	Notes
86	1	SW7	EMD	1950	Ex-Conrail 8869, Central Railroad of New Jersey 1081
Total	**1**				

SISSETON MILBANK RAILROAD

Reporting marks: SMRR   **Miles:** 38
Address: 405 W. Milbank Ave., Milbank, SD 57252
The Sisseton Milbank Railroad extends from Sisseton to Milbank, South Dakota. The enginehouse is at Milbank.

Nos.	Qty.	Model	Builder	Date	Notes
627	1	SW1200	EMD	1954	Ex-Milwaukee Road
Total	**1**				

SOO LINE RAILROAD

Reporting marks: SOO   **Miles:** 5,807
Address: Soo Line Building, Minneapolis, MN 55440

Nos.	Qty.	Model	Builder	Date	Notes
322, 325, 328	3	SW1200	EMD	1955	
375-378, 381-383	7	GP7	EMD	1952	
400-405, 408, 410-412, 414	11	GP9	EMD	1955-1957	1
532, 534, 543	3	SD10	EMD	1953-1954	2
730	1	GP35	EMD	1965	
737-756	20	SD40	EMD	1969-1971	
757-765, 767-789	32	SD40-2	EMD	1972-1975	
1200-1205	6	SW1200	EMD	1962-1965	3
1211, 1213, 1216, 1218, 1220, 1221, 1222	7	SW1200	EMD	1954	4
1400, 1401	2	SW1500	EMD	1966	5
1532-1563	30	MP15AC	EMD	1975	4
2002, 2003, 2008-2011, 2013-2016, 2018, 2025, 2026, 2031-2036, 2038, 2039, 2041, 2043, 2045, 2046, 2057-2059, 2064, 2066	30	GP40	EMD	1966-1968	4
2112-2115, 2117	5	SW9	EMD	1952-1953	1
2118	1	Slug	EMD	1953	6
2122, 2126	2	SW1200	EMD	1955	1
2401-2412	12	GP9	EMD	1954-1957	
2550, 2552, 2553, 2555	4	GP9	EMD	1954-1956	
4100-4106	7	GP9r	EMD	1952	7
4200-4204	5	GP9m	EMD	1954-1957	8
4225, 4227-4233	8	GP9	EMD	1944-1957	9
4300-4302	3	GP30m	EMD	1961-1963	10
4400-4406, 4408-4429, 4431-4452	51	GP38-2	EMD	1977-1983	
4506-4515	10	GP38-2	EMD	1973-1974	11
4598, 4599	2	GP39-2	EMD		12
4600-4603, 4612, 4648	6	GP40	EMD	1966, 1967	
6000-6057	58	SD60	EMD	1987, 1989	
6058-6062	5	SD60M	EMD	1989	
6240, 6241	2	SD39	EMD	1968	13
6400-6405	6	SD40	EMD	1970	14
6406-6410	5	SD40A	EMD	1970	15
6411	1	SD40	EMD	1966	16
6450	1	SD40B	EMD	1971	17
6491, 6492	2	SD45	EMD	1971	18
6601-6604, 6606-6623	22	SD40-2	EMD	1979-1984	
Total	**370**				

Notes:
1. Stored
2. SD9s rebuilt by Milwaukee Road
3. Ex-Minneapolis, Northfield & Southern 30-35

4. Ex-Milwaukee Road, same numbers
5. Ex-Minneapolis, Northfield & Southern 36, 37
6. Rebuilt from SW9 2118
7. Ex-Conrail units, rebuilt with Caterpillar engines
8. Ex-Conrail
9. Formerly numbered 550, 552-557
10. Rebuilt by Ziegler with Caterpillar engines, classified GP30C
11. Ex-Milwaukee Road 356-365
12. Ex-Kennecott Copper 793, 794
13. Ex-Minneapolis, Northfield & Southern 40, 41
14. Ex-Kansas City Southern 621-626
15. Ex-Illinois Central 6019-6023
16. Ex-Baltimore & Ohio 7500
17. Ex-Burlington Northern; rebuilt to B unit
18. Ex-Burlington Northern

SOUTH BRANCH VALLEY RAILROAD

Reporting marks: SBVR   **Miles:** 52
Address: P. O. Box 470, Moorefield, WV 26836
The South Branch Valley operates from Green Spring to Petersburg, West Virginia. The enginehouse is at Moorefield.

Nos.	Qty.	Model	Builder	Date	Notes
1	1	S1	Alco	1943	1
6135, 6240, 6506, 6600, 6604	5	GP9	EMD	1955-1957	2
Total	**6**				

Notes:
1. Ex-U. S. Navy
2. Ex-Chessie System, same numbers

For the last five units in an order of SD60s, Soo Line specified the safety cab and a solid red paint scheme.

Fred Radek

143

SOUTH BROOKLYN RAILWAY

Reporting marks: SBK   **Miles:** 2
Address: 990 Third Ave., Brooklyn, NY 11232
The South Brooklyn Railroad is a switching line in Brooklyn, New York.

Nos.	Qty.	Model	Builder	Date	Notes
N1, N2	2	50-ton	GE	1974	
Total	**2**				

SOUTH BUFFALO RAILWAY

Reporting marks: SB   **Track miles:** 12
Address: 2600 Hamburg Turnpike, Lackawanna, NY 14218
The South Buffalo Railway is a switching road at Lackawanna, New York.

Nos.	Qty.	Model	Builder	Date	Notes
10-14	5	Slug	SB	1972-1980	1
30-33	4	SW1200	EMD	1964-1965	
34-37	4	SW9r	EMD	1952	1
40, 45	2	S6	Alco	1955, 1957	1
50-53	4	SW1200	EMD	1957	
63, 78, 82-84, 87, 88, 90	8	S2	Alco	1945-1950	1
92, 93	2	S4	Alco	1951	2
94	1	S2	Alco	1951	1
95	1	S4	Alco	1951	2
97, 98, 100, 102-106	8	S2	Alco	1943-1952	1
Total	**39**				

Notes:
1. Stored unserviceable
2. Stored serviceable

SOUTH CAROLINA CENTRAL RAILROAD

Reporting marks: SCRF   **Miles:** 53
Address: P. O. Box 490, Hartsville, SC 29550

South Carolina Central operates from Florence to Bishopville, South Carolina, and Cheraw to Society Hill, S. C. Enginehouses are at Hartsville and Cheraw.

Nos.	Qty.	Model	Builder	Date	Notes
5805, 6097, 6187, 6440, 6550, 6555	6	GP9	EMD	1954-1957	1
Total	**6**				

Notes:
1. Ex-Chessie System, same numbers

SOUTH CAROLINA CENTRAL RAILROAD, CAROLINA PIEDMONT DIVISION

Reporting marks: CPDR   **Miles:** 34
Address: 268 E. Main St., Laurens, SC 29360
The Carolina Piedmont Division operates a former Charleston & Western Carolina (later Atlantic Coast Line) branch from Laurens, South Carolina, to East Greenville, 34 miles. The enginehouse is at Laurens.

Nos.	Qty.	Model	Builder	Date	Notes
8379, 8383, 8387	3	GP10	ICG	1974	
Total	**3**				

SOUTH CAROLINA CENTRAL RAILROAD, GEORGIA & ALABAMA DIVISION

Reporting marks: GAAB   **Miles:** 77
Address: 699 Main St., Dawson, GA 31742
The Georgia & Alabama Division operates from White Oak, Alabama, to Smithville, Georgia. The enginehouse is at Dawson, Ga. The road's motive power is also used on the Georgia Great Southern Division, which extends from Dawson to Albany, Ga., 22 miles.

Nos.	Qty.	Model	Builder	Date	Notes
21	1	GP9	EMD	1956	Ex-New Haven 1201
6541	1	GP9	EMD	1957	Ex-Baltimore & Ohio 6541
Total	**2**				

SOUTH CENTRAL TENNESSEE RAILROAD

Reporting marks: SCTR   **Miles:** 52
Address: P. O. Box 259, Centervillle, TN 37033
The South Central Tennessee Railroad operates between Colesburg and Hohenwald, Tennessee. The enginehouse is at Watson.

Nos.	Qty.	Model	Builder	Date	Notes
29	1	RS11	Alco	1959	1
30	1	RS11	Alco	1956	2
100, 200	2	H10-44	FM	1947	3
2062, 2070	2	GP9d	EMD	1957	4
5323	1	SW9	EMD	1953	5
5624	1	SW1200	EMD	1956	
Total	**8**				

Notes:
1. Ex-Southern Pacific 2910
2. Ex-Louisville & Nashville 950
3. Ex-Apache 100, 200
4. Ex-Santa Fe 2062, 2070; rebuilt by Santa Fe
5. Stored unserviceable

SOUTH CAROLINA CENTRAL RAILROAD, GEORGIA GREAT SOUTHERN DIVISION

Reporting marks: GGS   **Miles:** 22
Address: 699 Main St., Dawson, GA 31742
The Georgia Great Southern Division extends from Dawson to Albany, Georgia. The enginehouse is at Dawson.

Nos.	Qty.	Model	Builder	Date	Notes
2130	1	GP7r	EMD	1952	Ex-Santa Fe 2130
Total	**1**				

SOUTH CAROLINA CENTRAL RAILROAD, GEORGIA SOUTHWESTERN DIVISION

Reporting marks: GSWR   **Miles:** 258
Address: 908 Elm Ave., Americus, GA 31709
The Georgia Southwestern Division extends from Rhine, Georgia, to Mahrt, Alabama, and from Columbus to Bainbridge, Ga. Engine facilities are at Americus and Damascus, Ga.

Nos.	Qty.	Model	Builder	Date	Notes
2027, 2127, 2160, 2176, 2185	5	GP7r	EMD	1951-1952	1
Total	**5**				

Notes:
1. Ex-Santa Fe, same numbers; rebuilt 1979-1980

SOUTH CENTRAL FLORIDA RAILROAD

Reporting marks: SCFE   **Miles:** 104
Address: 900 South W. C. Owen Ave., Clewiston, FL 33440
The South Central Florida extends from Sebring to Lake Harbor, Florida. The enginehouse is at Clewiston.

Nos.	Qty.	Model	Builder	Date	Notes
9010-9012	3	GP7	EMD	1952-1953	
9013-9015	3	GP8	ICG		
9016	1	U30B	GE	1966	1
Total	**7**				

Notes:
1. Ex-Gettysburg 28, Conrail 2882, New York Central

SOUTHEASTERN PENNSYLVANIA TRANSPORTATION AUTHORITY

Reporting marks: SEPA   **Miles:** 276
Address: 200 W. Wyoming Ave., Philadelphia, PA 19140
Southeastern Pennsylvania Transportation Authority operates rail commuter service from Philadelphia to Trenton and West Trenton, New Jersey; Wilmington, Delaware; and Philadelphia International Airport, Newtown, Warminster, Doylestown, Chestnut Hill, Norristown, Parkesburg, and West Chester, Pa.

Nos.	Qty.	Model	Builder	Date	Notes
50-52	3	SW1200	EMD	1954	1
55	1	80-ton	GE	1953	2
90, 91	2	GP9	EMD	1955	3
2301-2307	7	AEM7	EMD	1987	4
CW6, CW7	2	45-ton	GE	1942, 1940	5
Total	15				

Notes:
1. Ex-Milwaukee Road 649, 626, 639
2. Ex-Standard Slag LM12
3. Ex-Conrail, Penn Central, Pennsylvania 7019, 7028
4. Electric locomotives
5. Ex-U. S. Army 7408 and 7330

SOUTHEAST KANSAS RAILROAD

Reporting marks: SEKR   **Miles:** 104
Address: 1230 West Walnut, Coffeyville, KS 67337
The Southeast Kansas Railroad operates from Bartlesville, Oklahoma, through Coffeyville, Kansas, to Nassau Junction Missouri. The enginehouse is at Coffeyville.

Nos.	Qty.	Model	Builder	Date	Notes
102, 103, 117, 123	4	GP7	EMD	1951-1952	1
142	1	RS3m	Alco	1951	2
Total	5				

Notes:
1. Ex-Missouri-Kansas-Texas, same numbers
2. EMD rebuild of Missouri-Kansas-Texas RS3

SOUTHERN ALABAMA RAILROAD

Reporting marks: SUAB   **Miles:** 16
Address: P. O. Box 1317, Shelbyville, TN 37160
The Southern Alabama operates from Troy to Goshen, Alabama. The enginehouse is at Troy.

Nos.	Qty.	Model	Builder	Date	Notes
7013	1	SW9	EMD	1952	
Total	1				

SOUTHERN INDIANA RAILWAY

Reporting marks: SIND   **Miles:** 5
Address: P. O. Box 312, Sellersburg, IN 47172
The Southern Indiana extends from Speed to Watson. The enginehouse is at Speed.

Nos.	Qty.	Model	Builder	Date	Notes
100-102	3	65-ton	GE	1946-1952	
Total	3				

SOUTHERN PACIFIC TRANSPORTATION CO.

Reporting marks: SP   **Miles:** 11,497
Address: One Market Plaza, San Francisco, CA 94105
This roster includes locomotives owned by and lettered for SP subsidiary Cotton Belt (St. Louis Southwestern). Number series are intermingled with those of SP proper, and SSW units roam all over the SP system.

Nos.	Qty.	Model	Builder	Date	Notes
1010-1013	4	Slug		1979	1
1500, 1502-1505, 1507, 1514, 1516-1525, 1527-1537, 1540, 1542					
	31	SD7r	EMD	1952-1953	2
1600-1613	14	TEBU	M-K	1980-1982	3

Nos.	Qty.	Model	Builder	Date	Notes
2251, 2254, 2257, 2259, 2261, 2270, 2271, 2277, 2282, 2286, 2293	11	SW1200	EMD	1964-1966	
2450-2454, 2457-2462, 2464-2475, 2477-2486, 2488-2504, 2506-2508, 2510, 2511, 2513-2517, 2519-2528, 2530-2540, 2542-2553, 2555-2568, 2570-2575, 2577, 2579-2593, 2595-2624, 2626-2639, 2641, 2642, 2644-2646, 2648-2651, 2653-2662, 2664, 2666-2674, 2676-2689	217	SW1500	EMD	1967-1973	
2690-2696	7	MP15	EMD	1974	
2697-2722, 2724-2759	62	MP15AC	EMD	1975	
2870, 2872, 2879	3	GP9	EMD	1959	
2887	1	GP9r	EMD	1956-1957	2
2889, 2893	2	GP9	EMD	1959	
2898, 2899	2	GP9r	EMD	1956-1957	2
2961-2970	12	SD35d	EMD	1964-1965	4
2971-2976	6	SD38-2	EMD	1973	

Nos.	Qty.	Model	Builder	Date	Notes
3102, 3103, 3105	3	SD35r	EMD	1965	2
3107, 3108	2	SD35	EMD	1965	
3186, 3188, 3190-3195	8	GP9r	EMD	1954-1955	2
3301, 3308, 3314, 3316, 3317, 3320, 3322, 3323, 3327-3329, 3333, 3336, 3338, 3339, 3342, 3343, 3347-3350, 3353, 3362, 3363, 3366, 3367, 3370, 3372, 3377-3380, 3382-3385, 3387, 3388, 3392, 3394-3396, 3398, 3399, 3402, 3403, 3408, 3412-3416, 3418, 3421, 3425, 3428, 3429, 3432, 3435, 3436, 3439, 3440, 3732, 3735, 3738, 3740, 3742-3745, 3748, 3750, 3751, 3753, 3755-3757, 3759-3761, 3763, 3764-3768, 3771, 3773, 3775, 3777, 3778, 3780, 3782-3785, 3787-3789, 3792-3795, 3800, 3802, 3808, 3811-3813, 3816, 3819, 3821, 3822, 3825, 3827-3831, 3833-3838, 3840-3844, 3846, 3848, 3850-3852, 3855, 3856, 3873, 3877, 3878, 3880, 3883	142	GP9r	EMD	1954-1959	2
4060	1	GP20	EMD	1962	

Southern Pacific GP40-2s sandwich TEBU 1602 at Benson, Arizona. The fourth unit is 6352, a rebuilt GP35.

Robert C. Del Grosso

Nos.	Qty.	Model	Builder	Date	Notes
4102-4105, 4107, 4112, 4115, 4117-4121, 4125, 4134, 4135, 4144, 4146, 4151-4153	20	GP20r	EMD	1962	2
4160, 4200, 4201, 4203	4	GP35d	EMD	1964	2
4301-4303, 4305, 4306, 4310, 4312-4315, 4317, 4319, 4320, 4322-4324, 4326, 4327, 4329, 4331-4334, 4336, 4338-4342, 4344-4347, 4349, 4351-4355, 4358, 4359, 4362, 4363, 4367, 4369-4375, 4377-4379, 4382-4384, 4389-4397, 4399-4403, 4406-4412, 4415-4418, 4420-4426, 4228-4432, 4434, 4436, 4438, 4440, 4451	99	SD9r	EMD	1955-1956	2
4800-4807, 4809-4844	44	GP38-2	EMD	1980	
5100-5114	14	B23-7	GE	1980	
5300, 5302, 5304, 5306, 5308-5310, 5314-5318	12	SD39	EMD	1968, 1970	
6300, 6302-6308, 6310-6320, 6322-6361	59	GP35r	EMD	1964-1965	2
6511, 6526, 6537, 6556, 6566, 6568, 6576, 6577, 6587, 6595, 6600, 6605, 6611, 6615, 6619, 6622, 6639, 6657, 6670, 6674, 6676, 6680	22	GP35	EMD	1964-1965	
6767-6832, 6834-6892	125	SD40T-2r	EMD	1974-1980	
7100-7138	39	GP40r	EMD	1966-1971	5
7200, 7201, 7230, 7231	4	GP40X	EMD	1978	6
7240-7257, 7259-7266, 7268-7273	32	GP40-2	EMD	1984	
7274-7299	26	GP40r	EMD	1966-1971	5
7300-7346, 7348-7381, 7383-7385	85	SD40-2r	EMD	1980-1981	
7399-7537, 7539-7548, 7550, 7552-7566	165	SD45T-2r	EMD	1973-1975	
7600-7602, 7608-7611, 7613-7617, 7619-7642, 7644, 7645, 7647-7649, 7651-7677	68	GP40-2	EMD	1974-1980	
7754-7769	16	B36-7	GE	1984	
7770	1	B36-7	GE	1980	
7771	1	B36-7B	GE	1980	7
7772, 7773	2	B36-7	GE	1980	
7774-7799, 7801-7809, 7811-7853, 7855-7867, 7869-7883	106	B30-7	GE	1978-1980	
7940-7956, 7958, 7959	19	GP40-2	EMD	1980	
7960-7967	8	GP40e	EMD	1982	
8000-8039	40	B39-8	GE	1987	
8040-8047, 8049-8081, 8083-8093	52	B40-8	GE	1988-1989	
8230-8259, 8261-8277, 8279-8300, 8301, 8303-8316, 8318-8341, 8350-8391, 8489-8506, 8508-8573	194	SD40T-2	EMD	1974-1980	
9189, 9194, 9195, 9197, 9201, 9206, 9207, 9213, 9215, 9218, 9226, 9231, 9232, 9234, 9238, 9239, 9242, 9243, 9245, 9250, 9254, 9256, 9257, 9259, 9260, 9262-9265, 9270, 9273, 9276, 9281, 9282, 9286-9289, 9291, 9294, 9295, 9298 9300, 9303, 9308-9311, 9313-9319, 9323, 9325, 9327-9330, 9333-9335, 9337, 9338, 9342-9349, 9351, 9353, 9355, 9358-9362, 9364-9368, 9371-9376, 9378, 9380, 9381, 9383-9388, 9391-9393, 9395, 9396, 9398, 9400-9404	114	SD45T-2	EMD	1972-1975	
9600-9618, 9620-9703, 9705-9769	168	GP60	EMD	1988-1991	
Total	2,067				

Notes:
1. Rebuilt from ex-Louisville & Nashville EMD switchers
2. Rebuilt by Southern Pacific 1970-1980
3. Slugs, rebuilt from GE U25Bs
4. Rebuilt 1973-1978; 2000 h.p.
5. Rebuilt 1991 by Morrison-Knudsen from secondhand units
6. GP50 prototypes
7. Cabless, rebuilt from wrecked unit

SOUTHERN RAILROAD OF NEW JERSEY

Reporting marks: SFL   **Miles:** 82
Address: P. O. Box 178, Winslow, NJ 08095-0178

The Southern Railroad of New Jersey (formerly named the Shore Fast Line) operates from Winslow Junction to Atlantic City, New Jersey; from Winslow Junction to Vineland; from Atlantic City to Linwood, and from Tuckahoe to Rio Grande, N. J. The enginehouse is at Winslow Junction.

Nos.	Qty.	Model	Builder	Date	Notes
1145	1	SW1200	EMD	1966	1
2875, 2876, 2884	3	U30B	GE	1967	2
Total	4				

Notes:
1. Ex-Missouri Pacific 1145
2. Ex-Conrail, same numbers

SOUTHERN RAILWAY OF BRITISH COLUMBIA

Reporting marks: BCE   **Miles:** 75
Address: New Westminster, BC V3M 4H3, Canada

Southern Railway of British Columbia (formerly B. C. Hydro Rail; originally British Columbia Electric) extends from New Westminster to Chilliwack, British Columbia.

Nos.	Qty.	Model	Builder	Date	Notes
151-153	3	MP15DC	EMD	1975	
381	1	SD38AC	EMD	1971	
382-384	3	SD38-2	EMD	1972, 1974	
900-909	10	SW900	GMD	1955-1964	
910, 911	2	SW900u	GMD	1967, 1969	1000 h.p.
Total	19				

SOUTHERN SAN LUIS VALLEY RAILROAD

Reporting marks: SSLV   **Miles:** 1
Address: 2255 Lava Lane, Alamosa, CO 81101

The Southern San Luis Valley operates freight service from Blanca to McClintock, Colorado.

Nos.	Qty.	Model	Builder	Date	Notes
1	1	20-ton	Plymouth	1941	
D500	1		SSLV	1955	Homebuilt center-cab
Total	2				

SOUTHWESTERN RAILROAD

Reporting marks: SW   **Miles:** 139
Address: P. O. Box 1876, Ogden, UT 84402

The Southwestern Railroad has two divisions, from Shattuck, Oklahoma, to Spearman, Texas, and from Whitewater, New Mexico, to Tyrone, Fierro, and Santa Rita. Engine facilities are at Perryton, Texas, and Hurley, N. M.

Nos.	Qty.	Model	Builder	Date	Notes
748	1	GP30r	EMD	1963	Ex-Union Pacific
2164, 2171, 2182	3	GP7r	EMD	1952	Ex-Santa Fe
2875, 2934	2	GP35r	EMD	1965	Ex-Santa Fe
5315	1	SD45r	EMD	1966	Ex-Santa Fe
Total	7				

SPENCERVILLE & ELGIN RAILROAD

Reporting marks: SPEG   **Miles:** 33
Address: 109 E. North St., Spencerville, OH 45887

The Spencerville & Elgin operates from Glenmore to Lima, Ohio. It is operated by Indiana Hi-Rail. The enginehouse is at Ohio City.

Nos.	Qty.	Model	Builder	Date	Notes
603, 606	2	RS3	Alco	1952	Ex-Vermont Railway
653	1	S4	Alco	1952	Ex-Penn Central 9764
Total	3				

STEELTON & HIGHSPIRE RAILROAD

Reporting marks: SH   **Miles:** 5
Address: Front and Swatara Streets, Steelton, PA 17113
The Steelton & Highspire is a switching road between Steelton and Highspire, Pennsylvania.

Nos.	Qty.	Model	Builder	Date	Notes
70	1	SW9	EMD	1952	1
71, 72	2	SW7	EMD	1950	2
73-75	3	SW1200	EMD	1956	3
77	1	SW1200	EMD	1957	4
Total	7				

Notes:
1. Ex-Conemaugh & Black Lick 119
2. Ex-Conemaugh & Black Lick 106, 117
3. Ex-Cambria & Indiana 38, 39, 42
4. Ex-Philadelphia, Bethlehem & New England 42

STERLING BELT LINE

Reporting marks: SBLN   **Miles:** 1
Address: 12th and Commerce Streets, Wellsville, OH 43968
The Sterling Belt Line is a switching operation at Wellsville, Ohio.

Nos.	Qty.	Model	Builder	Date	Notes
	1	Trackmobile			
Total	1				

STEWARTSTOWN RAILROAD

Reporting marks: STRT   **Miles:** 25
Address: P. O. Box 155, Stewartstown, PA 17363
The Stewartstown Railroad operates from York, Pennsylvania, south to New Freedom, then east to Stewartstown. The enginehouse is at Stewartstown.

Nos.	Qty.	Model	Builder	Date	Notes
9	1	35-ton	Plymouth	1943	
10	1	44-ton	GE	1946	
11	1	SW8	EMD	1953	
Total	3				

STOURBRIDGE RAILROAD

Reporting marks: SBRR   **Miles:** 25
Address: 356 Priestley Ave., Northumberland, PA 17857
The Stourbridge Railroad operates between Honesdale and Lackawaxen, Pennsylvania. The enginehouse is at Honesdale.

Nos.	Qty.	Model	Builder	Date	Notes
44	1	CF7	ATSF	1977	1
54	1	BL2	EMD	1950	2
Total	2				

Notes:
1. Ex-Santa Fe 2444
2. Ex-Bangor & Aroostook 54, used for excursions

STRASBURG RAIL ROAD

Reporting marks: SRC   **Miles:** 5
Address: P. O. Box 96, Strasburg, PA 17579
The Strasburg Rail Road extends from Strasburg to Leaman Place Junction, Pennsylvania. The enginehouse is at Strasburg. The road's primary business is excursion trains; it has several steam locomotives.

Nos.	Qty.	Model	Builder	Date	Notes
5	1	S1	Alco	1947	Ex-Alco plant switcher
33	1	44-ton	GE	1946	Ex-Pennsylvania 9330
Total	2				

Charles W. McDonald

Red-and-white SW9 No. 1201 is typical of EMD switchers found by the hundreds across the United States.

TACOMA MUNICIPAL BELT LINE RAILWAY

Reporting marks: TMBL   **Track miles:** 25
Address: P. O. Box 11007, Tacoma, WA 98411
Tacoma Municipal Belt is a switching road at Tacoma, Washington.

Nos.	Qty.	Model	Builder	Date	Notes
1200	1	SW9	EMD	1953	Ex-Union Pacific 1242
1201	1	SW9	EMD	1952	Ex-Chesapeake & Ohio 5089
1202	1	SW7	EMD	1951	Ex-Atlantic Coast Line 661
1203	1	SW1200	EMD	1965	Ex-Missouri Pacific 1271
1204	1	SW1200	EMD	1963	Ex-Missouri Pacific 1191
Total	5				

TENNESSEE SOUTHERN RAILROAD

Reporting marks: TSRR   **Miles:** 108
Address: P. O. Box 387, Pulaski, TN 38478
The Tennessee Southern extends from Godwin, Tennessee, to Florence, Alabama, with a branch from Columbia to Pulaski, Tenn. The enginehouse is at Pulaski.

Nos.	Qty.	Model	Builder	Date	Notes
706, 2329	2	GP7	EMD	1952, 1953	
5981, 6173, 6487, 6583	4	GP9	EMD	1955-1958	1
7314, 7317	2	SD18	EMD	1963	1
9424	1	GP18	EMD	1963	2
Total	9				

Notes:
1. Ex-Chessie System
2. Ex-Illinois Central

TENNKEN RAILROAD

Reporting marks: TKEN   **Miles:** 52
Address: 1200 E. Cherry St., Dyersburg, TN 38024
The Tennken Railroad operates from Dyersburg, Tennessee, to Hickman, Kentucky. Engine facilities are at Dyersburg.

Nos.	Qty.	Model	Builder	Date	Notes
9434, 9435	2	GP28	EMD	1964	Ex-Illinois Central Gulf
Total	2				

TERMINAL RAILWAY ALABAMA STATE DOCKS

Reporting marks: TASD   **Track miles:** 65
Address: P. O. Box 1588, Mobile, AL 36633

Terminal Railway, Alabama State Docks, is a switching railroad at Mobile, Chickasaw, and Theodore, Alabama.

Nos.	Qty.	Model	Builder	Date	Notes
761, 771, 772, 801-803, 821, 822	8	MP15DC	EMD	1976-1982	
Total	8				

TERRE HAUTE, BRAZIL & EASTERN RAILROAD

Reporting marks: TBER   **Miles:** 38
Address: P. O. Box 3271, Terre Haute, IN 47803

The Terre Haute, Brazil & Eastern extends from Terre Haute to Limedale, Indiana, with a branch at Staunton. The enginehouse is at Terre Haute.

Nos.	Qty.	Model	Builder	Date	Notes
100	1	SW1200	EMD	1965	Ex-Missouri Pacific 1268
101	1	SW8	EMD	1952	Ex-Rock Island 829
Total	2				

TEXAS CENTRAL RAILROAD

Reporting marks: TEXC   **Miles:** 24
Address: P. O. Box 458, Gorman, TX 76454

The Texas Central operates from Dublin to Gorman, Texas. The enginehouse is at Dublin.

Nos.	Qty.	Model	Builder	Date	Notes
1, 3, 4	3	S2	Alco	1946, 1948	
Total	3				

TEXAS CITY TERMINAL RAILWAY

Reporting marks: TCT   **Track miles:** 32
Address: P. O. Box 591, Texas City, TX 77592-0591

Texas City Terminal is a terminal road at Texas City, Texas. It is jointly owned by Union Pacific and Santa Fe.

Nos.	Qty.	Model	Builder	Date	Notes
35-37	3	MP15	EMD	1982	
Total	3				

TERMINAL RAILROAD ASSOCIATION OF ST. LOUIS

Reporting marks: TRRA   **Track miles:** 208
Address: 2016 Madison Ave., Granite City, IL 62040

TRRA is a switching railroad connecting all the major railroads in St. Louis and in the Illinois cities across the Mississippi River.

Nos.	Qty.	Model	Builder	Date	Notes
551, 553	2	Slug	TRRA	1975, 1976	1
560-562	3	Slug	TRRA	1984	1
1229-1231, 1235, 1238-1243	10	SW1200	EMD	1964-1965	
1501-1517	17	SW1500	EMD	1967-1972	
Total	32				

Notes:
1. Rebuilt from NW2s

Twin exhaust stacks are evidence of the 1983 repowering of Texas & Northern S4u No. 100.

TEXAS & NORTHERN RAILWAY

Reporting marks: TN   **Miles:** 8
Address: P. O. Box 300, Lone Star, TX 75688-0300
The Texas & Northern operates from Daingerfield to Lone Star, Texas. Engine facilities are at Lone Star.

Nos.	Qty.	Model	Builder	Date	Notes
1-3	3	Slug	Alco	1976-1983	1
24	1	S4r	Alco	1941	
41	1	S2r	Alco	1943	2
43, 44	2	RS1d	Alco	1945, 1950	3
45, 55, 56	3	S4d	Alco	1951-1952	3
100, 102	2	S4u	Alco	1951	4
992-995	4	CF7	ATSF	1973-1975	
996	1		TN	1982	5
997	1	GP7	EMD	1950	6
998, 999	2	MP15	EMD	1975	
Total	**20**				

Notes:
1. Rebuilt Alco S4s
2. Rebuilt 1978; 600 h.p.
3. Rebuilt 1979; 600 h.p.
4. Rebuilt 1983; 1200 h.p.
5. Homebuilt using Alco parts; 1800 h.p.
6. Ex- Great Northern 602

TEXAS, OKLAHOMA & EASTERN RAILROAD

Reporting marks: TOE   **Miles:** 40
Address: 412 E. Lockesburg Ave., De Queen, AR 71832

The Texas, Oklahoma & Eastern extends from Valliant, Oklahoma, to West Line, Arkansas. It is operated in conjunction with the De Queen & Eastern. Engine facilities are at De Queen, Ark.

Nos.	Qty.	Model	Builder	Date	Notes
D-4	1	SW900	EMD	1954	
D-12–D14	3	GP40	EMD	1971	
D-15, D16, D-20	3	GP40-2	EMD	1972-1974	Ex-Conrail
D-22	1	GP35	EMD	1965	Ex-SP
D-23–D-25	3	GP40	EMD	1968	
Total	**11**				

The principal item hauled by the GP40s and GP40-2s of Texas, Oklahoma & Eastern is forest products from the mills of the road's owner, Weyerhaeuser.

Charles W. McDonald

TEXAS MEXICAN RAILWAY

Reporting marks: TM   **Miles:** 157
Address: P. O. Box 419, Laredo, TX 78042-0419

The Texas Mexican Railway extends from Corpus Christi, Texas, to the Mexican border at Laredo. The road's shops are at Laredo.

Nos.	Qty.	Model	Builder	Date	Notes
850, 852	2	GP7	EMD	1950, 1951	1
853	1	GP9	EMD	1958	1
854, 855	2	GP18	EMD	1963	
856	1	GP28	EMD	1965	
857-860	4	GP38	EMD	1966-1971	
861-867	7	GP38-2	EMD	1974-1985	
868	1	GP35	EMD	1964	2
869, 870	21	GP60	EMD	1990, 1991	
Total	**20**				

Notes:
1. Short hood contains crew space
2. Ex-Burlington Northern, earlier Frisco

TEXAS NORTH WESTERN RAILWAY

Reporting marks: TXNW   **Miles:** 46
Address: Route 1, Box 9, Sunray, TX 79086-9702

The Texas North Western operates between Etter Junction and Morse, Texas, on a former Rock Island line. The enginehouse is at Sheerin.

Nos.	Qty.	Model	Builder	Date	Notes
88	1	SW7	EMD	1951	Ex-Conrail 9369
89	1	NW2	EMD	1948	Ex-Kansas City Southern 1221
Total	**2**				

Texas Transportation Company's entire locomotive roster waits with trolley poles down for work assignments.

TEXAS TRANSPORTATION COMPANY

Reporting marks: TXTC　　　　　　　　　**Miles:** 1
Address: P. O. Box 1661, San Antonio, TX 78296
Texas Transportation Company is an electrically operated switching railroad at San Antonio. It is owned by Pearl Brewing.

Nos.	Qty.	Model	Builder	Date	Notes
1	1	50-ton	BLW	1917	Electric
2	1	45-ton	Texas Electric	1921	Electric
Total	2				

TEXAS SOUTH-EASTERN RAILROAD

Reporting marks: TSE　　　　　　　　　**Miles:** 21
Address: P. O. Box 1661, Diboll, TX 75941
The Texas South-Eastern operates from Diboll to Lufkin, Texas, with a branch to Vair. The enginehouse is at Diboll.

Nos.	Qty.	Model	Builder	Date	Notes
22	1	70-ton	GE	1956	
301	1	DS44-7.5	BLW	1950	Ex-demonstrator 301
1007	1	VO1000	BLW	1944	
Total	3				

THERMAL BELT RAILWAY

Reporting marks: TBRY　　　　　　　　　**Miles:** 16
Address: P. O. Box 203, Bostic, NC 28018
Thermal Belt operates from Bostic, North Carolina, to Alexander Mills and Gilkey.

Nos.	Qty.	Model	Builder	Date	Notes
3	1	SW	EMC	1938	Ex-Pickens, St. Joseph Belt
Total	1				

TIOGA CENTRAL RAILROAD

Reporting marks: TIOC **Miles:** 26
Address: 25 Delphine St., Owego, NY 13827
Tioga Central extends from Owego to Harford, New York. The engine-house is at Owego.

Nos.	Qty.	Model	Builder	Date	Notes
14	1	S2	Alco	1947	
47, 62, 240	3	RS1	Alco	1945, 1950	
506	1	TE56	MK	1975	
Total	**5**				

TOLEDO, PEORIA & WESTERN RAILWAY

Reporting marks: TPW **Miles:** 290
Address: 1661 Route 22 West, Bound Brook, NJ 08805
The Toledo, Peoria & Western operates between Logansport, Indiana, and Fort Madison, Iowa. The enginehouse is at East Peoria, Illinois.

Nos.	Qty.	Model	Builder	Date	Notes
2001-2004	4	GP20r	EMD	1960	Ex-Santa Fe
3004, 3013-3016, 3019, 3022, 3034, 3040, 3051, 3053, 3059, 3063, 3068, 3073	15	GP20r	EMD	1960	Ex-Santa Fe
Total	**19**				

TONAWANDA ISLAND RAILROAD

Reporting marks: TIRL **Miles:** 5
Address: 3163 Buffalo Ave., Niagara Falls, NY 14303
The Tonawanda Island Railroad is a switching line between North Tonawanda and Tonawanda Island, New York. The enginehouse is at Tonawanda.

Nos.	Qty.	Model	Builder	Date	Notes
Unnumbered	1	50-ton	CLC	1950	
Total	**1**				

TOWANDA-MONROETON SHIPPERS LIFELINE

Reporting marks: TMSL **Miles:** 6
Address: RD 1, P. O. Box 18, Monroetown, PA 18832
Towanda-Monroeton Shippers Lifeline operates a former Lehigh Valley branch between the towns of its name. The enginehouse is at Monroeton.

Nos.	Qty.	Model	Builder	Date	Notes
26	1	SW1	EMD	1940	Ex-Richmond, Fredericksburg & Potomac 50
Total	**1**				

TRADEWATER RAILWAY

Reporting marks: TWRY **Miles:** 92
Address: P. O. Box 66, Sturgis, KY 42459
Tradewater Railway extends from Princeton to Waverly, Kentucky, with a branch from Blackford to Providence. The engine facilities are at Pyro Mine.

Nos.	Qty.	Model	Builder	Date	Notes
1338, 1342, 1347	3	GP18	EMD	1960-1961	1
1602	1	GP7	EMD	1951	
3419	1	GP10	ICG	1977	
Total	**5**				

Notes:
1. Ex-Rock Island

TRANSKENTUCKY TRANSPORTATION RAILROAD

Reporting marks: TTIS **Miles:** 50
Address: 105 Winchester St., Paris, KY 40361

Transkentucky Transportation operates between Mayville and Paris, Kentucky. The enginehouse is at Paris.

Nos.	Qty.	Model	Builder	Date	Notes
242-244, 246-248, 250-261	18	U28B	GE	1966	1
1074-1076	3	RS3	Alco	1946	
2074, 2075	2	Slug	Alco	1973	2
Total	**23**				

Notes:
1. Ex-Rock Island, Pittsburgh & Lake Erie, and Burlington Northern
2. Rebuilt from RS3s by Louisville & Nashville

TRI-RAIL (Tri-County Rail Authority)

Miles: 68

Address: 305 S. Andrews Ave., Fort Lauderdale, FL 33301

Tri-Rail operates commuter service on CSX rails between Miami and West Palm Beach, Florida.

Nos.	Qty.	Model	Builder	Date	Notes
801-805	6	F40PHL-2	M-K	1988	Rebuilt GP40s
Total	**6**				

TRONA RAILWAY

Reporting marks: TRC **Miles:** 31

Address: 13068 Main St., Trona, CA 93562

The Trona Railway extends from Searles to Trona, California. The enginehouse is at Trona.

Nos.	Qty.	Model	Builder	Date	Notes
52-54	3	AS616	BLW	1951-1954	
Total	**3**				

TULSA-SAPULPA UNION RAILWAY

Reporting marks: TSU **Miles:** 12

Address: 701 E. Dewey, Sapulpa, OK 74066

The Tulsa-Sapulpa Union Railway extends from Tulsa to Sapulpa, Oklahoma. The enginehouse is at Sapulpa.

Nos.	Qty.	Model	Builder	Date	Notes
101-103	3	SW1	EMD	1940-1947	
104	1	SW9	EMD	1952	Ex-Burlington Northern 157
107	1	SW7	EMD	1949	Ex-Davenport, Rock Island & Northwestern 107
Total	**5**				

TUSCOLA & SAGINAW BAY RAILWAY

Reporting marks: TSBY **Miles:** 537

Address: P. O. Box 550, Owosso, MI 48867-0550

The Tuscola & Saginaw Bay operates the former Ann Arbor main line from Ann Arbor to Thompsonville, Michigan, plus other lines in Michigan: Ashley to Middleton, Owosso to St. Charles, Cadillac to Petoskey, and Walton Junction to Grawn and Williamsburg. Engine facilities are at Owosso and Cadillac, Mich.

Nos.	Qty.	Model	Builder	Date	Notes
1	1	25-ton	GE	1953	
385, 387-394	9	GP35	EMD	1964	Ex-Ann Arbor, same nos.
1977	1	NW2	EMD	1942	Ex-Union Pacific 1034
Total	**11**				

TYBURN RAILROAD

Reporting marks: TYBR   **Miles:** 1
Address: P. O. Box 196, Penndel, PA 19047
The Tyburn Railroad operates switching services at Fairless, near Morrisville, Pennsylvania, and at Lancaster, Pa.

Nos.	Qty.	Model	Builder	Date	Notes
400	1	44-ton	GE	1947	
Total	1				

TYSON RAILROAD

Reporting marks: TSNR   **Miles:** 2
Address: 2210 Oaklawn Drive, Springdale, AR 72764
The Tyson Railroad is a switching road at Ivalee, Alabama.

Nos.	Qty.	Model	Builder	Date	Notes
293	1	SW900	EMD	1960	
Total	1				

UNION PACIFIC RAILROAD

Reporting marks: UP   **Miles:** 21,882
Address: 1416 Dodge St., Omaha, NE 68179

Nos.	Qty.	Model	Builder	Date	Notes
9-31	23	SL1	MP	1978-1981	1
61, 71	2	GP50	EMD	1981	2
100-135, 137-166, 170-184	81	B23-7	GE	1978-1981	2
200-254	55	B30-7A	GE	1981-1982	2
257-259	3	B23-7	GE	1980	2
300	1	SL1	MP	1982	1, 2
301-303	3	S5-2B	MK	1991	3
501-505, 507-512, 515	13	GP40	EMD	1968	4
551, 557, 558, 560, 561	5	U23B	GE	1974, 1976	2
581-585, 588-590, 592-594, 597	12	GP40	EMD	1969	4
651-665	15	GP40	EMD	1966-1971	5
667, 669-683	16	GP40	EMD	1968	4
782-795, 797-799	17	GP35	EMD	1965	5
850-862, 864-887	37	GP40	EMD	1968-1969	
906-914	9	GP40-2	EMD	1980	5
949, 951	2	E9A	EMD	1955	6
954-959	6	GP40X	EMD	1978	
960-970	11	GP50	EMD	1980-1981	2

Nos.	Qty.	Model	Builder	Date	Notes
970B	1	E9B	EMD	1955	6
972-980, 982-989	17	GP50	EMD	1981	2
1200-1236, 1238-1265, 1268-1270, 1272, 1273	70	SW10	UP	1980-1984	7
1300-1314	15	MP15	EMD	1974-1975	
1315-1317, 1320, 1321	5	SW1500	EMD	1972-1973	
1324, 1326, 1327	3	SW1500	EMD	1968	4
1329-1354, 1356-1376, 1379-1381, 1383-1392	60	MP15	EMD	1974-1982	
1393-1396	4	MP15AC	EMD	1980	4

Nos.	Qty.	Model	Builder	Date	Notes
1555-1589, 1591-1594, 1597-1617, 1619, 1621-1633, 1635-1646, 1648-1654, 1657-1684, 1686, 1688, 1689, 1692, 1696-1715, 1717-1721, 1723, 1725, 1728, 1729, 1731-1738, 1740-1743	166	GP15-1	EMD	1976-1982	2

Nos.	Qty.	Model	Builder	Date	Notes
1800--1828	29	GP38-2	EMD	1972	
1974-1978	5	GP38	EMD	1969-1970	4
1979-1989, 1991-1999	20	GP38AC	EMD	1969-1970	4
2000-2025, 2027-2059	59	GP38-2	ED	1974-1975	
2060-2063	4	GP38-2	EMD	1976	4

2077-2088, 2090, 2091, 2093, 2095-2102, 2104, 2105, 2107-2109, 2111, 2113-2117, 2119, 2110, 2121-2159, 2164, 2165, 2168, 2169, 2172-2176, 2178-2180, 2182-2186, 2188-2191, 2194-2200, 2202-2204, 2206, 2208-2211, 2215, 2216, 2220, 2222-2232, 2235, 2237-2240, 2243, 2244, 2247-2279, 2281-2289

Nos.	Qty.	Model	Builder	Date	Notes
174		GP38-2	EMD	1973-1980	2
2291-2295, 2298-2304, 2306, 2308, 2309, 2311-2314, 2316-2319					
	35	GP38-2	EMD	1981	
2335-2348	14	GP38-2	EMD	1973-1974	4
2350-2358	9	GP39-2	EMD	1977	8
2359-2378	20	GP39-2	EMD	1984	9
2400	1	C30-7	GE	1977	
2401, 2403-2414	13	C30-7rB	GE	1977	10
2415-2424, 2426-					
2431, 2433-2473	57	C30-7	GE	1978-1980	
2474	1	C30-7rB	GE	1980	10
2475-2488	14	C30-7	GE	1980	
2489	1	C30-7rB	GE	1980	10
2490-2535, 2537, 2538	38	C30-7	GE	1980-1981	
3000, 3001	2	SD40	EMD	1966	
3002-3007, 3123-3133, 3135-3138, 3140-3145, 3147-3165, 3167, 3168, 3170-3177, 3179-3194, 3196-3218, 3220-3265					
	140	SD40-2	EMD	1972-1980	
3266	1	SD40-2rB	EMD	1974	10
3267-3272, 3274-					
3308, 3310-3331	63	SD40-2	EMD	1974-1976	
3332	1	SD40-2rB	EMD	1976	10
3333-3335, 3337, 3338	5	SD40-2	EMD	1976	
3339	1	SD40-2rB	EMD	1977	10
3340-3430	91	SD40-2	EMD	1977-1978	
3431	1	SD40-2rB	EMD	1978	10
3432-3454	23	SD40-2	EMD	1978	
3455	1	SD40-2rB	EMD	1978	10
3456-3501, 3503-					
3533, 3536-3549	91	SD40-2	EMD	1978-1979	
3550	1	SD40-2rB	EMD	1979	10
3551-3555	5	SD40-2	EMD	1979	
3556	1	SD40-2rB	EMD	1979	10
3557-3616, 3618-3639	82	SD40-2	EMD	1979-1980	

Nos.	Qty.	Model	Builder	Date	Notes
3640, 3641	2	SD40-2rB	EMD	1980	10
3642-3782	141	SD40-2	EMD	1980-1981	
3783, 3786	1	SD40-2rB	EMD	1981	10
3785	1	SD40-2	EMD	1981	
3783, 3786	1	SD40-2rB	EMD	1981	10
3787-3808	22	SD40-2	EMD	1981	
3809-3843	35	SD40-2	EMD	1978-1981	4
3844, 3884	2	SD40-2	EMD	1985, 1981	
3900-3902, 3904-3973	73	SD40-2	EMD	1976-1980	2
4020-4029, 4031, 4033-4063, 4090-4104, 4106-4119, 4202-4217					
	87	SD40	EMD	1971	2
4218	1	SD40-2rB	EMD	1978	2, 10
4219-4265, 4267-4269,					
4271-4305, 4307-4312	91	SD40-2	EMD	1978-1980	2
4313	1	SD40-2rB	EMD	1980	2, 10
4315	1	SD40-2	EMD	1980	2
4316	1	SD40-2rB	EMD	1980	2, 10
4317-4321	5	SD40-2	EMD	1980	2
5000-5059	60	SD50	EMD	1984	2
6000-6013, 6015-6084	84	SD60	EMD	1987	
6060-6084	25	SD60	EMD	1987	
6085-6312	228	SD60M	EMD	1988-1991	
6936	1	DDA40X	EMD	1971	
9000-9059	60	C36-7	GE	1985	2, 11
9100-9479	380	C40-8	GE	1987-1991	
Locomotives still carrying Missouri-Kansas-Texas numbers					
53	1	SW1500	EMD	1968	
170, 173, 174,					
177, 180, 181	6	GP40	EMD	1966	12
194, 200	2	GP40	EMD	1968	
Locomotives still carrying Missouri Pacific numbers					
608, 622, 631, 632	4	GP40	EMD	1966, 1970	12
1377, 1378, 1382	3	MP15	EMD	1982	
1590, 1595, 1596, 1618, 1620, 1634, 1647, 1655, 1685, 1687, 1690, 1691, 1693-1695, 1716, 1722, 1724, 1726, 1727, 1730, 1739, 1744					
	23	GP15-1	EMD	1977-1982	

Union Pacific SD60M No. 6087 shows off the new look in diesels, the safety cab. The nose of the unit has extra reinforcement for collision protection.

Nos.	Qty.	Model	Builder	Date	Notes
2075, 2076, 2082, 2089, 2092, 2094, 2103, 2106, 2112, 2118, 2160-2163, 2166-2168, 2170, 2171, 2177, 2181, 2187, 2192, 2193, 2201, 2205, 2207, 2212-2214, 2217-2219, 2221, 2233, 2234, 2236, 2241, 2242, 2245, 2246, 2280, 2290, 2296, 2297, 2302, 2305, 2307, 2310, 2315, 2320, 2327, 2329, 2333	55	GP38-2	EMD	1973-1981	
2370	1	SD40-2	EMD	1976	
3306	1	SD40-2	EMD	1976	
3002, 3005-3008, 3010, 3012, 3015, 3017, 3023, 3028, 3042, 3060, 3064-3066, 3070, 3076, 3079, 3080, 3083, 3086, 3087, 3091, 3093, 3096, 3098, 3099, 3105, 3115, 3120	31	SD40	EMD	1966-1971	12
3132	1	SD40-2	EMD	1974	
4057, 4060-4062, 4066	5	SD40	EMD	1970	12
6071	1	SD40-2	EMD	1980	
Total	3,092				

Notes:
1. Slugs, rebuilt from EMD switchers
2. Ex-Missouri Pacific
3. Rebuilt from Kansas City Southern GP30s 4104, 4108, 4109
4. Ex-Missouri-Kansas-Texas
5. Ex-Western Pacific
6. For executive trains
7. Rebuilt EMD switchers
8. Ex-Missouri-Kansas-Texas 380-388, previously Kennecott Copper; rebuilt from high-cab configuration
9. Ex-MKT 360-379, in GP49 carbodies
10. Rebuilt to B unit, retaining the cab
11. Rated at 3750 h.p.
12. Leased to National Railways of Mexico

RENUMBERING OF MISSOURI-KANSAS-TEXAS UNITS

MKT Nos.	Model	UP Nos.
50-55	SW1500	1322-1327
56-59	MP15AC	1393-1396
211-221, 223-230	GP40	581-599
231-248	GP40	666-683
300-303	GP38	1975-1978
304-317	GP38-2	2335-2348
318-321	GP38-2	2060-2063
322-335, 337-343	GP38AC	1979-1999
360-379	GP39-2	2359-2378
380-388	GP39-2	2350-2358
600-612, 614-636	SD40-2	3809-3844

UNION RAILROAD

Reporting marks: Union Railroad   **Miles:** 234
Address: P. O. Box 68, Monroeville, PA 15146

The Union Railroad is a terminal railroad serving the steel industry in the Monongahela River Valley south of Pittsburgh. It leases locomotives to the McKeesport Connecting Railroad, a switching line at McKeesport, Pennsylvania. Union Railroad's shops are at Monroeville and Monongahela Junction.

Nos.	Qty.	Model	Builder	Date	Notes
1-9	9	SW1500	EMD	1972-1973	
10-33	24	MP15	EMD	1974-1977	
563-573, 580-583	16	SW9	EMD	1952-1953	
584, 585	2	SW1200	EMD	1954	
701, 702	2	TR5A	EMD	1951	
703	1	SW7	EMD	1950	
704	1	SW9	EMD	1951	
Total	55				

UNION RAILROAD OF OREGON

Reporting marks: UO   **Miles:** 3
Address: 957 Arch St., Union, OR 97833

The Union Railroad of Oregon operates freight service between Union and Union Junction, Oregon. The enginehouse is at Union.

Nos.	Qty.	Model	Builder	Date	Notes
1	1	25-ton	Plymouth	1928	
2	1	35-ton	Plymouth		
Total	2				

Air Force 1600 is an 80-ton diesel of 65-ton configuration weighted for extra tractive effort with a thick deck plate.

UNITED STATES AIR FORCE

Nos.	Qty.	Model	Builder	Date	Notes
1236+1246	8	44-ton	GE	1953	
1248, 1249, 1275, 1276	4	60-ton	Whitcomb	1953-1954	
1600-1602, 1604-1607, 1644, 1651, 1655, 1668-1673, 1686-1692					
	23	80-ton	GE	1952-1954	
1841, 1842, 1863, 1864	4	S12	BLH	1952	1
2000, 2007, 2021	3	SW8	EMD	1951	2
2104	1	MRS1	Alco	1953	2
4007, 4010, 4011,					
4013, 4041, 4044	6	RS4TC	BLH	1954	2
4900, 4901	2	GP40r	EMD	1967	3
4902, 4903	2	GP40r	EMD	1966	4
7277, 7370, 7460	3	S1	Alco	1943	2
7860	1	80-ton	GE	1943	2
Total	**57**				

Notes:
1. 1841, 1842 are in Alaska
2. Ex-Army
3. Ex-CSX; lettered for Boeing
4. Ex-CSX; being rebuilt by Air Force

Few Fairbanks-Morse locomotives remain in service anywhere. Most of the Army's H12-44s like No. 1849 are stored.

Charles W. McDonald

UNITED STATES ARMY

Nos.	Qty.	Model	Builder	Date	Notes
1011	1	5-ton	Brookville	1965	1
1200+1214	4	45-ton	GE	1942-1943	
1216+1246	9	44-ton	Davenport	1953	
1400+1414	3	65-ton	GE	1941	
1600-1607	8	80-ton	Davenport	1952	
1617+1694	42	80-ton	GE	1941-1952	
1808+1820	5	MRS1	EMD	1952	2
1821+1840	8	GP7L	EMD	1951	3
1843+1862	8	H12-44	FM	1953	4
1867-1881	14	GP10	ICG	1973-1974	5
2000, 2002, 2007-2012, 2027-2038	20	SW8	EMD	1951	6
2041	1	RSD4	Alco	1954	
2047+2079	3	MRS1	Alco	1954	2
4001-4012,					
4014-4040	39	RS4TC	BLH	1954	2
4600	1	GP10	VMV	1991	
4601-4603	3	GP10	ICG	1974	7
4604	1	GP10	VMV	1992	
7045+7051	4	45-ton	GE	1944	
7104	1	S2	Alco	1942	
7132+7142	6	S1	Alco	1941	
7186	1	DE44	Plymouth	1953	1
Total	**179**				

Notes:
1. Narrow gauge
2. Adjustable gauge
3. AAR Type A switcher trucks
4. Most stored unserviceable
5. Ex-Illinois Central Gulf 8144, 8149, 8011, 8105, 8164, 8329, 8205, 8027, 8176, 8245, 8157, 8293, 8204, 8289
6. 2000 and 2007 rebuilt 1964
7. Ex-Illinois Central Gulf 8343, 8418, 8088

163

UNITED STATES MARINE CORPS

Nos.	Qty.	Model	Builder	Date	Notes
112345, 112346	2	45-ton	GE	1943	1
152408	1	44-ton	GE	1943	
209483-209485	3	80-ton	GE	1953	2
248236	1	44-ton	GE	1941	3
248391	1	80-ton	GE	1943	4
250894	1	44-ton	GE	1942	5
262098	1	45-ton	GE	1943	
Total	**10**				

Notes:
1. 112345 is ex-Navy 65-00388
2. 209484 is ex-Navy 65-00487
3. Ex-Air Force 7310, Army 7310
4. Ex-Army 7285
5. Ex-Air Force 7508, Army 7508

UNITED STATES NAVY

Nos.	Qty.	Model	Builder	Date	Notes
65-00009	1	80-ton	GE	1945	
65-00027, 65-00060	2	VO1000r	BLW	1944	1
65-00068	1	65-ton	Whitcomb	1943	
65-00072, 65-00073, 65-00075, 65-00095, 65-00128, 65-00131, 65-00132	7	VO1000r	BLW	1942-1945	2
65-00201	1	65-ton	GE	1945	
65-00202	1	VO1000r	BLW	1945	
65-00205	1	45-ton	Vulcan	1942	
65-00255, 65-00257	2	80-ton	GE	1945	
65-00292, 65-00293	2	S12	BLW	1951	
65-00296, 65-00298, 65-00306–65-00310	7	80-ton	GE	1945-1952	
65-00315	1	S12	BLW	1951	
65-00328, 65-00329	2	50-ton	Porter	1942	
65-00342	1	80-ton	GE	1944	
65-00344, 65-00345	2	44-ton	GE	1945	
65-00347–65-00349, 65-00351, 65-00355, 65-00356, 65-00358–65-00360, 65-00364, 65-00365	11	80-ton	GE	1953	
65-00365–65-00371	7	S12	BLW	1952	
65-00367, 65-00368	2	VO1000r	BLW	1941	3
65-00383, 65-00385, 65-00386	3	80-ton	GE	1953	
65-00391	1	S12	BLW	1953	
65-00399	1	65-ton	GE	1942	
65-00411	1	80-ton	GE	1953	
65-00431	1	Mine loco	Brookville	1951	
65-00469	1	65-ton	GE	1943	
65-00476	1	Mine loco	GDMN		
65-00487	1	80-ton	GE	1953	
65-00495, 65-00496	2	Mine loco	Brookville	1942	
65-00498, 65-00499	2	65-ton	GE	1943	
65-00505	1	44-ton	GE	1942	
65-00510	1	65-ton	GE	1942	4
65-00513	1	44-ton	GE	1942	5
65-00514	1	61-ton	BLH	1954	6
65-00516	1	80-ton	GE	1952	7
65-00518	1	44-ton	GE	1944	8
65-00521	1	45-ton	GE	1942	9
65-00522, 65-00523	2	45-ton	GE	1944	10
65-00524	1	80-ton	GE	1951	
65-00526	1	VO1000r	BLW	1943	
65-00528	1	45-ton	GE	1944	11
65-00530, 65-00531	2	VO1000r	BLW	1942, 1943	

Navy 65-00554 is an example of the MRS1, a 1000-h.p. hood unit that was adaptable to difference track gauges and restricted clearance diagrams.

Charles W. McDonald

Nos.	Qty.	Model	Builder	Date	Notes
65-00537	1	45-ton	GE	1944	12
65-00540–65-00547,					
65-00554	9	MRS1	Alco	1953	13
65-00556	1	44-ton	Davenport	1953	14
65-00565–65-00567, 65-00569, 65-00576, 65-00578					
	6	65-ton	GE	1943	
65-00580	1	44-ton	GE	1942	15
65-00581	1	65-ton	GE	1943	16
65-00582	1	45-ton	GE	1941	17

Nos.	Qty.	Model	Builder	Date	Notes
65-00595, 65-00596, 65-00599, 65-00600, 65-00601, 65-00604, 65-00607					
	7	65-ton	GE	1941-1943	18
65-00608	1	44-ton	GE	1942	19
65-00610	1	VO1000	BLW	1943	20
65-00611–65-00614	4	MRS1	Alco	1954	21
65-00615	1	45-ton	GE	1941	22
65-00616, 65-00617	2	MRS1	Alco	1953	23

Nos.	Qty.	Model	Builder	Date	Notes
65-00618	1	80-ton	GE	1951	24
65-00619	1	80-ton	GE	1945	
65-00620–6500622	3	VO1000r	BLW	1941-1944	25
65-00623–65-00625	3	VO1000r	BLW	1943	26
65-00627–65-00631	5	SW1200	EMD	1954	
65-00632	1	44-ton	GE	1944	
65-00634, 65-00637	2	SW1200	EMD	1954	
Total	132				

Notes:
1. Ex-Seaboard Coast Line 38, 40
2. 65-00072 is ex-Frisco 205; 65-00095 is ex-Frisco 215
3. Ex-Frisco 201, 202
4. Ex-Air Force 7253
5. Ex-Air Force 7283
6. Ex-Air Force 1274
7. Ex-Air Force 1666
8. Ex-Air Force 8503
9. Ex-Air Force 7247
10. Ex-Air Force 8513, 8502
11. Ex-Army 8517
12. Ex-Army 8575
13. Ex-Army 2118, 2119, 2120, 2121, 2122, 2123, 2091, 2096
14. Ex-Army 1224
15. Ex-Army 7516
16. Ex-Army 7880
17. Ex-Army 7355
18. Ex-Army 1402, 7888, 7347, 7348, 7188, 7017, 7354
19. Ex-Marine Corps 250894, Air Force 7568, Army 7568
20. Ex-Louisville & Nashville 2150
21. Ex-Army 2048, 2066, 2063, 2058
22. Ex-Army 7091
23. Ex-Army 2057, 2064
24. Ex-Army 7393
25. Ex-Seaboard Coast Line 30, 39, 104
26. Ex-Frisco 210, 200, 206

UPPER MERION & PLYMOUTH RAILROAD

Reporting marks: UMP   **Track miles:** 15
Address: P. O. Box 404, Conshohocken, PA 19428
The Upper Merion & Plymouth is a switching road between Swedeland and Conshohocken, Pennsylvania.

Nos.	Qty.	Model	Builder	Date	Notes
9007	1	SW1	EMD	1941	Ex-19
9008, 9009	2	NW2	EMD	1948	Ex-1002, 1009
Total	3				

UTAH RAILWAY

Reporting marks: UTAH   **Miles:** 95
Address: 136 E. South Temple St., Salt Lake City, UT 84111
The Utah Railway operates between Provo and Hiawatha, Utah, partly by trackage rights on Denver & Rio Grande Western and partly by a paired-track arrangement with D&RGW; plus several branches in the coalfields at the south end of its line. The enginehouse is at Martin.

Nos.	Qty.	Model	Builder	Date	Notes
9001-9011	11	SD40u	EMD	1970	1
Total	11				

Notes:
1. Ex-CSX; rebuilt by MK to Dash 2 specifications

VALDOSTA SOUTHERN RAILROAD

Reporting marks: VSO   **Miles:** 10
Address: P. O. Box 1147, Valdosta, GA 31603-1147

The Valdosta Southern extends from Valdosta to Clyattville, Georgia. Engine facilities are at Clyattville.

Nos.	Qty.	Model	Builder	Date	Notes
184	1	GP7	EMD	1951	Ex-Western Ry of Alabama 709
473	1	SW1200	EMD	1954	Ex-Georgia Pacific Lumber
955	1	SW900	EMD	1958	
1284	1	GP10	ICG	1955	Ex-Illinois Central Gulf
Total	4				

VANDALIA RAILROAD

Reporting marks: VRRC   **Miles:** 3
Address: P. O. Box 190, Vandalia, IL 62471

The Vandalia Railroad is a switching line at Vandalia, Illinois.

Nos.	Qty.	Model	Builder	Date	Notes
2271	1	SW9	EMD	1953	Ex-Louisville & Nashville
Total	1				

VENTURA COUNTY RAILWAY

Reporting marks: VCY   **Miles:** 13
Address: P. O. Box 432, Oxnard, CA 93032

The Ventura County Railway connects Oxnard, California, with Port Hueneme. The enginehouse is at Oxnard.

Nos.	Qty.	Model	Builder	Date	Notes
7, 9, 11	3	S6	Alco	1955	1
Total	3				

Notes:
1. Ex-Southern Pacific 1251, 1226, 1277

VERMONT RAILWAY

Reporting marks: VTR   **Miles:** 129
Address: 1 Railway Lane, Burlington, VT 05401

The Vermont Railway operates from Burlington south through Rutland to Hoosick Junction, New York, on what used to be the main line of the Rutland Railway. It is affiliated with the Clarendon & Pittsford. Enginehouses are at Rutland and Burlington.

Nos.	Qty.	Model	Builder	Date	Notes
201, 202	2	GP38-2	EMD	1972, 1974	
501	1	SW1500	EMD	1966	
751	1	GP9	EMD	1954	1
801	1	GP18	EMD	1961	2
Total	5				

Notes:
1. Ex-Conrail, New York Central
2. Ex-Toledo, Peoria & Western 600

VIA RAIL CANADA

          Miles: 13,271
Address: 2 Place Ville-Marie, Montreal, PQ H3B 2G9, Canada

VIA Rail Canada operates intercity passenger service in Canada. The bulk of its trains operate in the Quebec CIty-Montreal-Toronto-Windsor corridor, but VIA reaches east to Halifax, west to Vancouver, and north to Hudson Bay at Churchill, Manitoba.

Nos.	Qty.	Model	Builder	Date	Notes
202-204	3	SW1000	EMD	1966-1967	1
6300-6314	15	FP9r	GMD	1954-1958	2
6400-6458	59	F40PH-2	GMD	1986-1989	
6506, 6514	2	FP9	GMD	1954, 1957	2
6902, 6903, 6905, 6907, 6912, 6916, 6917, 6919, 6920, 6921	10	LRC	BBD	1980-1983	
Total	89				

Larry Russell; collection of Louis A. Marre

The diversity that previously characterized VIA's locomotive roster — first-generation cab units from GMD and Montreal and lightweight LRCs from Bombardier — has been all but replaced by the standardized F40PH.

Notes:
1. Ex-Inland Steel 116-118
2. Ex-Canadian National

VIRGINIA RAILWAY EXPRESS
(Northern Virginia Transportation Commission)

Reporting marks: NVTC   **Miles:** 78
Address: 4350 N. Fairfax Drive, Suite 720, Arlington, VA 22203
At press time, Virginia Railway Express was scheduled to begin commuter service from Washington, D.C. to Manassas and Fredericksburg, Virginia.

Nos.	Qty.	Model	Builder	Date	Notes
V01-V10	10	RP39-2C	MK	1991	1
Total	10				

Note: 1. Rebuilt GP40s

WABASH & GRAND RIVER RAILWAY

Reporting marks: WGRY   **Miles:** 40
Address: 1 Elm St., Chillicothe, MO 64601
The Wabash & Grand River operates from Kelly to Chillicothe, Missouri, on a former Wabash route. The enginehouse is at Chillicothe.

Nos.	Qty.	Model	Builder	Date	Notes
2348	1	GP9	EMD	1952	1
7802	1	RS3md	Alco	1952	2
Total	2				

Notes:
1. Ex-CSX
2. Ex-Conrail, Central Railroad of New Jersey; rebuilt with 1200 h.p. EMD engine

WACCAMAW COASTLINE RAILROAD

Reporting marks: WCLR   **Miles:** 14
Address: P. O. Box 2022, Conway, SC 29526
The Waccamaw Coastline extends from Conway to Myrtle Beach, South Carolina. The enginehouse is at Red Hill, near Conway.

Nos.	Qty.	Model	Builder	Date	Notes
88C	1	F7A	EMD	1950	Ex-Milwaukee Road
502	1	E8A	EMD	1950	Ex-Chicago & North Western
943	1	GP18	EMD	1960	Ex-Norfolk & Western
1040	1	NW2	EMD	1947	Ex-Southern
2480	1	CF7	ATSF	1979	Ex-Santa Fe 2480
4257	1	E8A	EMD	1950	Ex-NJ Transit; Pennsylvania
Total	6				

WALKING HORSE & EASTERN RAILROAD

Reporting marks: WHOE   **Miles:** 8

Address: P. O. Box 1317, Shelbyville, TN 37160

The Walking Horse & Eastern extends from Shelbyville to Wartrace, Tennessee. The enginehouse is at Shelbyville.

Nos.	Qty.	Model	Builder	Date	Notes
1057, 1585	2	NW2	EMD	1948	Ex-Southern 2266, 2278
Total	2				

WARREN & SALINE RIVER RAILROAD

Reporting marks: WSR   **Miles:** 19

Address: P. O. Box 390, Warren, AR 71671

The Warren & Saline River extends from Warren to Hermitage, Arkansas. The enginehouse is at Warren.

Nos.	Qty.	Model	Builder	Date	Notes
538	1	SW7	EMD	1950	Ex-Illinois Central 420
539	1	SW1	EMD	1949	Ex-Rock Island 539
Total	2				

WASHINGTON CENTRAL RAILROAD

Reporting marks: WCRC   **Miles:** 400

Address: 6 East Arlington, Yakima, WA 98901

Washington Central operates from Pasco to Cle Elum, Washington, with branches to Naches, Moxee City, White Swan, and Granger; and from Connell to Airbase, with branches to Royal City and Schrag. The enginehouse is at Yakima.

Nos.	Qty.	Model	Builder	Date	Notes
201-203, 211, 212	5	SW1200	EMD	1965-1966	
301, 302	2	GP9	EMD	1957	
401, 402	2	CF7	ATSF	1973, 1971	
Total	9				

WASHINGTON COUNTY RAILROAD

Reporting marks: WACR   **Miles:** 14

Address: 84 South Main St., Barre, VT 05641

Washington County Railroad extends from Montpelier Junction, Vermont, to Barre, Websterville, and Graniteville. The enginehouse is at Montpelier.

Nos.	Qty.	Model	Builder	Date	Notes
25, 29	2	S1	Alco	1944, 1949	
Total	2				

WASHINGTON METROPOLITAN AREA TRANSIT AUTHORITY

Miles: 70

Address: 600 Fifth St. N. W., Washington, DC 20001

Washington Metropolitan Area Transit Authority operates the subway system in and around Washington, D.C.

Nos.	Qty.	Model	Builder	Date	Notes
814, 824	2	45-ton	GE	1944	Ex-U. S. Navy
846	1	44-ton	GE	1943	Ex-U. S. Army
854	1	44-ton	Davenport	1953	Ex-U. S. Army
865	1	44-ton	GE	1953	Ex-U. S. Army
Total	5				

WCTU RAILWAY

Reporting marks: WCTR   **Miles:** 14

Address: 7551 Crater Lake Highway, White City, OR 97503

WCTU is a switching line at White City, Oregon, north of Medford.

Nos.	Qty.	Model	Builder	Date	Notes
5117	1	70-ton	GE	1951	Ex-Southern Pacific 5117
5119	1	70-ton	GE	1955	Ex-Southern Pacific 5119
Total	2				

WESTERN RAIL ROAD

Reporting marks: WRRC   **Miles:** 2
Address: P. O. Box 311475, New Braunfels, TX 78131
Western Rail Road extends from New Braunfels to Dittlinger, Texas. Engine facilities are at Dittlinger.

Nos.	Qty.	Model	Builder	Date	Notes
534	1	SW1	EMD	1944	Ex-Rock Island
1007	1	SW1	EMD	1943	Ex-Southern Pacific
9617	1	SW1200	EMD	1957	
Total	3				

WEST JERSEY RAILROAD

Reporting marks: PWJ   **Miles:** 18
Address: 1 Elm St., Chillicothe, MO 64601-2729
The West Jersey Railroad extends from Swedesboro to Salem and Port Salem, New Jersey. The enginehouse is at Salem.

Nos.	Qty.	Model	Builder	Date	Notes
7803, 7804	2	RS3	Alco	1952	1
Total	2				

Notes:
1. Ex-Lamoille Valley, previously Delaware & Hudson

WEST TENNESSEE RAILROAD

Reporting marks: WTNN   **Miles:** 43
Address: 1 Depot St., Trenton, TN 38382
The West Tennessee operates from Lawrence to Kenton, Tennessee. The enginehouse is at Trenton.

Nos.	Qty.	Model	Builder	Date	Notes
9433	1	GP28	EMD	1964	Ex-Illinois Central Gulf
Total	1				

WHEELING & LAKE ERIE RAILWAY

Reporting marks: WE   **Miles:** 596
Address: 100 E. First St., Brewster, OH 44613
The Wheeling & Lake Erie extends from Connellsville, Pennsylvania, to Cleveland and Bellevue, Ohio. It has several branches and has trackage rights on CSX from Connellsville to Hagerstown, Maryland. The shops are at Brewster.

Nos.	Qty.	Model	Builder	Date	Notes
1765-1770, 1784, 1800	8	SD45	EMD	1970	1
2645, 2650-2657, 2660-2662, 2666, 2671-2673, 2675, 2676, 2679, 2680, 2682-2684, 2686, 2687, 2691, 2695, 2699, 2701-2703, 2705-2715	42	GP35	EMD	1965	2
9950, 9951	2	Slug	NW	1971, 1978	1
Total	52				

Notes:
1. Ex-Norfolk & Western, same numbers
2. Ex-Southern, same numbers

WHITE PASS & YUKON CORPORATION

Reporting marks: WPY   **Miles:** 110
Address: P. O. Box 4070, Whitehorse, YT Y1A 3T1, Canada
Service was suspended in 1983. Passenger service has since been restored on part of the line.

Nos.	Qty.	Model	Builder	Date	Notes
90, 91	2	84-ton	GE	1954	
92-94	3	86-ton	GE	1956	
95-97	3	84-ton	GE	1963	
98-100	3	85-ton	GE	1966	
101, 103, 104, 106, 107	5	DL535E	MLW	1969	
108-110	3	DL535E	MLW	1971	
Total	19				

WICHITA, TILLMAN & JACKSON RAILWAY

Reporting marks: WTJR   **Miles:** 78
Address: 4420 W. Vickery Blvd., Suite 110, Fort Worth, TX 76107

The Wichita, Tillman & Jackson operates from Wichita Falls, Texas, to Altus, Oklahoma, and from Waurika to Walters, Okla. The enginehouse is at Wichita Falls.

Nos.	Qty.	Model	Builder	Date	Notes
4364, 4367, 4370,					
4443, 4451, 4454	6	GP7	EMD	1950-1951	1
Total	6				

Notes:
1. Ex-Chicago & North Western, same numbers; originally Frisco 533, 514, 604, 590, 562, 517

WILLAMETTE VALLEY RAILROAD

Reporting marks: WVRD   **Miles:** 2
Address: 635 N. Walnut St., Independence, OR 97351

The Willamette Valley Railroad is a switching road at Independence, Oregon.

Nos.	Qty.	Model	Builder	Date	Notes
201	1	SW1200	EMD	1965	Ex-Southern Pacific 2273
2274	1	SW1200	EMD	1965	Ex-Southern Pacific 2274
2890	1	GP9	EMD	1959	Ex-Southern Pacific 2890
Total	3				

WILLAMINA & GRAND RONDE RAILWAY

Reporting marks: WGR   **Miles:** 5
Address: 635 N. Walnut St., Independence, OR 97351

The Willamina & Grand Ronde is a switching line between Willamina and Fort Hill, Oregon.

Nos.	Qty.	Model	Builder	Date	Notes
110	1	S2	Alco	1945	Ex-Longview, Portland & Northern
Total	1				

WILMINGTON & WESTERN RAILWAY

Reporting marks: WILM   **Miles:** 10
Address: P. O. Box 5575, Wilmington, DE 19808-5575

The Wilmington & Western Railway operates freight service between Landenberg Junction and Hockessin, Delaware. The enginehouse is at Greenbank. The route is used for excursion service by the Wilmington & Western Railroad.

Nos.	Qty.	Model	Builder	Date	Notes
3	1	S2	Alco	1949	1
114	1	SW1	EMD	1940	2
8408	1	SW1	EMD	1940	3
Total	3				

Notes:
1. Ex-Rohm & Haas 3, Chesapeake & Ohio 9115
2. Ex-Ocean City Western 114; Lehigh Valley 114
3. Ex-Baltimore & Ohio

WILMINGTON TERMINAL RAILROAD

Reporting marks: WTRY   **Miles:** 4
Address: 1717 Woodbine St., Wilmington, NC 28401

The Wilmington Terminal is a switching line at Wilmington, North Carolina.

Nos.	Qty.	Model	Builder	Date	Notes
1203-1205	3	SW1200	EMD	1966	1
Total	3				

Notes:
1. Ex-Missouri Pacific, same numbers; leased from North Carolina Ports Railway Commission

WINCHESTER & WESTERN RAILROAD

Reporting marks: WW   **Miles:** 102
Address: P. O. Box 264, Winchester, VA 22601

Winchester & Western's New Jersey Division operates from Millville to Dorchester, and from Vineland to Mauricetown and Seabrook, New

Jersey. The Virginia Division operates from Gore to Winchester, Virginia, then north to Williamsport, Maryland.

Nos.	Qty.	Model	Builder	Date	Notes
New Jersey Division					
459, 475, 498, 517, 520, 709, 732, 811	8	GP9	EMD	1955-1959	1
Virginia Division					
78, 80	2	S6	Alco	1955	2
351	1	RS11	Alco	1957	3
527	1	RS3	Alco	1950	
863	1	RS11	Alco	1959	4
2910	1	RS11	Alco	1959	5
3605, 3611	2	RS11	Alco	1956	6
Total	16				

Notes:
1. Ex-Norfolk & Western
2. Ex-Southern Pacific 1278, 1280
3. Ex-Norfolk & Western 351
4. Ex-Nickel Plate 863
5. Ex-Southern Pacific
6. Ex-Central Vermont 3605, 3611

WIREGRASS CENTRAL RAILROAD

Reporting marks: WGRC   **Miles:** 23
Address: 812 N. Main St., Enterprise, AL 36330

The Wiregrass Central extends from Waterford to Clintonville, Alabama. The enginehouse is at Enterprise.

Nos.	Qty.	Model	Builder	Date	Notes
1223	1	SW7	EMD	1952	Ex-Illinois Central Gulf 1223
2976	1	GP9	EMD	1959	Ex-Southern Pacific 2976
3832	1	GP9r	EMD	1959	Ex-Southern Pacific 3832
3872	1	GP9r	EMD	1958	Ex-Southern Pacific 3872
6084	1	GP9	EMD	1956	Ex-CSX; Baltimore & Ohio 6084
6226	1	GP9	EMD	1956	Ex-Chesapeake & Ohio 6226
Total	6				

WISCONSIN & CALUMET RAILROAD

Reporting marks: WICT   **Miles:** 143
Address: 203 S. Pearl St., Janesville, WI 53545-4521

Wisconsin & Calumet operates a network of lines in southern Wisconsin and nearby Illinois: Janesville to Waukesha, Janesville to Madison and Prairie du Chien, Janesville to Monroe, Janesville to Chicago (by trackage rights on Metra south of Fox Lake), and Madison to Freeport, Illinois. Engine facilities are at Janesville.

Nos.	Qty.	Model	Builder	Date	Notes
613-617	5	GP7	EMD	1950-1955	Ex-New York Central
618	1	GP9	EMD	1958	Ex-Baltimore & Ohio
Total	6				

WISCONSIN CENTRAL LTD.

Reporting marks: WC   **Miles:** 2,049
Address: P. O. Box 5062, Rosemont, IL 60017-5062

Wisconsin Central operates a network of former Soo Line and Milwaukee Road routes in Illinois, Wisconsin, and Michigan. It extends from Chicago to the Twin Cities, Superior, Wisconsin, and Sault Ste. Marie, Ontario. It is negotiating to purchase the Green Bay & Western and the Fox River Valley railroads. WC's principal shops are at North Fond du Lac and Stevens Point, Wis.

Nos.	Qty.	Model	Builder	Date	Notes
1	1	SW1	EMD	1942	1
582-590	9	SDL39	EMD	1969, 1972	2
700, 703, 704, 706-713, 715-719, 721 723, 724, 726,	17	GP30	EMD	1964	3
728, 731	5	GP35	EMD	1965	3
1230	1	SW1200	EMD	1963	4
1231	1	SW9	EMD	1951	5
1232	1	SW1200	EMD	1963	6
1233, 1234	2	SW1200	EMD	1966	7
1235-1237	3	SW1200	EMD	1965	8

172

Nos.	Qty.	Model	Builder	Date	Notes
1550, 1551, 1553-1557	7	SW1500	EMD	1968-1972	9
3000, 3002-3007, 3009, 3011, 3012, 3014, 3015, 3017, 3022-3025					
	17	GP40	EMD	1970-1971	10
4002, 4004-4013	11	GP35d	EMD	1965	11
6417, 6494, 6498, 6499, 6501, 6502, 6504-6508, 6510, 6511, 6517, 6522-6524, 6526, 6527, 6530-6535, 6537-6539, 6541, 6543, 6548, 6553, 6554, 6559, 6560, 6572, 6655, 6660, 6677, 6690					
	40	SD45	EMD	1966-1971	12
Total		114			

Notes:
1. Ex-Penn Central 8440, New York Central; named Francis J. Wiener
2. Ex-Milwaukee Road, same numbers
3. Ex-Soo Line, same numbers
4. Ex-Missouri Pacific 1107
5. Ex-Houston Belt & Terminal 31
6. Ex-Missouri Pacific 1278
7. Ex-Houston Belt & Terminal 33, 34
8. Ex-Southern Pacific 2260, 2287, 2288
9. Ex-Southern Pacific 2487, 2505, 2578, 2594, 2640, 2652, 2665
10. Ex-Union Pacific, Western Pacific
11. Ex-Union Pacific, Missouri Pacific 2603, 2608-14, 2616, 2602, 2605
12. Ex-Burlington Northern, same numbers

WISCONSIN & SOUTHERN RAILROAD

Reporting marks: WSOR   **Miles:** 147
Address: 511 Barstow St., Horicon, WI 53032

Wisconsin & Southern extends from North Milwaukee to Cambria, Markesan, Fox Lake, and Oshkosh, Wisconsin. The enginehouse is at Horicon.

Nos.	Qty.	Model	Builder	Date	Notes
1001	1	NW2	EMD	1948	1
1201, 1202	2	SW7	EMD	1951	2

Nos.	Qty.	Model	Builder	Date	Notes
2001, 2002	2	GP7r	EMD	1952	3
4490-4492	3	GP9	EMD	1957	4
4493	1	GP18	EMD	1960	4
Total	9				

Notes:
1. Ex-Upper Merion & Plymouth 1009
2. Ex-Milwaukee Road
3. Ex-Santa Fe 2001, 2002
4. Ex-Rock Island

WYOMING/COLORADO RAILROAD

Reporting marks: WTCO   **Miles:** 273
Address: 103 University, Laramie, WY 82070

The Wyoming/Colorado operates from Laramie, Wyoming, to Hebron, Colorado, and from Wolcott Junction to Saratoga, Wyo.

Nos.	Qty.	Model	Builder	Date	Notes
1510	1	FP7	EMD	1953	Ex-Alaska Railroad
1511	1	F7B	EMD	1952	Ex-Alaska Railroad
1512	1	FP7	EMD	1953	Ex-Alaska Railroad
Total	3				

YADKIN VALLEY RAILROAD

Reporting marks: YVRR   **Miles:** 100
Address: P. O. Box 1929, Laurinburg, NC 28352

The Yadkin Valley extends from Rural Hall to North Wilkesboro, North Carolina, and from Mount Airy to Brook Cove. The enginehouse is at Rural Hall.

Nos.	Qty.	Model	Builder	Date	Notes
201-203	3	GP7	EMD	1953-1953	
204-206	3	GP9	EMD	1957, 1955	
Total	6				

YOLO SHORTLINE RAILROAD

Reporting marks: YSLR   **Miles:** 11
Address: 3344 Braeburn St., Sacramento, CA 95821

The Yolo Shortline operates from West Sacramento to Clarksburg, California. The enginehouse is at Clarksburg.

Nos.	Qty.	Model	Builder	Date	Notes
50	1	50-ton	GE	1939	1
101	1	S1	Alco	1942	2
Total	**2**				

Notes:
1. Ex-Spreckels Sugar 1, Weirton Steel 69
2. Ex-Corn Products 511, Pacific States Steel 511, Western Pacific 511

YOUNGSTOWN & AUSTINTOWN RAILROAD

Reporting marks: YARR   **Miles:** 4
Address: P. O. Box 564, Sugar Creek, OH 44681

The Youngstown & Austintown operates freight service at Austintown, Ohio.

Nos.	Qty.	Model	Builder	Date	Notes
70	1	GP7	EMD	1953	Ex-Pittsburgh & Lake Erie 1500
Total	**1**				

YOUNGSTOWN & SOUTHERN RAILWAY

Reporting marks: YS   **Miles:** 50
Address: 7891 Southern Blvd., Youngstown, OH 44512

The Youngstown & Southern extends from Youngstown, Ohio, to Darlington, Pennsylvania, and from Negley, Ohio, to Smith Ferry, Pa. It is operated by the Pittsburgh & Lake Erie.

Nos.	Qty.	Model	Builder	Date	Notes
70, 71	2	SW7	EMD	1950	
Total	**2**				

YREKA WESTERN RAILROAD

Reporting marks: YW   **Miles:** 8
Address: P. O. Box 660, Yreka, CA 96097

Yreka Western operates from Yreka to Montague, California. The enginehouse is at Yreka.

Nos.	Qty.	Model	Builder	Date	Notes
20, 21	2	SW8	EMD	1953	
Total	**2**				

INDUSTRIAL LOCOMOTIVES

Industrial locomotives are listed alphabetically by state (Canadian provinces follow the U. S. listing) city, and company, then number. The data elements for the locomotives are number, model, builder, previous ownership, date built, and notes — either to the right or on the next line. Depending on the information available, some data elements may be absent.

ALABAMA

Albertville: Steel Processing Services

114	SW1200	EMD	N&PB 114	1956

Anniston: Monsanto Chemical

	25-ton	Whit	U. S. Air Force 1140	1952

Birmingham: American Cast Iron Pipe

102	NW2	EMD	BS 21, Lake Terminal	1948
103	HH600	Alco	BS 83	1937
104	NW2	EMD	BS 24, Lake Terminal	1948
201	SW1200	EMD	BS 201	1957

Birmingham: Birmingham Rail & Locomotive Co.

5252	SW9	EMD	1951

Birmingham: Empire Coke

206	SW	EMD	1937

Birmingham: Jim Walter Resources

51-54	SW1500	EMD		1972
55	NW2u	EMD	UP	1941	1200 h.p.
56	NW2u	EMD	UP	1946	1200 h.p.

Birmingham: Owens Illinois

8156	80-ton	GE

Birmingham: Stockham Valves & Fitting Co.

88	SW1	EMD	Georgia Marble, USP&F 46	1940

Brewton: Frit Car Service

2	44-ton	Dav	SLSF 1	1942
729	S2	Alco	South Carolina Ports Authority	1942

Chetopa: Alabama Byproducts

600	SW1	EMD	1953
900	SW900	EMD	1956
1000	SW1001	EMD	1971
	44-ton	GE	B&W	1948

Columbia: Alabama Power

31	65-ton	GE	1950

Ease Coulee: Stay Sales

1, 2	TWDT2	Plym	Century Coal	1956

Fountain: Alabama River Pulp

2	S3	Alco	Davenport, Rock Island & Northwestern 2	1952

Gadsden: Republic Steel

1074	NW2	EMD	Southern 1058	1948

Gantts Quarry: Alabama Marble

1	44-ton	GE	Georgia Marble	1950
2	35-ton	GE	Georgia Marble	1952
7	65-ton	GE	Georgia Marble	1946

Gantts Quarry: Georgia Marble

1	80-ton	Dav	Budd 3	1952

Greencastle: France Stone Co.

100	40-ton	Plym	Tuskeegee 100	1953

Huntsville: International Intermodal Center of Alabama

	SW8	EMD

Montgomery: Aesco Steel

2	45-ton	Vulcan	Vulcan Materials, Birmingham Slag	1947

Mount Vernon: Scott Paper Co.

3	44-ton	Whit	SLSF 3	1943

Muscle Shoals: Tennessee Valley Authority

100	DS44-660	BLW	1949
	Stored unserviceable		
200	S12	BLW	1952
936	65-ton	GE	

Muscle Shoals: TVA-Fertilizer Development Center

300	VO1000	BLW	U. S. Navy	1944

Stevenson: Meade Containerboard Co.

226	65-ton	GE	BRL 226, Lorton & Occoquan, U. S. Navy 65-00240	1943

Stevenson: Tennessee Valley Authority

315912	MRS1	Alco	U. S. Army	1951

Tarrant City: LPC Inc.

5	SW1	EMD	Lehigh Cement	1953

Tarrant: Poinsett Lumber & Manufacturing Co.

	25-ton	GE	1955

ALASKA

Sitka: Alaska Lumber & Pulp

1837	SW9	EMD	UP 1837	1953
	65-ton	GE	U. S. Navy 65000056	1943

ARIZONA

Sahuarita: Duval Sulphur

1208	S6	Alco	SP 1208	1955

ARKANSAS

Ashdown: Nekoosa Paper

2	S1	Alco

Ashdown: Georgia Pacific

259	RS1	Alco

Batesville: Arkansas Lime Co.

	50-ton	GE

Bearden: Glfford Hill Inc.

2	SW1	EMD	1942

Batesville: Arkansas Eastman

1511	SW1200	EMD	CV 1511	1960

East Camden: North American Car Co.

1	50-ton	GE	Goodpasture Grain,	
			U. S. Navy 65-00394	1942
2	45-ton	GE	Coffield Warehouse,	
			U. S. Army 7045	1942

East Camden: Arkansas Railway Supply

	80-ton	GE	Highland Corporation
	45-ton	GE	Foreman

East Camden: Arkansas Cement

514	NW2	EMD	BS 22, LT 1010	1948

Fort Smith: Planters Peanut Co.

1	25-ton	GE

Helena: Allied Chemical Co.

22	S12	BLW	SCL 220	1952
921	S8	BLW	U. S. Pipe & Foundry	1954

Helena: Riceland Foods

2	35-ton	Whit
3	25-ton	GE

Helena: Quincy Soybean Co.

1	25-ton	GE
2	25-ton	Plym

Little Rock: Big Rock Stone & Material

400	NW2	EMD	MP 1028	
1023	SW	EMD	MP 1023	1937
1272	SW1200	EMD	MP 1272	1965

Little Rock: Granite Mountain Quarries

4742	SW1	EMD	MP 6000, C&EI 98	1942

Little Rock: Little Rock Port Railway

1017	S2	Alco	Relco 1017

McGehee: Potlatch Forest Products

49	S2	Alco

North Little Rock: Koppers Co.

1	45-ton	GE	U. S. Navy 65-00081	1941

Newport: Razorback Steel

5970	45-ton	GE	American Bridge	1947

Process City: Weyerhaeuser Co.

	45-ton	GE	Omaha Power	1949

Stuttgart: Riceland Foods

1	25-ton	GE

Wilton: Braswell Sand & Gravel

10	00	Alco	CD&W 10	1952

Waldron: Tyson Feed Mill

1	45-ton	GE	North American Car,	
			Shamoon Industries	1956

CALIFORNIA

Anita: Jones Brothers Inc.

1251	45-ton	GE	Pacific States Steel,	
			U. S. Navy 65-00015	1942

Antioch: E. I. Du Pont

1	25-ton	GE	Hingham Shipyard	1943
2	44-ton	GE	ATSF 462	1943
2	S4	Alco	ATSF 1518	1951
3	S1	Alco	ATSF 2304	1943
4	S4	Alco	ATSF 1518	1943

Antioch: Fiberboard Products

	25-ton	Plym	1966

CALIFORNIA (continued)
Argus: Kerr-McGee

2962	RSD12	Alco	SP 2962	1961	

Aromas: Granite Rock Co.

801	65-ton	GE	Ex-100	1951	

Avon: Phillips Petroleum

1004	23-ton	GE	Tidewater Oil	1941	
1005	8-ton	Plym	Tidewater Oil	1944	
1006	25-ton	GE	Tidewater Oil	1947	

Benicia: Solano Rail Car Co.

8	VO1000	BLW	U. S. Army 7453	1942	

Castroville: Lone Star Industries

5110	15-ton	Plym			
5118	30-ton	Plym	U. S. Army 7574	1942	

Chico: Diamond International

870	25-ton	GE		1948	

Chowchilla: Certainteed Corp.

	25-ton	Plym			

Colton: General American Transportation

1013	S2	Alco	Relco 1013	1949	
6590	30-ton	Whit		1955	

Corona: Citrus Industry Supply

	S2	Alco	U. S. Navy 65-00603	1943	

Eloit: Pacific Coast Cement

5113	44-ton	BLW	U. S. Army 7489	1943	

Emeryville: Judson Steel

	35-ton	GE	Guy F. Atkinson 3	1953	

Fresno: E.B. Willis Co.

7	44-ton		Asarco 7, U. S. Navy 65-00442	1945	

Hamilton City: Holly Sugar Co.

2	25-ton	GE		1944	

Hatch: Guy F. Atkinson

3	25-ton	GE	Stone & Webster	1953	
4	25-ton	GE	McNary Dam Const.	1951	

Hayward: Ameron Pipe

661	20-ton	Whit			

Hayward: Oliver Bros. Salt

1			Home-built	1930	24" gauge
2	8-ton	Plym	Garfield Co.	1925	24" gauge
3	4-ton	Plym	California Clay Co.	1923	24" gauge

Irwindale: Consolidated Rock Products

1833	45-ton	GE	U. S. Army 7429	1941	

Keyes: A. L. Gilbert Co.

1204	S6	Alco	SP 1204	1956	

Livermore: Lone Star Industries

5104	45-ton	GE	Pacific Coast Cement	1943	

Livingston: Foster Farms

1210	S6	Alco	Bethlehem Steel, SP 1229	1956	
	S4	Alco	CCT 50, WP 563	1951	

Long Beach: Agrex Inc.

76	S12	BLW		1946	

Long Beach: Armco Steel Co.

1	25-ton	Porter		1948	
2	25-ton	GE		1945	

Long Beach: Koppel Bulk Terminal

1026	DRS66-15	BLW	Peabody Coal, Kaiser Steel 1026	1949	

Long Beach: Metropolitan Stevedore

2	S12	BLW		1945	
1002	RSD12	Alco	SP	1961	
2954	RSD12	Alco	SP 2954	1961	

Los Angeles: Martin-Marietta

4	45-ton	GE			
912	SW900	EMD	CRI&P 912	1954	
7589	RS11	Alco	CR 7589, NH 1402	1956	
7601	RS11	Alco	CR 7601, NYC 8001	1957	
7613	RS11	Alco	CR 7613, NH 1400	1956	

Manteca: Amstar-Spreckels Sugar Division

1	65-ton	GE		1951	

Martinez: Stauffer Chemical

10	25-ton	GE	Richmond Shipbuilders	1944	

Mendota: Amstar-Spreckels Sugar Division

5920	65-ton	GE		1962	
9	44-ton	GE	Pacific Terminal		

Mojave: Pabco Gypsum Inc.

2117	GP35	EMD	SP 6527	1964	

Newark: Leslie Salt Co.

1-18	4-ton	Vulcan		1936-1948	
19-22	5-ton	GE		1951	

CALIFORNIA (continued)

Newhall: Newhall Land & Farming Co.

22	SW1	EMD	Bethelem Steel	1940
3100	RS32	Alco	SP 4005	1962

Oakland: Port of Oakland

	45-ton	Whit	Howard Terminal, U. S. Army 7504	1943	36" gauge

Pine Creek: Union Carbide

4	8-ton	Plym	Cate Equipment	1957	36" gauge

Pittsburg: USX

12	80-ton	GE		1950
1214	S8	1	BLW	1952
112	DL535E	MLW	WP&Y 112	1982
113	DL535E	MLW	WP&Y 113	1982
1303	50-ton	GE		1956
1403	50-ton	GE		1956

Red Bluff: Diamond International

796	25-ton	GE	U. S. Navy 65-00320	1944

Richmond: J.L. Immel Co.

1	45-ton	GE	U. S. Maritime Commission	1942

Richmond: M. Lummis Co.

7	25-ton	Plym	Rockfield

Richmond: Lone Star Industries

5117	25-ton	Plym	Al. Drydock Co.	1942
4	8-ton	Plym	Grant Rock, Nevada County Narrow Gauge	1924

Richmond: Levin Richmond Terminal Co.

1	25-ton	Plym		
1195	SW900	EMD	SP 1195	1954
1402	NW2	EMD	SP 1402	1948
2285	SW1200	EMD	SP 2285	1965

Sacramento: Fruit Growers Express

6	50-ton	GE	

San Andreas: Calveras Cement

	45-ton	GE	Kaiser 3705	1952

San Francisco: Metal & Thermite Co.

1	25-ton	GE		1945

San Jose: Stauffer Chemical

2	44-ton	GE	PFE 1, Pine Flat Dam Co.	1950

San Pablo: Bay Area Rapid Transit

5003	7.5-ton	Plymouth		1969

San Pedro: American Bulk Loading Exports

500	S6	Alco	SP 1255	1956
501	VO1000	BLW		1944

San Pedro: Port of Los Angeles

1	VO1000	BLW	U. S. Navy 65-00249	1944

Santa Rita: Alameda County

	20-ton	Whit	U. S. Navy 65-00382	1926

Scotia: The Pacific Lumber Co.

101	80-ton	GE		1956
102	80-ton	GE		1956
103	80-ton	GE		1957

Stockton: Action Tank Co.

	10-ton	Plym	Dravo Corp.	1960

Stockton: Continental Grain Co.

102	20-ton	Plym	Corn Products Co.	
401	SW1	EMD	CC&P	
2598	CF7	ATSF	ATSF 2598	1974

Stockton: Delta Bulk Terminals

1001	C636	Alco	CR 6792	1968

Stockton: McCormick & Baxter Creosoting Co.

2	6-ton	GE	

Sunnyvale: Westinghouse Electric

1	25-ton	Plym	

Terminal Island: National Metals

505	65-ton	GE		
513	S12	BLW	SP 2131, SP 1497	1952
518	S6	Alco	SP	1955

Torrance: McDonnell-Douglas

	40	EMD		1943

Torrance: Pacific Railroad Contractors

	35-ton	GE	Armco Steel

Tracy: Holly Sugar Co.

1	SW1	EMD	SP 1000, EMC demonstrator 804	1939

Trona: Kerr-McGee

100	S12	BLW	SP 2150, SP 1543	1953
101	S12	BLW	SP 2157, SP 1550	1953
5321	S12	BLW	SP 2124, SP 1524	1952

178

CALIFORNIA (continued)
Trona: Kerr-McGee Soda Products

4419	SD9	EMD	SP 4419	1955

Turlock: Foster Farms

1213	S6	Alco	SP	1955
1218	S6	Alco	SP	1955

Tulare: J. D. Heiskell Co.

1886	S6	Alco	SP 1221	1956

Union City: Pacific States Steel

4	50-ton	GE	U. S. Navy 65-00017	1941
509	S1	Alco	UP 509	1942
511	S1	Alco	UP 511	1942

Vernon: Bethlehem Steel

18	SW1	EMD	SP 1016	1941

Victorville: Southwestern Portland Cement

412	GP40	EMD	PC 3084	1966

West End: Kerr-McGee

2284	SW1200	EMD	SP 2284	1965

Woodland: Amstar-Spreckels Sugar Division

1	50-ton	GE	Staley 97, Weirton Steel 97	1953

Wasco: Savage Coal Terminal

2156	GP7r	EMD	ATSF 2156	1952

COLORADO

Allen Mine: CF&I Steel

1	25-ton	Dav		1952	48" gauge
2	25-ton	Dav		1954	48" gauge

Boettcher: Ideal Cement

1845	S4	Alco	GN&A 1845, SP 1845	1955

Boulder: Public Service Company of Colorado

231	SW600	EMD		1962

Canon City: Colorado State Penitentiary

	13-ton	Whit		

Carbondale: Snowmass Coal Co.

997	SW8	EMD	Coors 997	1952

Denver: Cargill Inc.

	44-ton	GE	FJ&G 30, W&OD 47	1941

Denver: General Iron Works

142	25-ton	Whit	Hamm Brewing 1	1947

Denver: Koppers Co.

4	25-ton	GE		1948	36" gauge
5	25-ton	Porter		1948	

Denver: Public Service Company of Colorado

151	SW1001	EMD		1973
	12-ton	Bkvl		
	80-ton	GE		1965

Denver: U. S. General Services Administration

	80-ton	GE	U. S. Navy 65-00244	1941

Englewood: Railroad Builders Inc.

880	35-ton	Plym		
882	30-ton	Whit	UP, Boeing	1944
	15-ton	Bkvl		

Golden: Coors Brewing Co.

987	SW900m	EMD	CRI&P 553	1957
988	SW900m	EMD	CRI&P 550	1957
989	SW8	EMD	CRI&P 832	1952
990	SW8	EMD	CRI&P 831	1952
991	SW8	EMD	CRI&P 839	1953
998	SW1001	EMD		1980

Grand Junction: Public Service Company of Colorado

3	45-ton	GE		1949

Johnstown: Great Western Sugar Co.

40	44-ton	GE	D&RGW 40	1942

La Junta: City of La Junta

409	45-ton	GE	U. S. Navy 65-00409	1943

Longmont: Great Western Sugar Co.

23	35-ton	Dav	U. S. Army 7555	1942

Loveland: Great Western Sugar Co.

101	45-ton	GE	U. S. Gypsum	1942

Lyons: Martin-Marietta

1	20-ton	Plym	Dragon Cement	1955
3	45-ton	GE	Rocky Mountain Cement, Dewey Cement 102	1948

Portland: Ideal Cement

	NW2	EMD	SP 1911, SP1317	1941

Pueblo: CF&I Steel

1	44-ton	GE		1940
2	44-ton	GE		1947
3	50-ton	Whit		1946

179

COLORADO (continued)\
Pueblo: CF&I Steel (continued)]

4-11	25-ton	GE		1947
21, 22	44-ton	GE		1957
31	44-ton	GE	PC&F 3, NP 9	1946

Pueblo:

140	SW7	EMD	Great Western 140, BN 140, Great Northern	1950

Pueblo: U. S. Department of Transportation Test Center

1	U30C	GE		1971
3	GP40-2	EMD		1978
4, 5, 6	GP9	EMD	UP 205, 147, 162	1954
11	RSD1	Alco	U. S. Army 8004, ARR 1041, CRI&P 747	1941
12	RSD1	Alco	U. S. Army 8009, TC&I 601	1941
13	RSD1	Alco	ARR 1034, U. S. Army 8011, A&StAB 902	1941
14, 15	RSD1	Alco	U. S. Army 8016, 8018	1942
17, 18	RSD1	Alco	U. S. Army 8027, 8031	1942
8003	RSD1	Alco	U. S. Army 8003	1941
8017	RSD1	Alco	U. S. Army 8017	1942

Rangeley: Deseret-Western Railway

WFU1	E60C	GE		1983
WFU2	E60C	GE		1983
	SL125	GE		1982

Rocky Ford: American Crystal Sugar

64	F7A	EMD	D&RGW 5644	1946

Rocky Ford: Intermountain Transportation Service

171	GP9	EMD	BN 1803	1954
172	GP9	EMD	WM 5955	1954

CONNECTICUT

Groton: General Dynamics

3999	44-ton	GE	UP 1370	1947

Groton: Pfizer Chemical Co.

	SW8	EMD		1955
	80-ton	GE		1940

Hartford: Hartford Electric Light Co.

1	25-ton	GE		1949

Middletown: Hartford Electric Light Co.

	25-ton	GE		1952

New Britain: Stanley Inc.

873	45-ton	GE	Rohm & Haas RH-1	1942

New Haven: Connecticut Light & Power

2	25-ton	GE		1958

New Haven: New Haven Terminal

	25-ton	GE		1945

Waterbury: Connecticut Light & Power

1, 2	25-ton	GE		

Yantic: K & L Feed

8081	S4	Alco	Central Vermont 8081	1955

DELAWARE

Claymont: Phoenix Steel Corp.

989	NW5m	EMD	
DE08	45-ton	GE	
DE09	65-ton	GE	

Delmarva: Delmarva Power & Light

3, 4	SL100	GE		1977

Wilmington: Mechtron Corp.

7179	65-ton	GE	U. S. Army 7179	1943

DISTRICT OF COLUMBIA

Washington: Blue Plains Waste Water Plant

	23-ton	Vulcan	1953

Washington: District of Columbia

	45-ton	GE	1950

FLORIDA

Agricola: FD

1976	MP15	EMD		1976

Agricola: Mobil Chemical

8	SW8	EMD		1953
223	SW1	EMD		
226	SW1000	EMD		1966
227	SW900	EMD		1965
660	S4	Alco	Swift & Co.	
8	SCr	EMD	Dallas Union Terminal. 8, ATSF 651	1937

FLORIDA (continued)

Baldwin: David J. Joseph

	45-ton	GE	Southern Wood-Piedmont		
3	25-ton	Whit	SAL 1006	1950	36" gauge
4	40-ton	Plym	Eppinger & Russell	1952	

Bartow: Grace Agricultural Chemical

101, 102	SW1500	EMD	1969, 1968
121, 122	MP15	EMD	1975
131, 132	MP15DC	EMD	1980

Bartow: International Minerals & Chemical

103	65-ton			
200-202	S2	Alco		
205	RS1	Alco		
206	S2	Alco	B&O 9060	1949
207	RS1	Alco	Big Bend	

Bartow: Agrico

3	65-ton	GE		1947
4	70-ton	GE		1948

Bradley Junction: International Minerals & Chemical

102	65-ton	Whit		1947

Brewster: Amax Chemical Corp.

451	35-ton	Plym
452	35-ton	Plym

Brewster: American Cyanamid

13	DS44-660	BLW	1946
14	DS44-750	BLW	1950
16, 17	T6	Alco	1966
18	MP15DC	EMD	1976

Brewster: Borden Chemical

451, 452	TMDT	Plym	1972

Brewster: Brewster Phosphate Co.

19	MP15DC	EMD	1980

Brookville: Florida Portland Cement

2	S1	Alco	Long Island 405	1946

Bryant: U. S. Sugar

154	SW900	EMD		1955
156	SW900	EMD		1956
929	GP7	EMD	SCL 929	1950
968	GP7	EMD	DT&I 968	1953
1201	SW1200	EMD	Georgia Pacific 1201	1954

Canal Point: U. S. Sugar

100	Slug	BLW	SCL 54,SAL 1441	1950
968	GP7	EMD	SCL	1953

Century: International Paper

1	25-ton	GE	Alger-Sullivan	1953

Clear Springs: International Minerals & Chemical

52	35-ton	Whit		1947

Clewiston: U. S. Sugar

155	SW900	EMD		1955
158	SW1	EMD	URR 460	1949
159	SW900	EMD	Pickering Lumber	1959
756	GP7	EMD	SCL 756	1951
888	GP9r	EMD	SP 3727	1959
901	GP7	EMD	SCL 901	1950

Coronet: Smith-Douglas Phosphates

1, 2	35-ton	Plym		1959

Crystal River: Electric Fuels Corp.

9423	SW900	EMD	B & O 9423	1955
	25-ton	GE	Frit Car 2, Goodpasture Grain 94	1947

East Tampa: Gardiner Chemical

157	SW1	EMD	U. S. Sugar 157, WPBT 239	1941

Elfers: Stauffer Chemical

1	45-ton	GE	1946	
2	5-ton	BLW	1917	24" gauge
3	45-ton	GE	1951	

Foley: Buckeye Cellulose

230	35-ton	Porter	U. S. Navy	1942
235	DS44-1000	BLW	SCL 235	1949
1333	45-ton	GE	U. S. Navy	1942

Fort Green Springs: CF Mining Co.

	45-ton	Plym	1978

Gainesville: Koppers Co.

11	12-ton	Whit	Quaker Oats	1939

Jacksonville: C. G. Willis Co.

1450	25-ton	Plym

Jacksonville: Southeastern Specialties

9	35-ton	Plym		
	44-ton	GE	U. S. Navy 65-00138	1943

FLORIDA (continued)
Lakeland: General Power Systems

| 122 | S2r | Alco | SCL 122 | 1944 |

Manatee: Florida Power & Light

| 8505 | RSC2 | Alco | SCL 1105 | 1947 |

Medley: Lehigh Portland Cement

1	45-ton	GE	Ideal Cement, U. S. Army 7420	1942
3	80-ton	GE	Lone Star Industries	
501	SW8	EMD		
991	44-ton	GE	Maule Industries 991, U. S. Navy 65-00044	1944
1004	S2	Alco	Maule Industries 1004	1942
9186	RS1	Alco	Maule Industries, B&O 9186	
	45-ton	GE	MIddle Fork 2, W&OD 49	1942

Medley: Vulcan Materials

| 2121 | 45-ton | GE | Dardanelle & Russellville 12 | |

Medley: Maule Industries

| | S1 | Alco | B&O 9185 | 1945 |

Mulberry: Agrico

181	NW2r	EMD		
182	NW2r	EMD		
1581	GP10	ICG		

Mulberry: Estech

8	SW900m	EMD		1957
223	SW1	EMD		1940
660	S3m	Alco		1949
1976	MP15DC	EMD		1976

Mulberry: Grace Agricultural Chemical

| 2000 | VO1000 | BLW | SCL 89 | 1943 |

Mulberry: International Minerals & Chemical

| 501 | 25-ton | GE | Jackson County Grain | 1963 |

North Miami: General Portland Cement

| 6 | S3 | Alco | DRI&NW 6 | 1953 |

Naranja: Naranja Rock Co.

| | 8-ton | Plym | USA | 1940 |

Nichols: Mobil Chemical

22	65-ton	Whit	U. S. Army 7969	1943	Stored
23	65-ton	Whit	U. S. Army 8416	1944	Stored
220	50-ton	Porter	U. S. Army 7060	1965	

222	65-ton	Whit		
227	SW900	EMD		1965
228	SW1000	EMD		1966
994	80-ton	GE		1971
1	45-ton	Porter		1942
2	25-ton	Whit	Lake Ontario Steel	1944

Noralyn: International Minerals & Chemical

| 101 | 65-ton | Whit | | 1947 |
| 104 | 65-ton | Whit | | 1948 |

Orlando: Florida Light & Power

| | 25-ton | Plym | | |

Orlando: Orlando Utilities

| | 44-ton | GE | U. S. Air Force 1242 | 1953 |

Pace: Air Products & Chemical

| 3102 | 44-ton | GE | L&N 3102 | 1950 |
| | 44-ton | GE | L&N | 1949 |

Pace: Air Products Co.

| 1803 | SW7 | EMD | Chrome Crankshaft, UP 1803 | 1950 |

Palataka: Georgia-Pacific Corp.

| 3 | 45-ton | GE | Cherokee Brick 4 | 1951 |

Panama City: Port of Panama City

| | 65 ton | GE | International Paper, U. S. Army 7307 | 1940 |

Panama City: Cove Contractors

| | 45-ton | GE | U. S. Navy 65-00013 | 1943 |

Pensacola: Air Products Co.

| 1010 | S1 | Alco | L&N 31 | 1941 |

Pensacola: Lone Star Cement Co.

| | 25-ton | Plym | Lone Star Cement, Houston, Texas | |

Perry: Turner Lumber Co.

| 3 | 25-ton | GE | | 1955 |

Pierce: Agrico

6	70-ton	GE	Humble Oil 998	1954
8	70-ton	GE	AD&N 170	1948
10	S2	Alco	TP-MP Terminal 10	1948
11	S2	Alco	TP-MP Terminal 11	1948
1515	GP7	EMD	Precision National 1500, BN 1529	1951
1525	GP7	EMD	Precision National 1501, SLSF 527	1950

182

FLORIDA (continued)

Pierce: Agrico (continued)

1535	GP7	EMD	Precision National 970, DT&I 970	1953	
1545	GP7	EMD	Precision National 971, DT&I 971	1953	
1555	GP7	EMD	Precision National 969, DT&I 969	1953	
1586	GP10	ICG			
2010	GP20	EMD	UP 499	1960	

Piney Point: Borden Chemical

1	40-ton	Plym		1966
2	40-ton	Plym		1972

Plant City: Central Farmers Co-op

622	VO1000	BLW	Auto Train 622	1944

Port Mayaca: Florida Power & Light

8723	RSC02	Alco	SCL 1102	1947

Port St. Joe: Basic Magnesia

115	23-ton	GE	Duke Power 115	1941

Port Sutton: Tampa Electric Co.

38	45-ton	Plym		1963

Port Tampa: Hardaway Construction Co.

	30-ton	Dav	CRI&P 347	1941

River Junction: Gulf Power Co.

	80-ton	GE	1952

Suwanee: Occidental Chemical Co.

6660	GP9	EMD	B&O 6660	

Tampa: Cargill Inc.

881	SW1	EMD	MILW 881	1940
	44-ton	GE	MILW 991, 1700	1941

Tampa: Florida Portland Cement

1	S3	Alco		1951

Tampa: Gardiner Chemical

1004	NW2	EMD	International Minerals & Chemical	
1-3	100-ton	GE		

Tampa: Royster Inc.

	45-ton	GE	Iowa Terminal	1944

Tampa: Tampa Electric Co.

1206	NW2	EMD	Relco 1206	

Tampa: Tampa Port Authority

300	65-ton	Porter	Swift 300, ARR 1102, U. S. Army 7033	1942
S8	45-ton	GE	U. S. Navy 65-00485	1943

Tarpon Springs: Stauffer Chemical

1	45-ton	GE	

Telogia: Reichhold Chemicals

2	10-ton	Plym		
	20-ton	GE	Cherry River Lumber	1947

West Miami: Lehigh Portland Cement

4	45-ton	GE		1970 Remote control

West Palm Beach: Port of Palm Beach

238	SW1	EMD	EJ&E 238	1941

White Springs: Occidental Petroleum Co.

1030	NW2	EMD	Southern 1030	1949
1091	NW2	EMD	Southern 1091	1950

GEORGIA

Athens: Seaboard Grains

9416	SW900	EMD	B&O 9416	1955

Atlanta: Atlantic Steel

16	SW8	EMD	SCL 16, ACL 56	1952
17	SW8	EMD	SCL 17, ACL 57	1952
1206	SW1200	EMD	IT 1206	1955
1947	65-ton	GE		1947
1948	80-ton	GE		1948
1952	80-ton	GE		1952

Atlanta: Evans Railcar

	25-ton	GE	Philadelphia Gas	1953

Atlanta: MARTA

1001	SL50	GE		1983

Atlanta: Martin-Marietta

65	65-ton	GE	U. S. Army 7175	1941
7172	65-ton	GE	U. S. Army 7172	1943

Augusta: Columbia Nitrogen

1042	S2	Alco	SCL 46, SAL 1430	1946
	44-ton	GE	Fort Worth Sand & Gravel, B&SE 199	1953

Augusta: Southern Wood-Piedmont

1	16-ton	Plym	Piedmont Wood	1947
2	30-ton	Plym	J. M. Huber	1940

Bainbridge: Kaiser Agricultural Chemical Co.

44	45-ton	Whit	M&B 44, U. S. Navy 65-00218	1942

183

GEORGIA (continued)

Baldwin: Fieldale Farms Feed Mill

	25-ton	GE	Hillyer-Deutsch, Edwards 2	1962

Baxley: Georgia Power

1401	SW1500	EMD		1969

Baxley: Hercules Powder

61	SW900	EMD		1961
63	SW1	EMD	EL 350	1940
68	SW1	EMD	L&N 14	1941

Camak: Martin-Marietta

7589	RS11	Alco	CR 7644	1957
8547	40-ton	Whit	Bowdon Railroad	1953
2	S2	Alco	B&O 9053, 509	1948

Cartersville: Atlantic Steel

1951	S4m	Alco	L&N 2326, L&N 2226	1951
	SW7	EMD	L&N	1950

Clinchfield: Medusa Cement

	25-ton	Plym		1953

East Point: Southern Wood-Piedmont

494	25-ton	Whit		
	25-ton	GE	Oregon Shipbuilding	1941
6	40-ton	Plym	Southern Wood Preservatives	1953

Eden: Dawes Silica Mining

4	25-ton	GE	U. S. Army 7775	1944

Fitzgerald: Fitzgerald Rail Service

606	HH660	Alco	Relco 606, BCK 43	1940
9548	NW2	EMD	B&O 9548	1948

Fortson: Vulcan Materials

4061	80-ton	GE	Jerita, C&O, United Road Machinery 5	1943
4386	80-ton	GE	Birmingham Slag, UM&P 53	1942

Garden City: National Gypsum

	35-ton	Plym	G&MCT, U. S. Army 7627	1943

Gray: Martin-Marietta

7163	S4	Alco	EL 528	1952

Kennesaw: Vulcan Materials

3	65-ton	GE	Continental Foundry, U. S. Army 7353	1942
4842	45-ton	GE	U. S. Air Force 7417, U. S. Army 7417	1942
3	45-ton	GE	Worthington Pump	1952

Macon: Florida Rock Inc.

7943	65-ton	GE	U. S. Army	1943

Macon: Georgia Power

1201	45-ton	GE		1957

Macon: Southern Wood-Piedmont

5	35-ton	Whit	CofG 152	1949

Macon: Transco

	25-ton	GE	Ford Motor Co. 152	1949

Marble Hill: Georgia Marble

77	SW1	EMD		1947
81	NW2	EMD	Southern 1065	1948

Martinez: Martin-Marietta

1	S2	Alco	B&O 9115, C&O 5026	1949

Milledgeville: Georgia Power

1601	100-ton	GE		1960
1602	125-ton	GE		1964

Monticello: Georgia-Pacific Corp.

1	70-ton	GE	Meadow River Lumber 8	1957

Newnan: Georgia Power

1402	SW1500	EMD		1971
1405	SW1500	EMD		1973

Norcross: Western Electric Co.

1	50-ton	GE		1971

Port Wentworth: Atlantic Creosoting

661S	25-ton	GE	Pineland Timber	1955

Putney: Georgia Power

2	50-ton	GE		1962

Rome: Florida Rock

9095	S4	Alco	B&O 9095	1956

Rome: General Electric

2	45-ton	GE		1956

Rome: Georgia Power

5	50-ton	GE		1959
1502	SW1500	EMD		1971
1503	SW1500	EMD		1971

Savannah: Catham Iron & Metal

504	80-ton	GE	Louisiana Southern 504, NEO 253	1937
505	80-ton	GE	Louisiana Southern 505, NEO 254	1937
2	50-ton	GE	U. S. Navy 65-00250	1941

Savannah: Georgia Power

1	45-ton	GE		1952
3	45-ton	GE		1951

GEORGIA (continued)

Savannah: Georgia Power (continued)

4	25-ton	GE		1951
2101	125-ton	GE		1967

Savannah: Kaiser Agricultural Chemical Co.

	44-ton	GE	U. S. Army 7930	1943

Savannah: Savannah Machine & Shipyard

	45-ton	GE	U. S. Navy 65-00185	1943

Stone Mountain : Stone Mountain Scenic Railroad

	SW1	EMD	Groveton Paper, B&M 1114	1946

Thomasville: Dawes Silica Mining

1	6-ton	Plym	DE. Ordnance Plant	1941
2	30-ton	Vulcan	U. S. Navy 4	1941
3	25-ton	GE	Cherokee Brick 9	1951

Thomasville: Georgia Marble

3	25-ton	GE	Alberene Stone 3	1953

Toccoa: Westinghouse Air Brake Co.

	40-ton	Plym	Castalia Quarry	1954

Valdosta: ADM Corp.

140	SW9	EMD	SCL	1952

Valdosta: Gold Kist Inc.

140	SW1200	EMD	SCL	

Warrenton: Martin-Marietta

14	44-ton	GE	EL 537, DL&W 53	1948

Wrens: J. M. Huber

1	SW1500	EMD	Georgia Power	
2	15-ton	Plym		1966

IDAHO

Caldwell: J. M. Simplot Co.

	CR16	Whit	Sumpter Valley.100	1930

Dry Valley: Western Railroad Builders

500	SD9	EMD	MILW 500	1954
508	SD9	EMD	MILW 508	1954
511	SD9	EMD	MILW 511	1954

Nampa: Amalgamated Sugar

32	65-ton	GE	Simplot 32	1943
1	25-ton	GE		1948
2	65-ton	Porter	U. S. Air Force 7159, U. S. Army 7159	1943

Soda Springs: Beker Industries

	80-ton	GE	U. S. Army 7862	1943

Soda Springs: Washington Construction Co.

1200	40-ton	GE	Cargill	1955

ILLINOIS

Alton: Alton Box Board

SE01	44-ton	GE	SLSF 4	1943

Alton: Archer Daniels Midland

1	Slug	GE	Rebuilt from 45-ton diesel	
2	Slug	GE	Rebuilt from 45-ton diesel	
8	45-ton	GE		
110	44-ton	GE	Laclede Steel	
229	45-ton	GE	Almont Shipping	1945

Baldwin: Peabody Coal Co.

1	Slug	BLW		
104	SW1	EMD		1940
600	GP7	EMD		Rebuilt by MK
601	SW8	EMD		1951
607	GP7	EMD		1952
616	GP7	EMD		1952
1201	SW1200	EMD		1957

Barberton: Luntz Corp.

	S3	BLW	B&W 1776	1954

Bartonsville: Keystone Steel & Wire

453	SW8	EMD	Precision National 1112	1953

Bloomington: Cargill Grain Inc.

5	NW2u	EMD	North Carolina Ports Railway 5, Southern 1036	1946

Bureau: West Virginia Pulp & Paper

4800	SW1	EMD	CRI&P 4800; IC 600, 9014	1939

Cairo: Bunge Grain

1052	S4	Alco	Relco 1052	1946

Chicago: Association of American RailroadS

	9-ton	Bkvl		1950
	S1	Alco	Morrell Meats 7	1944

Chicago: Corn Products Co.

327	SW1200	EMD	SOO 327	1955

ILLINOIS (continued)

Chicago: Federal Enamel Stamping Co.

| 1 | 30-ton | Plym | Briggs & Turivas, A. M. Byers | 1940 |

Chicago: Interlake Inc.

| 14 | 35-ton | GE | | 1953 |

Chicago: Interstate Terminal Warehouse

| 343 | 45-ton | GE | Republic Steel 343, | |
| | | | GE-Cummins demonstrator | 1937 |

Chicago: Keystone Steel & Wire

| 450, 451 | SL85 | GE | | 1975 |

Chicago: Midwest Dock

| 877 | SW1m | EMD | MILW 877, 946, 1630 | 1940 |

Chicago: Morton Thiokol

| 101 | SW1200 | EMD | N&PB 101 | 1955 |

Chicago: Republic Steel

| 163 | NW2 | EMD | RT 20, Southern 1066, 2275 | 1948 |
| 816, 817 | S6 | Alco | | 1957 |

Chicago: Sanitary District of Chicago

| 4 | MP15DC | EMD | | 1983 |

Chicago: Standard Oil

| 1 | SL110 | GE | Republic Steel | 1976 |

Chicago: Stolt Creek Coal

| 5 | SW1 | EMD | | 1947 |

Chicago: Thrall Car Manufacturing Co.

| 1 | 45-ton | GE | ACF 70 | 1942 |
| 2 | 45-ton | GE | St. Regis Paper 2 | 1954 |

Chicago: Verson All-Steel Press Co.

| 8 | 65-ton | GE | Nekoosa Paper | 1941 |

Chicago: Wisconsin Steel

| 10 | 45-ton | Dav | U. S. Army 7192 | 1942 |

Chicago Heights: Calumet Steel

| 107 | 40 | EMD | | 1942 |
| 108 | 65-ton | GE | U. S. Army 1440 | 1941 |

Chillicothe: Koppers Co.

| 58 | 35-ton | GE | Martin-Marietta | 1958 |

Chillicothe: Martin-Marietta

| 17 | 25-ton | GE | Concrete Materials | 1955 |

Chillicothe: McGrath Sand & Gravel

| 6 | 30-ton | Dav | Martin-Marietta | 1930 |

Cicero: Rescar Inc.

| 4 | 25-ton | GE | Gifford Hill, Texas Gravel 210 | 1953 |

Cisco: Cisco Grain Co-op

| 7300 | GP9 | EMD | Prairie Trunk 7300 | 1956 |

Coffeen: Central Illinois Public Service

| 1362 | GP7m | EMD | | |

Cora: Cora Dock & Terminal

| 2-22 | NW2 | EMD | Precision National Corp. | |

Crystal Lake: Vulcan Materials

3	44-ton	GE	Consumers Co.,	
			Union Freight RR 3	1946
6	45-ton	GE	Colorado Springs Utilities,	
			U. S. Department of the Interior	1955

Danville: General Motors-Central Foundry

| 1979 | SW9 | EMD | | |
| 1323 | MP15 | EMD | UP 1323 | 1975 |

Decatur: Central Soya

| 495 | NW2 | EMD | BN 495 | 1949 |

Dillsburg: Cargill Grain Inc.

| 903 | SW900 | EMD | Iowa RR, CRI&P 903 | 1959 |
| 6009 | SD7 | EMD | BN 6009 | 1952 |

Dixmoor: National Railway Equipment

| 9610 | SW9 | EMD | B&O 9610 | 1952 |

Dixon: DBM Corp.

| | 20-ton | Plym | Lee County Electric, | |
| | | | Case Brothers | 1949 |

Dixon: Green River Development Corp.

| | 35-ton | Plym | | |
| | 40-ton | Plym | | |

East Dubuque: Dubuque Sand & Gravel

| 537 | SW1 | EMD | CRI&P 537 | 1949 |

Dunfermline: United Electric Coal

17	SW1	EMD	Monon 50	1942
115	S2	Alco	D&RGW 115	1944
	45-ton	Whit	Barber Asphalt	1935

DePue: Mobil Chemical

| RE1009 | S2 | Alco | KCT 59 | 1949 |

East St. Louis: Shippers Car Line

| 1 | 45-ton | GE | U. S. Army 7935 | 1942 |
| 2 | 45-ton | GE | | |

186

ILLINOIS (continued)

East St. Louis: Union Carbide

| 161 | 45-ton | GE | Marquette Cement, American Cement 2 | 1946 | |
| 2 | 30-ton | Dav | Linde Air Co. | 1942 | |

Federal: Laclede Steel

| 3 | 45-ton | GE | Union Carbide, Columbia Southern Chemical 2 | 1956 | |

Freeburg: Peabody Coal Co.

| 1616 | AS416 | BLW | NS 1616 | 1955 | |
| 4489 | GP9 | EMD | CRI&P 4489, 1328 | 1959 | |

Fulton: Agri-Industries

| | SW1 | EMD | | | |

Fairview Heights: Construction & Mining Inc.

| 7704 | SW9 | GMD | CN 7704 | 1952 | |

Gibson City: Cargill Grain Inc.

| 18 | 80-ton | GE | U. S. Navy 65-00506, U. S. Army 7286 | 1943 | |

Gibson City: Central Soya

| 5 | 25-ton | GE | | | |

Gilman: Continental Grain Co.

| 1528 | S4 | Alco | ATSF | 1952 | |

Granite City: Granite City Steel

| 1241 | SW9 | EMD | UP 1827 | 1953 | |

Granite City: Respondek Railroad Corp.

| 103 | SW1200 | EMD | N&PB 103 | 1956 | |

Galesburg: Koppers Co.

| 1 | 25-ton | GE | Allegheny Ludlum Steel | 1951 | |

Henning: Cargill Grain Inc.

| | 45-ton | Whit | | | |

Joliet: Commonwealth Edison

| 8-15 | SW1 | EMD | | 1941-1956 | |
| 16 | SW1200 | EMD | | 1964 | |

Joliet: Relco

707	S3	Alco	GATX 46707, Ann Arbor 5	1950	
1010	S4	Alco	D&H 3037	1950	
1028	S2	Alco			
1031	S2	Alco			
1056	S4	Alco	GTW 8296	1956	
1060	S2	Alco	Savannah State Docks, B&O 9132, C&O 5038	1949	
1203	NW2	EMD		1949	
1204	NW2	EMD	BN 593	1948	
1205	NW2	EMD	CR 9213	1948	
1209	NW2	EMD	BN 402	1947	

Joliet: Amoco Chemical

| RE801 | S2 | Alco | ERIE 505 | 1946 | |

Joliet: U. S. Railcar

| 1 | 25-ton | GE | | 1960 | |

Joliet: Farmers Grain

| 1 | SW900 | EMD | B&O 9402 | 1955 | |

Joppa: Joppa Steam Plant

| 1-3 | 44-ton | GE | | 1951 | |

Kaskaskia: Kaskaskia Regional Port District

| 1321 | GP9 | EMD | CRI&P 1321 | 1957 | |

Kellog: Consolidated Coal Co.

| 5803 | GP7 | EMD | CR 5803 | 1953 | |

La Grange: Electro-Motive Division, General Motors

117	SW1001	EMD		1979	
4975	GM7C	EMD		1975	Electric
4976	GM10B	EMD		1976	Electric
5740	SD45X	EMD		1970	4200 h.p.
9500-9502	SD45X	EMD		1970	No engine
9503, 9504	SD45X	EMD		1972	No engine

Lemont: Thomas Steel

| 2479 | CF7 | ATSF | ATSF 2479 | 1979 | |

Mapleville: C.F. Industries

| 1 | S6 | Alco | SP 1202, CM&N 1202 | 1955 | |

McCook: Pace Engineering

| 29 | 35-ton | GE | Marble Cliff Quarries | 1957 | |

McCook: Vulcan Materials

| 3782 | SW1200 | EMD | BN 258, FW&D 609 | 1959 | |

Metropolis: AEP-Ohio Power

| 3 | SL110 | GE | | 1975 | |
| 4, 5 | SL144 | GE | | 1976 | |

ILLINOIS (continued)
Minooka: Relco

902	SW8	EMD	CRI&P 816	1952
903	SW8	EMD	CRI&P 825	1952
1064	S2	Alco		1946
	Leased to Keokuk Junction Railway			
1260	SW1200	EMD	TRRA 1227	1964
1261	SW9	EMD	TRRA 1216	1952

Mound City: Behimer & Kissnor Inc.

454	S2	Alco	Consolidated Grain & Barge Co.
	S2	Alco	

Naples: Naples Terminal Co.

3	RS1	Alco	AWW 3, SOO 101	1945
5	RS1	Alco	AWW 5, SOO 104	1947

Newton: Central Illinois Public Service

2000	GP35	EMD	SP 6548	1964

Paxton: Illinois Grain Co.

1019	S2	Alco	Relco 1019	

Peoria: Archer Daniels Midland

1	SW1200	EMD	IT 1204	1955

Peoria: CF Industries

1202	S6	Alco	CM&N 1202

Pinckneyville: Consolidated Coal Co.

	S1	Alco		1944
7339	RS1	Alco	Truax-Traer Coal, M&StL 213	1946

Pontiac: Bunge Grain

1007	S4	Alco	D&H 3035	

Quincy: Kansas Gas & Electric

	35-ton	Plym		1943

Quincy: Quincy Soybean Co.

1	20-ton	Plym	Illinois Brick	1957
2	40-ton	Plym		1962

Riverdale: ACS

78	92-ton	GE	Armco Steel B78	1962

Roberts: Pillsbury Inc.

1021	S4	Alco	Relco 1021

Robinson: Union Carbide

1	50-ton	GE	PA&M 50, A.E. Byers D-2	1948
3	SL110	GE		1975
1117	SW8	EMD	SP 1117	1953

South Chicago: USX

1	VO660	BLW	American Steel & Wire	1941
12	VO1000	BLW	American Steel & Wire	1946
21	VO1000	BLW	American Steel & Wire	1943
922	VO1000	BLW	Oliver Iron Mining 933	1951

Sheldon: Early & Daniels

2	80-ton	GE

Sparta: Ziegler Coal

222	NW2	EMD	TRRA 558	1947

Springfield: Central Illinois Light Co.

	SL110	GE		1975

Sterling: Northwestern Steel & Wire Co.

1-3	SW1200	EMD		1980

Taylorville: Allied Mills

602	HH660	Alco	Relco 602, EJ&E 210	1940

Washington: U. S. Railway Equipment Co.

1	65-ton	GE	Interlake Steel	1942
2	65-ton	GE	U. S. Air Force 1215	1951

Worthington: Allis Chalmers

5	SW1	EMD		1948
6	SW1	EMD		1947

Worthington: American Steel Foundries

909	SW1001	EMD		1979
9010	SW1001	EMD		1979

INDIANA

Boonville: Peabody Coal Co.

9842	RSD15	Alco	ATSF 9842	1960
9843	RSD15	Alco	ATSF 9843	1960

Boonville: Squaw Creek Coal

	U33C	GE	Southern	1970

Burns Harbor: Bethlehem Steel

14	SW9	EMD	BAR, P&LE 8938, NYC 8938	1951
15	SW9	EMD	P&LE 1237	1952
16	SW9	EMD	P&LE 1241	1952
20	SW1200	EMD	CRI&P 922	1965
21	SW1200	EMD	CRI&P 923	1965
22	SW900	EMD	CRI&P 902	1959

INDIANA (continued)

Burns Harbor: Bethlehem Steel (continued)

23	SW900	EMD	CRI&P 905	1959
24	SW900	EMD	CRI&P 906	1959
25	SW900	EMD	CRI&P 909	1959
26	SW900	EMD	CRI&P 910	1959
27	SW900	EMD	CRI&P 913	1959

Burns Harbor: Burns Bailing

603	HH660	Alco	EL 324, DL&W 409	1940

Carlisle: Con-Agra

6252	SD24	EMD	MMID 6252, BN, CBQ 512	1959

Cayuga: Public Service of Indiana

1	SW1000	EMD		1969
1	45-ton	GE		1947

Clare: Public Service of Indiana

2	65-ton	GE		1948

Clymer: Farmers Co-op Elevator

	45-ton	GE	U. S. Air Force 7245,	
			U. S. Army 7245	1941

Colfax: Anchor Grain

91	SW1200	EMD	MP 1155	1964

Chandler: Squaw Creek Coal

1002	GP7	EMD	Peabody Coal 1002, SLSF 542	1951

Decatur: Central Soya

1	35-ton	GE	Marble Cliff Quarries	1957
2	35-ton	GE	Hamm Brewing	1958

East Chicago: Amoco Oil Co.

555	SW1	EMD		

East Chicago: Blaw Knox Foundry Mill

26	65-ton	GE	TVA 26	1944

East Chicago: General American Tank Line

602	HH660	Alco	Relco 602, EJ&E 210	1940
1	65-ton	Whit	U. S. Army 8477	1944
2	65-ton	Whit		1948

East Chicago: Inland Steel

61, 62	S1	Alco		
70, 72,				
76-81	SW1	EMD		1948-1951
88-98,				
103-114	SW1200	EMD		1956-1965

115-118	SW1000	EMD		1966-1967
119-125	SW1500	EMD		1968-1973
126, 127	SW1001	EMD		1976
132-137	SW9	EMD		1952
S1-S7	Slug		Rebuilt Alco S1s	

East Chicago: Youngstown Sheet & Tube

1-5	35-ton	GE		1959, 1966
423	80-ton	GE		1942
900-904	SW900	EMD		1956, 1965
905-914	SW1000	EMD		1967-1969
915	NW2	EMD	Precision National 8711,	
			P&LF 8711	1947
916-918	NW2	EMD		
919	SW9	EMD	CSL 26, Mississippi Central 208	1953
920	SW9	EMD	CSL 27, Mississippi Central 204	1953

East Mount Carmel: Amax Coal Co.

WG01	GP38-2	EMD		
WG02	GP38-2	EMD		

Edgar: USX

119	NW2	EMD	CRI&P 4906	1949

Emporia: Emporia Grain Co.

1	25-ton	GE		
2	30-ton	Plym		
3	S2	Alco		
611	S2	Alco		

Evansville: Cargill Grain Inc.

1134	G8	GMD	Columbia Cellulose	

Frankfort: A. E. Staley Co.

	45-ton	Plym	Swift Grain Co.	1962

Fort Wayne: International Harvester

201	25-ton	GE		

Fort Wayne: USX

11	45-ton	GE		1942

Gary: Gary Slag Co.

476, 477	VO1000	BLW	EJ&E 476, 477	1949
9403	S3	Alco	PC 9403	1950

Gary: USX

2	S2	Alco	GTW 8091	1942
30	NW2	EMD	P&LE 8707	1947
31	S2	Alco	BN 908, NP 708	1943

INDIANA (continued)
Gary: USX (continued)

32	NW2	EMD	KCS 1218	1948	
33	NW2	EMD	GTW 7901	1941	
34-37	SW900	EMD	American Steel & Wire	1958	
38	NW2	EMD	P&LE 8714	1947	
39	NW2	EMD	GTW 7910	1942	
40	SW8	EMD	CWP&S 40, CUT 33	1951	
41	SW8	EMD	International Harvester 42, CWP&S 42	1951	
42	SW8	EMD	NL&G 38	1953	
43, 44	SW1	EMD	Am. Sheet & Tube; EJ&E 237, 246	1941	
45-60	80-ton	GE		1947-1957	
61	S2	Alco	BS 535, URR 535	1948	
62, 63	S2	Alco	OIM 921, 927	1943, 1944	
65	S2d	Alco	OIM 926	1944	
66, 67	S2	Alco	OIM 917, 916	1942	
68	S2	Alco	EJ&E 458	1944	
69	S2d	Alco	EJ&E 453	1940	
70	S2	Alco	EJ&E 455	1944	
71	S2d	Alco	EJ&E 462	1948	
72, 73	S2	Alco	EJ&E 459, 460-	1944	
74	S2d	Alco	EJ&E 452	1940	
75	S2	Alco	EJ&E 454	1941	
76	S2d	Alco	EJ&E 456	1944	
77	S2	Alco	BS 525, URR 525	1945	
78	S2	Alco	EJ&E 461	1944	
79, 80	S2	Alco	Youngstown & Northern 214, 218	1942, 1940	
81	S2	Alco	N&SS 1008	1946	
82	NW2	EMD	LT 1008	1947	
83	S2	Alco	N&SS 1010	1946	
84	S2	Alco	Hannibal Connecting 603	1940	
85	S2	Alco	MSS 607	1942	
86, 87	S2	Alco	Youngstown & Northern 219, 216	1949, 1945	
88-90	SW1001	EMD		1970, 1971	
2589	SW8	EMD	CRI&P 837	1953	
9110	SL115	GE		1975	
9111	SL115	GE		1976	
9597	NW2	EMD	Precision National Corp.		
9667	S2	Alco	PC 9667, NYC 8594	1952	

Gary: Vulcan Materials

4	50-ton	GE	West Palm Beach Terminal, U. S. Navy 65-00380	1942	

Hammond: Amaizo Rail Switching Service

202, 608	SW1200	EMD	MILW 630, 608	1954	
4907	NW2	EMD	CRI&P 4907, P&LE 4907	1949	

Hammond: CF&I Steel

101, 102	GP7			1951	

Hammond: Commonwealth Edison of Indiana

3	SW1	EMD		1947	
4	SW9	EMD		1948	

Hammond: Fruit Growers Express

6	45-ton	GE		1952	

Hammond: La Salle Steel

	25-ton	GE	U. S. Army 7763	1943	

Hammond: Standard Railway Equipment

	80-ton	GE	U. S. Navy 65-000362	1953	

Indianapolis: Central Soya

9256	NW2m	EMD	CR 9256	1948	
3	65-ton	GE			

Indianapolis: Citizens Gas & Coke Utility

2, 3	80-ton	GE			

Indianapolis: Farm Bureau

	80-ton	Whit	PPG 402	1954	

Indianapolis: Illinois Cereal Mills—Evans Division

2308	SW9r	EMD	SP 2308	1953	
8418	SW1	EMD	B&O 8418	1942	Stored

Indianapolis: Indianapolis Coke

7	SW1200	EMD		1955	

Indianapolis: Spencer Industries

2335	65-ton	Whit	American Aggregates	1944	Stored
6089	65-ton	Whit	American Aggregates	1944	

Jeffersonville: Jeffboat Shipyard

102	S5	Alco	Transfer Terminal	1954	

Keeport: Indiana-Ohio Stone Co.

	40-ton	Plym		1957	

Kokomo: Kokomo Grain Co.

102	45-ton	GE	Laona & Northern, U. S. Navy 65-00246	1942	
	45-ton	GE	Purdue Univ., U. S. Army 7066	1942	

190

INDIANA (continued)
Kokomo: Penn-Dixie Steel

108	70-ton	GE	MKT 1651	1949
109	70-ton	GE	Kingan 5	1949

La Paz: Merchants Grain Co.

1207	SW9	EMD	TRRA 1207	1952

Lafayette: Eli Lilly Corp.

494	NW2	EMD	BN 494	1949
	80-ton	GE	G&W 22	1945

Leesburg: Western Indiana Gravel

2	25-ton	GE	1946

Linden: Cargill Grain Inc.

	65-ton	GE	U. S. Army 7270	1941

Logansport: Bunge Grain

713	S1	Alco	Relco 713, L&N 59	1949

Logansport: Transcar Service Co.

	44-ton	GE	MC&SA 17, T&NO 17, PE 1654	1944

Lynnville: Peabody Coal Co.

1	H16-66	FM	Squaw Creek Coal	
230	RS3	Alco	GN 230	1953
343	AS16	BLW	USP&F 43, MP 936,4327	1954
413	RS3	Alco		1953
416	GP7	EMD	MK 510, SLSF 510	1950
470	SW1	EMD	DL&W 358	1940
1001	GP7	EMD	SLSF 594	1951

Lynnville: Squaw Creek Coal

1001	H16-66	FM

Merom: Hoosier Energy

1	SL115	GE	1980

Michigan City: Joy Manufacturing Co.

2	12-ton	Whit	1951

Michigan City: Northern Indiana Public Service

12	SW1	EMD	GN 78	1941

Mishawaka: Indiana & Michigan Electric Co.

6	65-ton	GE	1967

Monon: Monon Crushed Stone

	20-ton	Dav	U. S. Army 7668	1941

Mount Vernon: Consolidated Grain & Barge

1155	RS1	Alco	H. O. Forgy, GM&O 1055, IT 755	1950
7300	GP7	EMD		

Peru: Kickapoo Sand & Gravel Co.

1	25-ton	GE		1946
2	25-ton	GE	Warren Slag	1949

Petersburg: Indianapolis Power & Light

	110-ton	GE	1966

Princeton: Cargill Grain Inc.

	45-ton	Whit	U. S. Army	1942
	65-ton	BLW	Hercules Powder	1953

Reynolds: TAB

2	45-ton	GE	Lone Star Industries, Lehigh Portland Cement	1945

Roachdale: Merchants Grain Co.

1233	SW1200	EMD	TRRA 1233	1964

South Bend: Kuert Concrete Inc.

	20-ton	Plym	Ball Glass Jar	1926

South Bend: Notre Dame University

5332	65-ton	Porter	U. S. Army	1942

South Bend: Western Indiana Gravel

1	45-ton	GE	Hass Gravel, U. S. Army 7074	1941

Switz City: B&LS Contracting

	S2r	Alco
	Caterpillar engine	

Terre Haute: Graham Grain

12	SW1	EMD

Tipton: Cargill Grain Inc.

	80-ton	GE	U. S. Air Force 7448	1941

Valparaiso: Northern Indiana Public Service

	SW1	EMD	DE 210	1949

Walton: Erny's Grain Elevator

9531	NW2	EMD	B&O 9531	1949
9539	NW2	EMD	B&O 9539	1949

Warrrick: Southern Indiana Gas & Electric

	80-ton	GE	1951

Waterloo: DeKalb County Co-op

103	DS44-1000	BLW	DT 103	1947

Watson: Merchants Grain Co.

1210	SW9	EMD	TRRA 1210	1952

Whiting: Standard Oil

555	35-ton	Whit	1953
2	20-ton	Dav	1946

INDIANA (continued)
Whiting: Union Tank Car Co.

1	50-ton	Whit		1947	
	45-ton	GE	U. S. Army 7190	1942	

Whiting: International Mill Services

7901	SW900	GMD	CN 7901	1953	

Yankeetown: Yankeetown Dock Co.

20	SD38-2	EMD		1972	
21	SD38-2	EMD		1972	
22	SD38-2	EMD		1978	

IOWA

Alta: Cargill Grain Inc.

	80-ton	Whit	Beech Mountain 100	1945	
1	44-ton	Dav	U. S. Army 1218	1953	

Avon: Farmers Grain Dealers Association

1	40-ton	Dav		1954	
2	40-ton	Dav		1954	

Beaver: Cargill Inc.

2138	30-ton	Plym		1951	

Belmond: Central Soya

6	45-ton	GE	General Electric 2	1942	

Belmond: Iowa-Illinois Gas & Electric Co.

1	25-ton	Dav		1946	
2	50-ton	GE		1948	Remote control

Bettendorf: Bettendorf Inc.

	45-ton	Dav	Wisconsin Steel, U. S. Army 7219	1942	

Bettendorf: J. I. Case Co.

1	45-ton	Whit	U. S. Navy 65-000167	1941	

Bondurant: Bondurant Grain

5202	NW2	EMD	C&O 5202	1949	

Britt: Continental Grain Co.

207	RS3	Alco	Lasco 1609, LS&I 1609	1955	
1054	S2	Alco	Relco 1054		

Buffalo: Cargill Grain Inc.

43	S1	Alco	Nekoosa Paper	1949	

Buffalo: Martin-Marietta Cement Division

	45-ton	GE		1975	

Burt: Big Six Grain Co.

LO604	HH660	Alco	Relco 604, MILW 983	1939	

Carlisle: Agri-Industries

151	GP7	GMD	AC 151	1951	

Carter Lake: Paxton & Vierling Steel

1	50-ton	GE	Sheffield Steel	1942	

Cedar Rapids: Corn Sweeteners

	SW1	EMD			

Cedar Rapids: Diesel Supply Co.

507	SW1	EMD	IC 607	1946	

Cedar Rapids: Iowa Electric Light & Power

52	SW1	EMD	PB&NE 212, P&BR 155	1940	
59	65-ton	GE	U. S. Navy 65-00054	1943	

Clinton: Chemplex

1002	S4	Alco	D&H 3049	1950	

Clinton: Corn Processing Co.

701	S1	Alco	Minnesota Transfer 61	1941	

Clinton: Interstate Power Co.

150	20-ton	Dav	U. S. Army 7705	1941	

Council Bluffs: Cargill Inc.

	35-ton	Plym		1933	

Council Bluffs: Nucor Steel

1272	SW1	EMD	C&NW 1272	1953	
2307	SW1200r	EMD	SP 2307	1954	
2313	SW1200r	EMD	SP 2313	1954	

Davenport: Alter Inc.

	65-ton	GE	Massachusetts Electric 2, Worcester County 2	1950	

Davenport: Occidental Chemical Co.

715	S1	Alco	Relco 715		

Denison: Cook Industries

1031	S2	Alco			

Des Moines: Agri-Industries

1	44-ton	Dav	Arco Petroleum Products Co.		
8313	SW1	EMD	B&O 8413	1942	

Des Moines: Central States Warehouse Co.

8421	SW1	EMD	B&O 8421	1942	

Des Moines: Iowa Power & Light

2	45-ton	GE		1952	Remote contro

IOWA (continued)

Des Moines: Monarch Cement
1	65-ton	GE	Olin Chemical, U. S. Army 7018	1942
2	45-ton	GE	Marquette Cement,	
			U. S. Army 7797	1941

Des Moines: Pillsbury Inc.
| 2 | 45-ton | Plym | Martin-Marietta | 1940 |

Des Moines: A. E. Staley Co.
| 185 | NW2 | EMD | MP 1010 | 1948 |

Des Moines: Swift Edible Oil Co.
| | NW2 | EMD | | 1948 |

Dubuque: N-Ren Nitrogen Corp.
| 4 | 25-ton | | St. Pauls, N.H., 2 | |

Eagle Grove: Agri-Processing Corp.
| 4 | SW900 | EMD | Waterloo 4 | 1958 |

Fort Madison: Chevron Chemical Co.
| 1028 | S2 | Alco | Relco 1028, TRRA 569 | 1948 |

Fort Madison: First Mississippi Co.
| 1006 | S2 | Alco | Relco 1006, KCT 57 | 1945 |

Hartley: Albert City Elevator
| 1 | SW900 | EMD | Waterloo 1 | 1957 |

Jefferson: Farmers Co-op Association
| 3 | 45-ton | Plym | | 1969 |

Jordan: West Central Co-op
| | 45-ton | Plym | | 1979 |

Keokuk: Foote Minerals Co.
1	25-ton	Whit	Keokuk Metals, Walsh Construction	1941
2	45-ton	GE	Kemco 2	1952
	Stored unserviceable			
3	50-ton	GE	Kemco 3	1948
4	45-ton	GE	U. S. Army 1212	1941

Keokuk: Union Electric Co.
| | Electric | | AMCR | 1917 |

Keota: Keota Washington Transportation
104	VO1000	BLW	Pacific Lumber 104,	
			U. S. Army V-1800	1945
105	VO1000	BLW	Pacific Lumber 105	1945

Lake Manawa: Iowa Power & Light
| 1 | 45-ton | Plym | | 1953 |

Lakota: Farmers Co-op Association
| 1 | 40-ton | Plym | Rohm & Haas | 1958 |

Linwood: Ferruzi Grain Terminal
| | 44-ton | GE | Bayside Elevator, PRR 9324 | 1949 |

Linwood: Pillsbury Inc.
| 144 | 65-ton | Whit | Material Service, U. S. Army 7970 | 1943 |

Marshalltown: Iowa Electric Light & Power
| | 25-ton | GE | Tama & Toledo 2 | 1944 |

Mason City: Lehigh Portland Cement
| 1 | 23-ton | GE | | 1939 |
| 2 | 65-ton | Whit | U. S. Army 7989 | 1943 |

Mason City: Northwestern States Portland Cement
| 1 | 65-ton | GE | U. S. Army 7395 | 1941 |
| 2 | 80-ton | GE | Ohio River 5 | 1959 |

Missouri Valley: Loveland Elevator
| | 20-ton | Whit | Giddeon Grain | |

Oelwein: Transco
| 101 | S2 | Alco | Mattoon Service, KCT 51 | 1940 |
| 543 | SW1 | EMD | CRI&P 543 | 1942 |

Onawa: Langren Grain Co.
| 1 | 20-ton | Vulcan | Midwest Const. 101, | |
| | | | U. S. Army 2083 | 1942 |

Palo: Iowa Electric Light & Power
| 50 | 65-ton | GE | U. S. Army 7152 | 1942 |

Pickering: Continental Grain Co.
| 205 | RS1 | Alco | Calumet & Helca, LS&I 1002 | 1951 |

Port Neal: Iowa Public Service Co.
| | 45-ton | GE | U. S. Navy 65-00011 | 1942 |

Port Neal: Terra Chemical Co.
| | 45-ton | GE | Port Utilities Comm. | 1955 |

Ralston: Farmers Co-op Association
| 2 | 45-ton | Plym | | 1957 |

Waverly: Koehring Co.
| 65 | 35-ton | Plym | | |
| 101 | 35-ton | Plym | Standard Slag 25 | 1938 |

Winterset: Penn Dixie Cement
3	70-ton	GE	MSV 3	1950
36	Model 40	EMD	U. S. Rubber, U. S. Army	1941
37	80-ton	GE		1956

KANSAS

Augusta: Mobil Oil

| | 30-ton | Whit | | |

Chanute: Ash Grove Cement

| 3 | 45-ton | GE | | 1944 |
| 4 | 65-ton | GE | | 1953 |

Independence: Lehigh Portland Cement

| 7 | 45-ton | Whit | Atlas Cement 7 | 1949 |
| 8 | 45-ton | GE | Atlas Cement 8, U. S. Gypsum 103 | 1949 |

Independence: Universal Atlas Cement

| 5 | 45-ton | Whit | Southeastern Shipbuilding | 1943 |

Iola: Iola Industries

| 1 | 20-ton | GE | Lehigh Cement | 1938 |

Kansas City: Bartlett Grain

| 1 | S3 | Alco | | 1956 |
| 9539 | NW2 | EMD | B&O 9539 | 1949 |

Kansas City: Board of Public Utilities

| | 80-ton | GE | GE, U. S. Army 1660 | 1952 |

Kansas City: Cedrite Technologies (plant closed)

| 9428 | SW900 | EMD | B&O 9428 | 1955 |

Kansas City: General American Transportation

| 705 | S1 | Alco | Relco 705, Minnesota Transfer 62 | 1941 |

Lawrence: Cooperative Farm Chemical Association

32	S3	MLW		
1244	TR4Au	EMD	ATSF 1244	1951
1245	TR4Bu	EMD	ATSF 1245	1951

Lawrence: Kansas Power & Light

| 2 | SL110 | GE | | 1976 |

Olathe: Johnson County Industrial Railway

| | 80-ton | GE | U. S. Army 7891 | 1943 |

Pittsburg: Gulf Oil Co.

| 3 | 44-ton | Whit | U. S. Army 7044 | 1942 |

Pittsburg: Thermex Energy Co.

| 1 | 25-ton | GE | Gillies Bros. | 1949 |

Portland: Portland Flour Mills

| 401 | ML-06 | Plymouth | | |

Riverton: Empire District Electric Co.

| 6 | S4 | Alco | Gulf Oil 6, VTR | 1953 |
| | 35-ton | Whit | | |

Topeka: PLM Corp.

| 421 | C415 | Alco | CRI&P 421 | 1968 |

Turner: Erman Howell

| 629 | DS44-750 | BLW | ATSF 629 | 1949 |

Wichita: Vulcan Materials

| 1 | 50-ton | GE | Chevrolet Motor Division | 1965 |

KENTUCKY

Ashland: Armco Steel Co.

| 1202 | S6 | Alco | | 1956 |
| 1215 | SW1001 | EMD | | 1981 |

Ashland: Mansbach Metals

1	80-ton	GE	Lukens Steel 40, Sharon Steel 18	1952
2	70-ton	GE	Detroit Steel 110, Steel Corp. of West Virginia, Conners Steel	1946
113	70-ton	GE	Empire Steel	1956

Beaver Dam: Peabody Coal

| 1 | S6 | Alco | South Hopkins Coal | |

Calla: Southeast Coal

| 100, 200, 300, 400, 500 | S2 | Alco | | |

Central City: Peabody Coal Co.

| 44, 45 | AS16 | BLW | U. S. Pipe & Foundry 44, 45 | 1950 |

Charleston: Roberts Brothers Coal

| 1 | MRS1 | Alco | W&W | 1952 |

Colson: Red Fox Coal

| 600 | S2 | Alco | | |

Corbin: Glover Corp.

| 871 | SW1 | EMD | DRI&NW 871 | 1940 |

Corbin: Shamrock Coal

| 9950-9957 | SD40-2 | EMD | | 1979 |

Corbin: USX

| 1218 | S6A and S6B | Alco | | 1956f |

DeCoursey: Corbin Railroad Service Co.

| 33 | SW9 | EMD | BAR 33 | 1951 |

194

KENTUCKY (continued)

Evanston: Consolidated Coal Co.

6199	GP9	EMD	C&O 6199	1956	

Evanston: Kentucky Criterion Coal

675	GP9	EMD	NW 6759	1959	

Ghent: Clean Coal Terminal

1000	VO1000	BLW	L&N 2161	1943	EMD engine
1001	VO1000	BLW	L&N 2160	1943	EMD engine

Ghent: Kentucky Utilities

1, 2	45-ton	GE		1949, 1956

Grahamville: Tennessee Valley Authority

22	H12-44	FM		1954
30	S2	Alco	U. S. Navy 65-00076	1942
39	RSD1	Alco	ARR 1032, U. S. Army 8024	1942
40	RSD1	Alco	U. S. Army 8028	1942

Guthrie: Koppers Co.

14	35-ton	Plym	1953	30" gauge

Isom: Enterprise Coal

1503	S2	Alco	Remote control

Jackson: Kentucky May Coal Co.

6	NW2	EMD	NCPR 6, Southern 1058, 2267	1948

Jamboree: Pikeville Coal

9415	SW900	EMD	B&O 9415	1955

Kimper: Berwind Coal Co.

6010	GP9	EMD	C&O 6010	1958

Kimper: McCoy Elkhorn Coal Co.

8101	GP10	ICG	ICG 8101	1975

Kite: Knott County Coal Co.

2	S2r	Alco	1948

Leatherwood: Blue Diamond Coal

1	S2	Alco

Lexington: R. J. Reynolds

4	25-ton	GE	Hazelbrook Coal	1956

Louisville: American Coal Terminals

1234	S6	Alco	SP 1234	1956

Louisville: Big River Electric Co.

1	SL144	GE	1983

Louisville: Du Pont

	80-ton	GE	U. S. Steel 55	1953

Louisville: General Electric

1-3	80-ton	GE		
44	25-ton	GE	Louisiana Dock Co.	
1234	S6	Alco	SP 1234	1956

Louisville: Louisville Gas & Electric

	SL110	GE	1980

Louisville: Riverway Inc.

1	40-ton	Plym	Martin-Marietta 18547, Bowdon Railway 5639	1953

Louisville: Rohm & Haas

	50-ton	GE	GE 18	1953

Louisville: Tube City Iron & Metals

4	43-ton	GE	Du Pont 1, General Electric	1940

Lynch: USX

101	S2	Alco
102	S2	Alco

Madisonville: Midwest Coal Handling Co.

2492	CF7	ATSF	ATSF 2492	1974
2495	CF7	ATSF	ATSF 2495	1974
2508	CF7	ATSF	ATSF 2508	1974
2511	CF7	ATSF	ATSF 2511	1974
2627	CF7	ATSF	ATSF 2627	1972

Maysville: Eastern Kentucky Power

1	SL110	GE		1975
2	SW1001	EMD		1980

Millard: Clark-Elkhorn Mining Co.

1	S2	Alco	1944

Oven Fork: Blue Diamond Coal

2	RS3	Alco

Pilgrim: Pontiki Coal

60	SD45	EMD

Rockport: Pyramid Mining Co.

1	S6	Alco	1956

St. Charles: Charolais Corp.

1246	SW9	EMD	ICG 1246	1950
		Stored unserviceable		
1249	SW9	EMD	ICG 1249	1952
1250	SW9	EMD	ICG 1250	1951

KENTUCKY (continued)
Silver Grove: Midstates Grain Terminal
530	RS3	Alco		1951

West Henderson: Agrico
4	65-ton	GE		1949

LOUISIANA

Addis: Copolymer Inc.
	35-ton	GE	Cargill	1959

Alsen: Schuylkill Metals Co.
80	80-ton	GE	Ethyl I80	1948

Amelia: Intercoastal Terminal Inc.
1	25-ton	Whit		
2	25-ton	GE		

Amite:
Gifford-Hill Inc.
359	SW1	EMD	EL 359	1940
7012	65-ton	GE	U. S. Army 7021	1942

Avondale: Avondale Shipyards
1	80-ton	GE	U. S. Air Force 7291, U. S. Army 7291	1942
1055	NW2	EMD	MP 1055	1948
1651	S12	BLW	NOBP 61	1955
2372	S8	BLW	NOBP 51	1951
	45-ton	GE		

Avondale: Continental Grain Co.
1037	S2	Alco	Relco 1037, Intra-Plant Switching Corp.	
1236	SW1200	EMD	MP 1236	1952

Ama: Archer Daniels Midland
2	GP7	EMD	Farmer's Export 2, SLSF 581	1951

Baton Rouge: Alcoa Aluminum
9	SW1500	EMD		1970

Baton Rouge: Allied Chemical Co.
705	S2	Alco	Relco 705	1944

Baton Rouge: Ethyl Corp.
120	SW9	EMD	ICG 461	1951

Baton Rouge: Exxon-Coker
3	SL144	GE		1981

Baton Rouge: Ideal Cement Co.
9	80-ton	GE		1944

Baton Rouge: Olin Chemical
1	SW1001	EMD		1982

Baton Rouge: Relco
1049	S2	Alco	SCL 48, SAL 1432	1948

Bogalusa: Crown-Zellerbach
1, 2	RS1	Alco		

Bossier City: Kerr-McGee
2	25-ton	GE	Diamond Alkali	1948

Bridge City: Continental Grain Co.
1056	NW2	EMD	Southern 1056	1948

Convent: Peavy Inc.
1280	SW600	EMD	C&NW 1280	1954

Destrahan: Bunge Grain
202-204	SW1	EMD		1949
1039	S2	Alco	SCL 76	1944

Donaldsonville: Agrico
4	S3	Alco	DRI&NW 4	1952
5	70-ton	GE		1952
9	70-ton	GE	AN&N 172	1948

Gibsland: Willamette Industries
	45-ton	GE	Pacolet Manufacturing Co., LS 501	1941

Georgetown: Gulf South RailCar Service
1	25-ton	Plym	Crown-Zellerbach, Linton-Summit Coal, Simmons & O'Connor Coal	1941

Harvey: Intercoastal Terminal Inc.
	45-ton	GE	Avondale Shipyards 3, U. S. Air Force 7325, U. S. Army 7325	1941

Houma: Imco Inc.
1	50-ton	Plym	Gardinier Chemical, Continental Grain, American Cyanamid	1962

Isabel: Louisiana Industries
3	23-ton	Porter	Celotex 3, United Fruit 1	1929

Joyce: Crown-Zellerbach
3	45-ton	GE	Tremont Lumber, Lone Star Cement	1955

Lacombe: Coastal Sand & Gravel
	40	EMD	American Creosote, U. S. Army 7953	1943

Lake Charles: Amoco Chemical
SE03	SW1001	EMD		1978

Lake Charles: Olin Mathieson Chemical
2	SW1	EMD		1949

LOUISIANA (continued)
Laplace: Bayou Steel
47   SW9   EMD   WP&S 47   1952
Marerro: Celotex Corp.
7   S3   Alco   DRI&NW 7   1953
Mathews: South Coast Corp.
1   45-ton   GE   U. S. Army 8534   1944
New Orleans: Amaco Chemical Co.
SE08   SW1001   EMD   1978
New Orleans: American Cyanamid
1060   NW2   EMD   1948
          S2   Alco   SP 1754, T&NO 66   1949
New Orleans: Bulk Terminal Co.
5   SW1   EMD   1949
New Orleans: Judice Armature
69   SW1   EMD   Hercules Powder   1942
New Orleans: New Orleans Public Bulk Terminal
1   80-ton   GE   1951
2   80-ton   GE   1951
          Stored unserviceable
New Orleans: New Orleans Sewerage & Water Board
L-4   12-ton   Plym   1935
L-5   8-ton   GE   1910
          Rebuilt from electric, 1947
L-6   16-ton   Plym   1977
New Orleans: Port of New Orleans Public Grain Elevator
          80-ton   GE   1962
New Orleans: Southern Scrap Metals
1   50-ton   Whit
2   65-ton   GE   U. S. Navy
3   45-ton   Whit
          65-ton   GE   U. S. Navy   1943
Plaquemine: Copolymer Corp.
          45-ton   GE
Pollack: Witte Gravel
100   44-ton   Dav   Martin-Marietta   1941
Port Allen: Cargill Inc.
601   SW8   EMD   Bamberger 601, CSL 601   1952
Raceland: South Coast Sugar Co.
          10-ton   Plym

Reserve: Bayside Grain Elevator
811   NW2   EMD
1045   S2   Alco   Relco 1045, Louis Dreyfus Co.
Shreveport: Tri States Concrete Co.
1256   SW1200   EMD   MP 1256   1966
St. Gabriel: Relco
901   S2   Alco   SCL 116, ACL 35, 626   1944
Superior: Superior Tie & Timber
7   SW1   EMD   MILW 879   1940
Taft: Union Carbide
1207   NW2   EMD   Relco 1207
Taylor: Woodward Walker Lumber
463   44-ton   GE   Midwest Grain, ATSF 463   1943
Westwego: Continental Grain Co.
724   S2   Alco   South Carolina Ports   1945
1037   NW2   EMD   Southern 1037   1947
Westwego: Edward Levy Metals
          80-ton   GE   Southwest Steel 7394, U. S. Air
                                Force 7394, U. S. Army 7394   1945
Westwego: Ideal Cement & Metals
          25-ton   GE   1951
Westwego: Lone Star Industries
5   45-ton   GE   1953

MARYLAND

Appliance Park: General Electric
2   80-ton   GE   Coors 998, Sharon Steel 16   1952
Baltimore: Baltimore MTA
801   60-ton   Plym   1981
Baltimore: Central Soya
204   RS3   Alco   SJ&LC 204, L&HR 4   1950
Baltimore: Kerr-McGee
3   45-ton   GE   U. S. Navy   1941
Baltimore: Striegel Supply
          RSC2   Alco   Florida Power & Light, SAL
Baltimore: Marquette Cement
2   45-ton   GE   AM Cement 60   1948
4   45-ton   GE   1942

MARYLAND (continued)
Mount Airy: Mount Airy Cold Storage Co.

8411	SW1	EMD	WW 8411, B&O 8411	1940

Sparrows Point: Vulcan Materials

2	25-ton	GE	M&T Chemical Co., Everett Foundry 1956

MASSACHUSETTS

Chicopee: General Electric

9	50-ton	GE	Continental Steel	1941

Lee: Seaview Transportation

2	65-ton	Vulcan	Warwick 105, U. S. Navy 65-0019	1943

Ludlow: Ludlow Industrial Park

11	25-ton	GE	Ludlow Manufacturing	1944

Lynn: Perini Corp.

1	SW1200	EMD		

Mount Tom: Holyoke Water Power Co.

1849	SW1	EMD	B&M 1113	1941

Palmer: Quabaug Transfer

3606	RS11	Alco	CV 3606	1956
3611	RS11	Alco	CV 3611	1956

Pittsfield: General Electric

6	S1	Alco		
8	110-ton	GE		1967
8	50-ton	GE		1941

Watertown: Atlas Railroad Contractors

454	S2	Alco	LI 454	1949

West Cambridge: West End Iron Works

10	44-ton	GE	G&U 10	1946

MICHIGAN

Alabaster Junction: National Gypsum — Gold Bond Division

9	65-ton	Whit	URR, U. S. Army 7972	1943
16	65-ton	GE	Dragon Products, GE 16, U. S. Army 7254	1942

Albion: Anderson's Hartung Elevator

1	35-ton	Plym		

Alpena: Huron Portland Cement

	44-ton	Porter	U. S. Army 7446	1942

Brighton: Great Lakes Steel

206	SW900	EMD	Boke Trading 206, Hanna Furnace 17, Peaker Services	
1557	SW1500	EMD	P&LE 1557	1972
1565	SW1500	EMD	P&LE 1565	1972
1568	SW1500	EMD	P&LE 1568	1972

Cadillac: Mitchell Corp.

1001	DCL	Plym	Chris-Craft Co.	

Carrollton: Michigan Elevator Exchange

	25-ton	Plym		1962

Conners Creek: Detroit Edison Co.

213	SW600	EMD		1954	
214	SW900	EMD		1956	Remote control

Detroit: Buick Motor Division

792, 1776	SW900	EMD		1954, 1955

Detroit: Detroit Edison Co.

211	SW1	EMD		
217	SW1	EMD	Dominion Terminal Associates	
1535	SW1500	EMD		1973
1539	SW1500	EMD		1973
1540	SW1500	EMD		1973

Detroit: Ford Motor Co.

2003	SW9	EMD		

Detroit: S. G. Keywell

1	25-ton	GE	Evans Products.	1967

Detroit: Marrow Steel

4	80-ton	GE	U. S. Navy 65-00501	1944

Ecorse: Great Lakes Steel

6, 7	SW900	EMD		1956
11, 12	SW1	EMD		1940
16	SW1200	EMD		1962
17, 18	SW1	EMD		1945
22	SW1	EMD		1949
27-29	SW9	EMD		1951-1952
30	SW1	EMD		1941
33, 34	SW1	EMD		1948, 1949
36, 38	SW1	EMD		1953
39-53	SW1200	EMD		1956-1965
56, 57	SW9	EMD		1953
58	SW1000	EMD		1971

MICHIGAN (continued)

Ecorse: National Steel Co.

62	SW1200RS	GMD	CN 1297	1958

Ecorse: Nicholson Terminal & Dock Co.

15	DS44-660	BLW	Wyandotte Southern	1948
16	1000	LH	TP&W 300, Lima demonstrator 1000	1949
22	750	LH	Cincinnati Union Terminal	1950
25	35-ton	Lima	Stored unserviceable	1942

Erie: Consumers Power

2	80-ton	GE	G&W 21	1945

Essexville: Consumers Power

401	RS1	Alco	VTR 401, Rutland 401	1951

Filer City: Packaging Co. of America

	45-ton	Plym		1967

Gaylord: Champion International

75	44-ton	GE	WM 75	1943

Green Oak: Weirton Steel

302	SW1500	EMD	P&LE 1568	1972
			Rebuilt by Peaker Services	

Hemlock: Hemlock Farmers Co-op

2	25-ton	GE	Memphis Terminal, U. S. Navy 65-00221	1942

Holland: Consumers Power

1	SL125	GE		1979

Holland: Louis Pandos Iron & Metal

16	20-ton	Dav	U. S. Army 2049	1941
50	35-ton	GE	Bucyrus-Erie	1952

Ishpeming: Ishpeming Steel Co.

7	65-ton	GE	Cliffs-Dow	1957

Lansing: Board of Water & Light

7232	SW900	EMD	GTW 7232	1956
799	GP7	EMD		1951

Lansing: Summit Steel

720	S1	Alco	Relco 720, C&NW 620	1948

Manistique: Manistique Pulp & Paper

16	45-ton	Whit	Union Carbide	1941

Marysville: Detroit Edison Co.

208	65-ton	GE		1942

Melvindale: Supreme Warehouse Corp.

999	35-ton	Plym	Toledo Port Authority	1943

Monroe: North Star Steel

864	S5	Alco	BN 864	1954

Monroe: Detroit Edison Co.

1, 2, 5	SD40	EMD		1970
7-11	U30C	GE		1972
13, 15, 16	SD40	EMD		1972
18-22	U30C	GE		1975

Newaygo: Westmac Inc.

7	44-ton	GE	Northwest Oklahoma	1946

Newport: Detroit Edison Co.

L1	15-ton	Bkvl		1957
	15-ton	Bkvl		1958

Oxford: Spencer Industries

47	65-ton	Whit	American Aggregates, C&BL, U. S. Army 7978	1943
6816	65-ton	Porter	American Aggregates, ARR 1106, U. S. Army 7313	1941

Ottawa Lake: Merchants Grain Terminal

62	S2	Alco	DC 62	1946

Petoskey: Penn Dixie Cement

1	25-ton	GE		1953

Pontiac: Chevrolet Motor Division

96	SW1001	EMD		1972

Pontiac: General Motors-Central Foundry

2	SW1200	EMD	MP 1297	1966
818	SW900	EMD		1953
1971	SW1001	EMD		1971
1972	SW1000	EMD		1972

River Rouge: RailCar Repair & Alterations Co.

	SW1	EMD	Welded Tube Co., Great Lakes Steel	

MICHIGAN (continued)

Saginaw: Oglebay Norton Co.

8	44-ton	Porter	Saginaw Dock, U. S. Navy 65-00188	1943	

Sunfield: American Bean & Grain

719	S1	Alco	Relco 0719	

Trenton: Detroit Edison Co.

216	SW1000	EMD		1967	Remote control

Trenton: McLouth Steel

55	SW1200	EMD	Peaker Services 55 MP	1964

Trenton: Detroit Edison Co.

212	SW1	EMD		1950

Warren: Jones & Laughlin Steel

81, 82	45-ton	GE	Rotary Electric.Steel	1948
83	45-ton	GE	Rotary Electric.Steel	1950
84	45-ton	GE	American Viscose	1947

White Pine: White Pine Copper Co.

200	S12	BLW	Copper Range 200	1951

Weirton: Weirton Steel Co.

312	SW1500	EMD	P&LE 1544	1972

MINNESOTA

Alpha: Cargill Grain Inc.

	35-ton	GE		1962

Babbitt: Reserve Mining

1200, 1201	SW8	EMD		1952
1211	SW9	EMD		1953
1212	SW1200	EMD		1962
1220-1225	SD9	EMD		1955-1959
1226-1232	SD18	EMD		1960-1962
1233-1236	SD28	EMD		1965

Duluth: American Steel & Wire

2	VO660	BLW		1942

Duluth: International Multifoods

	25-ton	GE		1950

Duluth: Michigan Elevator Co.

7350	65-ton	GE	Peavy Grain	
9044	S2	Alco	B&O 9044	1944

Fairfax: Fairfax Farmers Elevator

	23-ton	GE	

Guckeen: Blue Earth Farmers Co-op

651	65-ton	GE	U. S. Navy 65-00598, U. S. Army 7382	1943
	44-ton	GE		

Guckeen: Louis Dreyfus Co.

13	SW1	EMD		1941
1056	S4	Alco	Relco 1056	

Hoyt Lakes: Erie Mining

4210-4214	F9A	EMD	100-104	1956
4215, 4216	GP38	EMD	7250, 7251	1967
4217-4219	C420	Alco	7220-7222	1965
4220-4225	F9B	EMD	200-205	1956
4317	C424	Alco	500	1964
7200-7214	RS11	Alco	300-314	1956
7244	S12	BLW	Monongahela 415	1953
7247	S12	BLW	GN 26	1953
7248	S12	BLW	Monongahela 422	1954

Hoyt Lakes: LTV Steel

7215	RS11	Alco	BN 4190, NP 910	1958
7216	RS11	Alco	BN 4186, NP 906	1958
7222	C420	Alco		1965
7223	C424	Alco		1964
7241, 7243	S12	BLW		1955, 1956

Minneapolis: General Electric

1257	S6	Alco	SP 1257	1956

Minneapolis: North Star Steel

387	SW8	EMD		1953

Rosemount: CF Industries

712	S1	Alco	Relco 712	
9027	S2	Alco	B&O 9027	1943
9077	S2	Alco	B&O 9077	1948

Savage: Cargill Inc.

	65-ton	GE	U. S. Army 7278	1943

MINNESOTA (continued)
St. Hilaire: St. Hilaire Co-op Elevators

303	SW7	EMD	SLSF 303		1951

St. Paul: Alter Inc.

1747	GP9	EMD	C&NW 1747		1957

St. Paul: Cargill Grain Inc.

7	NW2	EMD	B&O 9549		1948
38	S6	Alco	U. S. Pipe & Foundry		1955
1004	SW1	EMD	Diesel Services		

St. Paul: Koppers Coke

3	S1	Alco			1946

St. Paul: Northern Waterways

100	44-ton	Dav	Martin-Marietta		1946

St. Paul: Pillsbury Inc.

1	80-ton	GE	U. S. Navy 65-00242		1944

MISSISSIPPI

Bayou Castole: Chevron Oil

117	NW2	EMD			

Bude: Shippers Car Line

3	45-ton	GE	Metro Track 1, Du Pont Chemical	1950

Columbus: Kerr-McGee

1	25-ton	GE	Pennsylvania Forge, Public Service Electricity & Gas	1947
3	35-ton	GE	BR&L	1953

Crosby: Masonite Corp.

2	25-ton	Plym	Swift Grain	1965

Greenville: Delta Oil Mills

1214	SW7	EMD	ICG 1214	1950

Hattiesburg: Hercules Powder

46	50-ton	Whit		1946

Jonestown: Delta Oil Mills

99	80-ton	GE	TVA 9	1952

Laurel: Masonite Corp.

1	S2	Alco		

Marks: Cook Industries

716	S1	Alco		

Moss Point: Dreyfus Grain Co.

9476	S4	Alco		

Moss Point: Mississippi Power Co.

63	C420	Alco	Mississippi Export 63	1965

Pascagoula: Chevron Chemical

117	NW2	EMD	U. S. Navy 65-00117	1942

Pascagoula: Jackson County Grain Terminal

13	NW1	EMD	L&N 13	1941
791	NW2	EMD	PC&N 791, UP 1052	1944

Pascagoula: Montgomery Maintenance

791	NW2	EMD	Tie Plant	

Pascagoula: Koppers Co.

3101	15-ton	Porter		24" gauge
3102	23-ton	GE		24" gauge

Vicksburg: Vicksburg Chemical Co.

1	70-ton	GE	East St. Louis Junction	

Vicksburg: Cedar Chemical Co.

9054	70-ton	GE		1948
9101	SW9	EMD	ICG 1240, Mississippi Central 205	1953

Wiggins: International Paper

9	25-ton	GE	Chicago Bridge & Iron	1947

MISSOURI

Clarkesville: Dundee Cement

	50-ton	GE		1950

Crystal City: Pittsburgh Plate Glass

911	80-ton	Whit		1952
1247	S6	Alco	SP 1247	1956

Crystal City: USS Agri Chem

870	SW1	EMD	MILW 870	1940

Glover: American Smelting & Mining

1	S12	BLW	CRI&P 758	1953

Kansas City: Armco Steel Co.

12	SL115	GE		1976

Kansas City: Cedrite Technologies

9428	SW900	EMD	B&O 9428	1955

Kansas City: Continental Grain Co.

2401	35-ton	Plym	MP 2401	1961

Kansas City: Intercon

1	80-ton	Whit	Kansas City Power & Light	
1237	SW9	EMD	MP 1237	1951

MISSOURI (continued)
Kansas City: Mid America Car Co.

79	S6	Alco	Kansas City Power & Light, SP 1279	1956

Kansas City: Watco

9, 10	SW9	EMD	

Mexico: Missouri Farmers Association

420	S1	Alco	LIRR 420	1949
	50-ton	Plym	Republic Steel 824, Plymouth demonstrator	1957

Montrose: Kansas City Power & Light

91	SW1	EMD	BN 91	1940

Montrose: Peabody Coal Co.

801	SW8	EMD	Poplar Ridge Coal	1951

North Kansas City: Shippers Car Line

7	80-ton	GE	

Sikeston: City of Sikeston

1	SL85	GE	1980

Sikeston: Sikeston Power

1	110-ton	GE	1980

Springfield: City Utilities

683	SW1	EMD	McLouth Steel 4	1951

St. Joseph: St. Joe Zinc

6	SW7	EMD	MP 1205	1950

St. Joseph: Quaker Oats

	45-ton	GE	Lange-Stegmann Fertilizer	
1209	SW9	EMD	TRRA 1209	1952
1216	SW9	EMD	TRRA 1216	1952
1217	SW9	EMD	TRRA 1217	1952
1232	SW1200	EMD	TRRA 1232	1955
1236	SW1200	EMD	TRRA 1236	1965

Thomas Station: Associated Electric Co-op

73	70-ton	GE	East St. Louis Junction	1949
425	C424	Alco	N&W 425, N&W 3905	1964

MONTANA

Billings: Billings Grain Terminal

84	SW1	EMD	BN 84, CB&Q 9139	1939

Billings: Montana Power

101	S4	Alco	Minnesota Transfer 101	1951

Butte: Montana Resources

111	NW2	EMD	GN	1949
1010	SW900	EMD	Pickering Lumber	1957

Columbia Falls: Anaconda Aluminum

100	SW1200	EMD	Tooele Valley 100	1955
900	NW3	EMD		

East Helena: Asarco

8	SW1200	EMD	SOO 2123	1955

East Helena: Watco

55	SW1	EMD	Henningsen Cold Storage

Laurel: Long Construction Co.

1101	110-ton	GE	Foley Brothers	1929

Miles City: PLM Corp.

423	C424	Alco	BN 4004, SPS 104	1968
424	C415	Alco	Brandon Corp., CRI&P 424	1968
509	S1	Alco	WP 509	1942

Miles City: Trancisco

143	SW7	EMD	BN 143, GN 11	1950

Silver Bow: Stauffer Chemical

101	GP7	EMD	BA&P 101	1952

Silver Bow: Port of Montana

1166	S4	Alco		
7902	SW900	GMD	CN 7902	1953

NEBRASKA

Alliance: PLM Corp.

420	S6	Alco	SP 1245	1956

Southwest Electric Power

3	88-ton	GE	Diamond Shamrock 22, Winifrede RR	1957

Avon: Agri-Industries

1269	SW1	EMD		1953

Beatrice: Co-op Grain

707	S1	Alco	Relco 707

Beatrice: Southeastern Nebraska Co-op

1	S3	Alco	Relco

Fremont: Consolidated Blenders

	25-ton	Plym

Grand Island: Platte Generating Station

1	S6	Alco	SP 1206	1955

NEBRASKA (continued)
Grand Island: Cornhusker Railcar Service

36	SW9	EMD	BAR 36, P&LE 1244	1951	

Hartwell: Cargill Grain Inc.

1990	NW2	EMD		1948	

Kearney: Peavy Grain

352	RS1	Alco	SOO 352	1954	
1281	SW600	EMD	C&NW 1281	1954	

Norfolk: Nucor Steel

69	DS44-1000	BLW	SCL 69	1950	
864	SW1	EMD	MILW 1186	1940	

Omaha: American Smelting & Refining

1	10-ton	Plym		1967	36" gauge

Omaha: Omaha Public Power

1200	S6	Alco	SP 1200	1955	

Omaha: Southwest Electric Power

4	SL136	GE	C&NW 1199, BN 1101	1978	

Omaha: Unarco Transportation Equipment

	50-ton	GE	Union Carbide, Weirton Steel	1950	

Ravenna: Cargill Grain Inc.

89	NW2	EMD	SP 1923	1941	

South Omaha: Scoular Grain Co.

	S2	Alco	Wilson Co.		

Sidney: Nebraska Western Technical College

60A	FP7A	EMD	MILW 60A	1950	
5057	U25B	GE	MILW 5057	1965	

Sidney: Sidney Warehousing & Terminal

244	S6	Alco	SP 1222	1956	

Superior: Ideal Cement

5	45-ton	Whit		1943	

Sutherland: Nebraska Public Power District

100	S6	Alco	SP 1239	1956	

NEVADA
Mercury: U. S. Atomic Energy Commission

1502	MRS1	Alco	U. S. Navy 65-00549, U. S. Army 2116	1953	
1503	MRS1	Alco	U. S. Navy 65-00550, U. S. Army 2117	1953	

NEW JERSEY
Atlantic City: Atlantic City Electric

8960	NW2	EMD	PDC	1948	
	15-ton	Bkvl		1948	

Bayonne: American Cyanamid

4	65-ton	GE		1946	
7	44-ton	GE	Terra Chemical, U. S. Navy 65-00346	1943	

Bayonne: East Jersey Terminal

18	65-ton	GE		1948	
19	80-ton	GE	BR&L 157, U. S. Steel	1948	

Bayonne: River Terminal Development Co.

1	44-ton	Dav			
2	80-ton	GE	CNJ		

Camden: Linco Stone & Supply

106	S2	Alco	Jersey Southern 106, Wyandotte Terminal 106	1948	

Camden: Port Authority Transit Corp.

404	60-ton	BLW	Niagara Junction 9	1937	Electric
405	60-ton	BLW	Niagara Junction 8	1938	Electric

Camden: Public Service Electric & Gas

D1	45-ton	GE		1947	

Carneys Point: E. I. Du Pont

105	80-ton	GE		1946	
106, 107	SW1000	EMD		1967	
108	SW1	EMD		1941	
109	NW2	EMD	MP 1027, KO&G 1001	1949	
110	NW2	EMD		1949	

Deepwater: Atlantic City Electric Co.

2	25-ton	GE		1958	
1200	SW7	EMD			

East Camden: Transitank Corp.

	45-ton	Porter	Solvay Process	1943	

Elizabeth: Vulcan Materials

1	25-ton	GE	M&T Chemical, U. S. Army 7757	1942	

Florence: Griffin Pipe Co.

1	8-ton	Bkvl		1954	
2	45-ton	GE		1945	

203

NEW JERSEY (continued)

Gloucester City: Holt Hauling & Warehousing

	50-ton	GE	U. S. Navy 65-00125	1941
	50-ton	Vulcan	Westinghouse 1	1942

Harrison: Guyon Pipe

992	44-ton	GE	Coors 992, Alexander 2	1946

High Bridge: Taylor Wharton Iron & Steel

7	50-ton	GE		1947

Kearny: River Terminal Development Co.

4	80-ton	GE	Eastern Gas 1, LS&BC 7	1941
5	44-ton	Dav	U. S. Army 1226	1953
6	80-ton	GE	MS Test Site, U. S. Army 7441	1943

Kearny: Tropicana

98	70-ton	GE	L&N 98	1948
	Alco 6-251 engine			
5281	NW2	EMD	C&O 5281	1949

Linden: Linden Chlorine Products

14	S4	Alco	M&E 14	1952
43	RS1	Alco	CNJ	1950

Mahwah: American Brake Shoe

200	SW1	EMD	UTLX 200, CSL 200	1942

Merchantsville: Cook Industries

	65-ton	Whit	Quality Supply, U. S. Navy 65-00219	1943

Millville: Armstrong Cork

	30-ton	Plym	Warwick 103, B&M 101	1938

Millville: Kerr Glass Co.

5	45-ton	Plym	Armstrong Cork	1968

Newark: Naporano Iron & Metals

8319	S12	BLW	PC 8319	1952

Parlin: E. I. Du Pont

	6-ton	Vulcan		1948	36" gauge

Perth Amboy: Raritan River Steel

8629	SW900	EMD	CR 8629, NYC 9629	1954

Phillipsburg: Ingersoll Rand

90, 91	45-ton	GE	U. S. Army 7927, 7928	1944

Phillipsburg: Jersey Central Power & Light

1	45-ton	GE		1941
2	35-ton	Plym		1949

Phoenix: National Lead

	25-ton	GE		1951

Port Newark: Koppers Co.

13	4-ton	Whit		1948
9	15-ton	Whit		1948

Ridgefield: Public Service Electric & Gas

	100-ton	GE		1958

Roebling: Roebling Steel Co.

3-5	45-ton	GE		1947
7	100-ton	GE	Wickwire Spencer	1960
8	80-ton	GE	CF&I Steel, Wickwire Spencer	1947

Sayreville: New Jersey Steel Structure Co.

3	35-ton	GE		1959
7	65-ton	GE		

South Amboy: McCormick Sand & Gravel

1268	S4	Alco	NMI	1951

South Camden: South Jersey Port Corp.

1199	35-ton	GE	New York Shipbuilding	1952
3111	35-ton	GE	New York Shipbuilding	1958

Trenton: General Electric

1	25-ton	GE		1948

NEW MEXICO

Carlsbad: Duval Corp.

86	S2	Alco	SB 86	1950

NEW YORK

Albany: Albany Port Authority

1	65-ton	GE	Wyandotte Terminal	1941
	Stored unserviceable			
3	RS3	Alco		
11	SW900	EMD	B&O 9405	1955
12	SW900	EMD		
13	SW900	EMD	Cargill Grain Inc.	

Alsen: Marquette Cement

4	80-ton	GE	U. S. Steel	1952

Bainbridge: New York State Electric & Gas

3	25-ton	Plym		1945

Binghamton: New York State Electric & Gas

2	25-ton	Plym		1945

NEW YORK (continued)
Buffalo: Hanna Furnace
| 17 | SW900 | EMD | | 1954 |

Buffalo: International Milling
| | 45-ton | GE | Lehigh Portland Cement | 1943 |

Castleton-on-Hudson: Fort Orange Paper Co.
| | S1 | Alco | LI 417 | 1949 |

Cortland: Wickwire Bros. Inc.
| 7 | 25-ton | GE | | 1940 |

Stored unserviceable

Deferiet: Champion Paper Co.
| 7249 | GP9 | EMD | St. Regis Paper | 1956 |
| 1 | RS1 | Alco | St. Regis Paper | 1950 |

Depew: Dresser Industries
| P3 | 45-ton | GE | | |

East Syracuse: Speno Rail Service
| 4 units | GP38R | EMD | Ex-Conrail | 1984 |

Rebuilt to cab units
| 4 units | F40PH-2M | EMD | | 1982-1985 |

Units are renumbered as assignments change.

Hornell: VIC Industries
| 404 | RS1 | Alco | St. Regis Paper | 1949 |

Hudson: Independent Cement
| | S2 | Alco | UAC, McKeesport Connecting | 1948 |

Jamesville: General Crushed Stone
| 5, 6 | 80-ton | GE | Allied Chemical 5, 6 | 1950, 1951 |

Leroy: General Crushed Stone
| | 65-ton | GE | Whitten Machinery 3 | 1943 |

Leroy: Genesee Leroy Stone Corp.
| 27 | 44-ton | GE | | |

Milliken: New York State Electric & Gas
| 8 | S2d | Alco | LI 453 | 1949 |

New York: New York Dock Railway
| 1026 | NW2 | EMD | Southern 1026 | 1946 |
| 1044 | NW2 | EMD | Southern 1044 | 1947 |

New York: New York Transit Authority
10, 11	35-ton	GE		1959
50-82	45-ton	GE		1965-1977
71-82	45-ton	GE		
EL1-EL10	SL50E	GE		1983 Electric
L883-L902	SL50	GE		1983

New York: Staten Island Rapid Transit Operating Authority
| 407 | S1 | Alco | LI 407 | 1946 |

Newburgh: Steel Style Shipyard
| | 25-ton | Whit | Atlantic Steel, U. S. Navy 65-00330 | 1952 |

Olean: Dresser Industries
| | 45-ton | GE | U. S. Army 7210 | 1941 |

Port Ivory: Proctor & Gamble
| 36 | S4 | Alco | | |

Port Ivory: Atlantic Cement Co.
| | 25-ton | BLW | | |
| | 45-ton | GE | | 1961 |

Rochester: Eastman Kodak
8	SW1000	EMD		1968
10	MP15AC	EMD		1981
1521	GP7	EMD		1952

Rochester: Rochester Gas & Electric Co.
| | 45-ton | GE | | 1941 |

Roseton: Central Hudson Gas & Electric
| | 25-ton | GE | | 1951 |
| | 65-ton | GE | Briggs Manufacturing Co. | 1951 |

Schenectady: General Electric
| 20, 21 | SL85 | GE | | 1976, 1978 |

Solvay: Allied Chemical Co.
1A	S4	Alco		1955
2A	S4	Alco	D&H 3042	1950
2B	S4	Alco	B&O 9085	1956
4A	S4	Alco	C&O 5109	1955
5A	S4	Alco	NY	
	S6	Alco	SP 1231	1956

Solvay: Marley Junk Yard
| | 25-ton | GE | | |

Somerset: New York State Electric & Gas
| 1 | SL144 | GE | | 1983 |

Syracuse: Crucible Steel
| 40 | 45-ton | GE | | 1948 |

Tonawanda: Dunlop Tire Co.
| | 50-ton | GE | | |

Watervliet: Allegheny Ludlum Steel
| 156 | 65-ton | GE | Reynolds Aluminum, U. S. Army 7262 | 1942 |
| 102 | 80-ton | GE | | 1970 |

NORTH CAROLINA

Allen: Duke Power Co.

90	45-ton	GE		1956

Aurora: Texas Gulf Sulphur

101	70-ton	GE	Laurinburg & Southern 101	1947
114	SW1	EMD	Laurinburg & Southern 114	1948
146	SW900	EMD		1956

Beaufort: Grace Agricultural Chemical

2900	80-ton	Whit		1941

Belews Creek: Duke Power Co.

104	80-ton	GE		1973
105	80-ton	GE		1973

Bridgeton: Frit Car Service

1	44-ton	GE	Southern 6010	1945
	35-ton	GE		

Brook Cove: R. J. Reynolds

1	65-ton	GE		

Canton: Champion Paper Co.

110	NW2	EMD	BR&L 183, DT	1947
112	S2	Alco	SCL 112, ACL 31	1944
2081	S4	Alco	NW 2081	1953
6050	S2r	Alco	WM 141	1944
6056	SW1001	EMD		1979

Castle Hayne: Diamond Shamrock Chemical

101	45-ton	GE		1955
6029	45-ton	GE	Whitehead & Kale	1950

Cliffside: Duke Power Co.

60	25-ton	GE		1949
103	80-ton	GE		1972

Clover: Duke Power Co.

113	25-ton	GE		1950

Cove: R. J. Reynolds

2	65-ton	GE	U. S. Navy 65-00548, U. S. Army 7876	1943

Cowans Ford: Duke Power Co.

112	25-ton	GE		1950

Darlington: Nucor Steel

1	DS44-1000	BLW	SCL 66, SAL 1454	1950
2	DS44-1000	BLW	SCL 93, K&IT 54	1953
3	S12	BLW	SCL 222, SAL 1481	1953

East Laurinburg: Libbey Owens Ford Glass Co.

108	35-ton	Plym	LS 108, Magor Car Co.	1952

Eden: Duke Power Co.

10	25-ton	GE		1950
50	25-ton	GE		1949

Enka: Vulcan Materials

7	65-ton	GE	U. S. Army 7181, U. S. Navy 65-00058	1943

Goldsboro: Carolina Power & Light

1	35-ton	GE		1959

Kernersville: United Metal Recyclers

1	80-ton	GE	North Carolina Ports L-5, U. S. Army 7896	1943
2	80-ton	GE	North Carolina Ports L-6, U. S. Army 7363	1941

Lillington: Becker Sand & Gravel

10	S2	Alco	N&SS 1012	1942
12	S2	Alco		1949
14	S4	Alco	Minnesota Transfer 105	1951
15	S4	Alco	South Buffalo 111	1953
16	S4	Alco	C&O 5102	1953
17	S4	Alco	C&O 5112	1953
18	S4	Alco	N&W 3322	1953
9059	S2	Alco	Savannah State Docks	1945

Lumberton: Carolina Power & Light

5	35-ton	GE		1964

Mocksville: R. J. Reynolds

3	45-ton	GE	Ethyl Corp.	1950

Moncure: Carolina Power & Light

2	35-ton	GE		1951

Morehead City: North Carolina Ports

4	65-ton	GE	GE 2001	1942
1036	NW2	EMD	Southern 1036	1946

Morganton: Great Lakes Carbon Co.

	45-ton	GE	BR&L 160, Yancey 1	1955

Mount Airy: North Carolina Granite Co.

2	50-ton	GE		1949

Mount Holly: Duke Power Co.

11	45-ton	GE		1955

Navassa: Swift & Co.

	VO1000	BLW	SCL 95	1944

Navassa: USX

76	S3	Alco	DRI&NW 1	1951

NORTH CAROLINA (continued)

New Hill: Carolina Power & Light

3	25-ton	Whit		1953

Riegelwood: Federal Paperboard Co.

	45-ton	GE		1950

Roxboro: Carolina Power & Light

63	RS1	Alco	WT 63	1951
4	SL110	GE		1979

Salisbury: Fiber Industries

2	45-ton	GE	Goodpasture Inc., U. S. Army 7048	1942

Skyland: Carolina Power & Light

111	S2	Alco	L&S 111	1950

Southport: Pfizer Inc.

	45-ton	GE	Electric Boat Co.	1942

Spencer: Duke Power Co.

20	25-ton	GE		1951
40	40-ton	GE		1960

Stoneville: Vulcan Materials

8000	RS1	Alco	GM&O 1130	1945
8001	RS1	Alco	GM&O 1132	1944

Terrell: Duke Power Co.

101	80-ton	GE		1966

Wilmington: Almont Shipping Co.

110	S6	Alco		1955
1032	GP7	EMD	North Carolina Ports 1032, Georgia 1032	1952

Wilmington: Grace Agricultural Chemical

	45-ton	Whit	U. S. Navy 65-00161	1944
	80-ton	Whit	U. S. Navy 65-00102	1944

Wilmington: Horton Iron & Metal

1	50-ton	Porter	Mechanical Pipe	
2	50-ton	Porter	Ohio Stone Co., U. S. Navy 65-0	1942

Wilmington: Kaiser Agricultural Chemical Co.

14	44-ton	GE	PA&M 13, NKP 90	1949
72	70-ton	GE	U. S. Air Force 7930, U. S. Army 7930	1943

Wilmington: Southern Wood-Piedmont

1	50-ton	Whit	U. S. Navy 65-00270	1943

Winston-Salem: Corn Products Co.

69	SW1000	EMD	Corn Industries Co.	1969

Winston-Salem: R. J. Reynolds

12	80-ton	GE	MD&W	1949

NORTH DAKOTA

Bowbells: Farmers United Co-op Elevator

2	SW1	EMD	St. Joseph Terminal 2	1950

Churchs Ferry: Farmers Co-op

8414	SW1	EMD	B&O 8414	1940

Dickinson: Cargill Grain Inc.

1	50-ton	GE		

Enderlin: National Sun Industries

1	50-ton	GE	Dayton Power & Light	1959

Fargo: North Dakota Wheat Growers

865	SW1	EMD	DRI&NW 865, MILW 865	1940

Ladish: Ladish Malting Co.

LM2	SW9	EMD	BN 161, CB&Q 9270	1951
LM3	SW1200	EMD	N&PBL 102	1956

Langdon: Cargill Grain Inc.

1	NW2	EMD		

McVille: Sheyenne Valley Grain Co-op

1214	SW1200	EMD	SOO 2124	1955

Minot: Cargill Grain Inc.

856	NW2	EMD	BN 586, NP 99, NYO&W 115	1948

Mooreton: Southeast Grain Co.

9534	NW2	EMD	B&O 9534	1949

Spiritwood: Ladish Malting

9	65-ton	GE	BRW 7079, U. S. Army 7079	1944
LM1	SW1	EMD	BN	

OHIO

Arcanum: Continental Grain Co.

RE604	HH660	Alco	MILW 980	1940

Briar Hill: Syro Steel

23	S4	Alco	Youngstown Sheet & Tube	

Campbell: North Star Steel

1010	S1	Alco	VEPCO	

Canton: Timken Inc.

1985	S4	Alco	B&O 9100	1957
4845	SL85	GE		1977
5703	25-ton	GE	METR 4, U. S. Navy 65-00390	1942
5704	25-ton	GE		

OHIO (continued)
Canton: Timken Inc. (continued)

5705	25-ton	GE	Canadian Stone 12	1952
5706	25-ton	GE	G&U 10, Thurso & Nation 6	1952
5911	S2	Alco	LI 453	1949
7948	80-ton	GE		1953
8628	SW8	EMD	CR 8628	1953

Carey: National Lime & Stone

1	50-ton	GE	Weirton Steel 88	1948
2	20-ton	Plym		
3	20-ton	Plym		
4	35-ton	Plym		
74	45-ton	GE	American Viscose	1953

Carey: Wyandot Dolomite

	35-ton	Plym

Carthage: Fredrick Steel Co.

7097	45-ton	GE	U. S. Air Force 7097	1942

Cincinnati: Cargill Grain Inc.

109	S4	Alco	IHRC 108	1951

Cincinnati: Hatfield Terminals

1	45-ton	GE	Jeffboat

Clarksfield: Sunrise Grain

1	30-ton	Plym	Dayton Malleable Iron	1954

Cleveland: Atlas-Lederer Co.

2	S1	Alco	Studebaker Corp.	1945
9190	NW2m	EMD	CR 9190	1948

Cleveland: Cleveland Electric Illuminating

15	SL110	GE		1984
18	SL110	GE		1978
100-108	GP38-2	EMD		1975

Cleveland: Cleveland RTA

3	65-ton	GE	General Electric, U. S. Navy 65-00056	1943

Cleveland: Federal Steel & Wire

	65-ton	GE	Republic Steel	
360	25-ton	GE		1959
361	25-ton	GE		1959

Cleveland: Interstate Terminal & Warehouse Co.

343	80-ton	GE	Republic Steel 343	1937

Cleveland: Zen-Noh Grain

1	S4	Alco

Coalburg: Midwest Steel & Alloy

8508	SW1	EMD	CR 8508, PRR	1949

Columbus: Buckeye Steel Casting

6	SW1	EMD	Ford Motor Co.	1947
9	SW900	EMD	B&O 9404	1955
7	SW600	EMD		1955
10	SW900	EMD	B&O 9403	1955

Columbus: Continental Grain Co.

1215	SW9	EMD	TRRA 1215	1952
3	50-ton	Dav	Glidden Co.	1955
	Stored unserviceable			

Cumberland: Muskingum Electric

100	E50C	GE		1968 Electric
200	E50C	GE		1968 Electric
	100-ton	GE		1967

Fairborn: Southwestern Portland Cement

18	SW1200RS	GMD	CN 1255	1956

Findlay: Blanchard Valley Farmers Co-op

1	35-ton	Plym

Grafton: Academy Iron Co.

	25-ton	GE

Grafton: Landmark Grain Co.

2	45-ton	GE		1945

Greenfield: Blue Rock Transportation Co.

1235	65-ton	Whit	U. S. Army 8122	1944

Huron: Amber Milling

1060	S2	Alco	Relco 1060	1950

Huron: Pillsbury Inc.

154	50-ton	GE

Hamilton: Great Miami Inc.

1371	SW1	EMD		1946
1372	SW7	EMD		1950

Ivorydale: Proctor & Gamble

1828	GP9m	EMD	BN 1828	1954

Kenton: Landmark Grain Co.

66	44-ton	GE	Ford Motor Co.
	S4	Alco	Ford Motor Co.

Lima: Proctor & Gamble

	50-ton	GE	Cincinnati Milling	1948

OHIO (continued)

Lordstown: Transco

101	RS1	Alco	Morton Salt 101, ATSF 2397	1947
333	SW1	EMD	CR 8597	1948

Lynn Tipple: Cravatt Coal Co.

500	70-ton	GE	North American Coal, Kellys Creek Coal	
501	70-ton	GE	North American Coal, Kellys Creek Coal	
	35-ton	Plym	Ohio Ferro Alloy, Griscom-Russell	1941

Marion: Central Soya

2	25-ton	Plym		
4	25-ton	Plym		
9189	NW2m	EMD	CR 9189	1941

Marion: Union Tank Car Co.

9752	S4	Alco	CR 9752	1953
12226	SW1	EMD		

Masury: General American Transportation

6575	SL85	GE		1978
6576	SL85	GE		1981

Middletown: Armco Steel Co.

162-165	SW1001	EMD		1978-1979
168, 169	SW1001	EMD		1981

Maumee: The Andersons

900203	50-ton	GE	
900210	25-ton	Plym	
900211	25-ton	Plym	
902420	35-ton	Plym	
902451	30-ton	Plym	

Monroeville: Monroeville Grain Co-op

| 1 | 25-ton | GE | Western Massachusetts Electric, Connecticut Light & Power | |

North Bend: Pillsbury Inc.

| 3 | 23-ton | GE | Lehigh Cement | 1941 |

South Point: South Point Ethanol

1	GP9	EMD	N&W 890	1959
2	GP18	EMD	N&W 932	1960

Toledo: The Andersons

| 18-ton | Plym | Larson Bros. Sand 3 | 1952 |

Toledo: Cargill Grain Inc.

| 10 | 44-ton | GE | PC&N 2, AS 629 | 1948 |

Toledo: Koppers Co.

63	S1	Alco	Minnesota Transfer 63	1946
64	S1	Alco	Minnesota Transfer 64	1946

Troy: Continental Grain Co.

| RE601 | HH660 | Alco | MILW 981 | 1940 |

Woodville: Martin-Marietta

| 504 | SW1 | EMD | Relco 504 | 1949 |

Woodville: Steetley Resources Co.

| 45-ton | GE | Ohio Lime, U. S. Air Force 8536, U. S. Army 8536 | 1944 |

Worthington: Allis Chalmers

| 45 | 35-ton | GE | |

Youngstown: McDonald Steel Corp.

| O777 | S2 | Alco | U. S. Steel 73 | |

Youngstown: Syro Steel Co.

12211	S2	Alco	
12079	S1	Alco	
12080	S4	Alco	

OKLAHOMA

Muskogee: Port of Muskogee

5683	S12	BLW	Metro Stevedore, SP 2154	1953
	45-ton	GE		

Sand Springs: Armco Steel Co.

| 4802 | SW1 | EMD | CRI&P 4802 | 1940 |

Sand Springs: Armco Steel Co., Sheffield Division

1	80-ton	GE	U. S. Navy 65-00026	1943
2	25-ton	GE		1956
3	80-ton	GE	Laclede Steel 13, U. S. Army 7387	1943
4	80-ton	GE	Laclede Steel 14	1943

Tulsa: Port of Catoosa

100	NW2	EMD	BN 555	1949
101	65-ton	GE		

Tulsa: Port of Tulsa

1	65-ton	GE	U. S. Air Force 1408, U. S. Army 1408	1941
2	65-ton	GE	U. S. Army 7016	1942

Williams: Agrico

2	NW2	EMD	UP 1022	1941
7	65-ton	GE	U. S. Army 7169	1954

OREGON

Gaston: Stimpson Lumber Co.

| | 25-ton | GE | Oregon Technical Institute | 1941 |

Glendale: Gregory Timber Co.

| | 45-ton | GE | Robert Dollar Co., U. S. Air Force 7488, U. S. Army 7488 | 1942 |

Grants Pass: Southern Oregon Plywood Co.

| 1 | 12-ton | Plym | Pacific Portland Cement | 1927 |

Lake Oswego: Oregon Portland Cement

| 852 | 25-ton | GE | | 1956 |

Linnton: Georgia-Pacific Corp.

| 1 | 45-ton | GE | Omaha Power | 1952 |

Milwaukie: Samuels Steel Products Inc.

18	18-ton	Plym	Alaska Steel, Kaiser Corp.	1948
20	20-ton	Dav	Alaska Junk Co., U. S. Army 7706	1941
25	25-ton	GE	Alaska Junk Co., Columbia River Construction	1942

Newberg: Publishers Paper Co.

1	80-ton	GE	U. S. Air Force 7265, U. S. Army 7265	1943
3	80-ton	GE	Spaulding Pulp	1952
4	80-ton	GE	Hercules Powder, U. S. Army 7360	1941
51	DS44-000	BLW	Alaska Terminal	1940
777	SW9	EMD	CRI&P 777	1953

Nyssa: Amalgamated Sugar

| 100 | 45-ton | GE | | 1948 |
| 500 | 25-ton | GE | | 1949 |

Portland: Columbia Grain Co.

1029	S2	Alco	Schnitzer Steel, USMC, U. S. Army	1941
1206	SW1200	EMD	Relco 1206, Tampa Electric	1963
	NW2	EMD	BN 487	1945

Portland: International Terminal

| 1548 | S1 | Alco | Schnitzer Steel | 1945 |

Portland: McCormick & Baxter Creosoting Co.

| 1 | 25-ton | GE | International Paper | 1942 |

Portland: Oregon Steel Mills—Gilmore Steel

650	50-ton	Porter	U. S. Navy 65-00113	1944
700	45-ton	GE	Inter City Materials, U. S. Army 7795	1940
765	T6	Alco	Portland Terminal.	1968

Portland: Portland Public Docks

1	25-ton	GE	Maritime Commission	1942
2	44-ton	GE	Howard Lumber, GN 5200	1940
3	45-ton	GE	U. S. Air Force 7499, U. S. Army 7499	1943
404	45-ton	GE	U. S. Marine Corps 112346	1943

Portland: Schnitzer Steel Products

1	25-ton	GE	Inter City Materials, Kaiser Corp.	1942
2	25-ton	GE	Inter City Materials, ARR 51	1941
1531	44-ton	GE	St. Regis Paper, Chattanooga Traction	1941
1546	45-ton	GE	Willamette Western, U. S. Air Force 7316, U. S. Army 7316	1942
1547	45-ton	Plym	American Ship Dismantling, U. S. Navy 65-00208	1945
1548	S1	Alco	International Terminal 603, NP 131, 603	1945
1549	50-ton	GE	American Ship Dismantling, Asarco	1952
1566	45-ton	Whit	Cargill Terminal, USMC, U. S. Navy	1944
2457	45-ton	GE	American Ship Dismantling, U. S. Army 7209	1941

Riddle: Hanna Nickel Smelting

| 551 | 50-ton | GE | Bechtel Corp. | 1953 |
| 553 | 65-ton | GE | QNS&L 90, East Erie Commercial | 1948 |

Springfield: U. S. Railway Manufacturing Co.

| 1 | 45-ton | Whit | Oregon-Washington Plywood, U. S. Navy 65-00153 | 1945 |

The Dalles: J. H. Baxter Co.

| | 25-ton | GE | UP MW8 | 1967 |

The Dalles: U. S. Corps of Engineers

| 75 | 25-ton | GE | U. S. Army | 1942 |

Tigard: Fought & Co. Inc.

| | 8-ton | Bkvl | Portland Dock Yard, U. S. Navy 65-00287 | 1943 |

Tillamook: Port of Tillamook Bay

| 110 | 80-ton | GE | U. S. Navy 65-00285 | 1943 |
| 111 | 80-ton | GE | LP & N 80 | 1943 |

Toledo: Georgia-Pacific Corp.

| 2 | 50-ton | GE | Johnson Lumber | 1951 |

Toledo: Simpson Timber

| 1202 | SW1200 | EMD | BN 259 | 1959 |

PENNSYLVANIA

Aliquippa: Jones & Laughlin Steel

1	60-ton	Plym		1960	
2	65-ton	Whit		1944	
2-6	80-ton	GE		1956-1960	
6, 7	65-ton	Whit	U. S. Army 8413, 8417	1944	
7-9	50-ton	GE		1947, 1950	
10, 11	25-ton	GE		1946, 1948	
32	110-ton	GE		1966	
101	SW8	EMD	NOPT 2	1953	
102, 103	SW1000	EMD		1967, 1969	
107	S4	Alco	B&O 9092	1956	
146-148	S2	Alco	MCRR 146-148	1948-1949	
801	SW8	EMD		1953	

Aliquippa: Koppers Coke

1, 2	44-ton	GE		1945

Allenport: Sharon Steel

5	SW900	EMD		1955

Allentown: Lehigh Structural Steel

	45-ton	GE		1948

Allentown: Lone Star Cement Co.

	45-ton	GE	Greenville Manufacturing, American Aggregates	1952

Ambridge: Armco Steel Co.

1	100-ton	GE	National Supply 1	1944
B77	100-ton	GE		1960
E107, E108	S1	Alco	Armco Middletown Works	1950

Ambridge: Babcock & Wilcox

8705	NW2	EMD	P&LE 8705	1947
8743	NW2	EMD	P&LE 8743	1949

Ambridge: USX

18, 19	65-ton	GE		1942
39	65-ton	Porter		

Apollo: Apollo Fabricators

	MDT	Plym		

Annville: Wimpy Minerals Co.

3606	RS11	Alco	W&W, CV, DW&P 3606	1956

Bakersfield: EEC

21, 22	SL85	GE		1980

Bethlehem: Bethlehem Steel

36" GAUGE

5, 6	50-ton	Whit		1940	
7	20-ton	Whit		1940	
13, 15, 17, 18	50-ton	Whit		1941-1943	
19	10-ton	Whit		1940	
20	25-ton	Whit	OH 1	1940	
21	50-ton	Whit	OH 3	1941	
25A-27A	50-ton	Whit		1952, 1953	Cow
25B-27B	50-ton	Whit		1952, 1953	Calf
28A, 29A	50-ton	Plym		1956, 1957	Cow
28B, 29B	50-ton	Plym		1956, 1957	Calf
34A-37A	50-ton	Whit		1940-1941	Cow
34B-37B	50-ton	Whit		1940-1941	Calf

STANDARD GAUGE

1-3	SC	EMD		1937	
4	SW	EMD		1937	
5, 6	SC	EMD		1937, 1936	
7	SW8m	EMD		1953	
8-10	SWu	EMD		1937	
11	SW900	EMD		1955	
22	20-ton	GE	121	1949	
60, 61, 65	SW1	EMD	C&BL 60, 61, 65	1950	Remote control
66	65-ton	Whit	BSCX 142, U. S. Army 8440	1944	Remote control
70	SW1	EMD	PC 85062, PRR 9206	1949	Remote control
75	SW1	EMD	P&BR 156, PB&NE 214	1940	Remote control
80	SW1	EMD	RDG 19	1940	Remote control
401, 402	40-ton	Whit		1956	Remote control
902, 905, 906, 909, 910, 913	SW900	EMD		1959	
1245	SW9	EMD		1951	

Brackenridge: Allegheny Ludlum Steel

1	25-ton	GE		1948	
3	25-ton	GE		1956	
12	65-ton	Whit	U. S. Navy	1943	
14-18	65-ton	GE		1950-1953	
19	80-ton	GE	Ford 500	1946	
20	65-ton	GE	Consumers Power	1949	

PENNSYLVANIA (continued)
Braddock: USX

102	SW8	EMD	Donora Southern 807	1951
109	SW8	EMD		1951
110	S2r	Alco	P&LE 8546	1949
114	NW2	EMD		
1	Slug	URR	Union Railroad Alco S2 528	
2	S2	Alco	URR 534	1948

Brownsville: Hillman Barge Co.

69	8-ton	Plym	U. S. Army 818	1937
	25-ton	GE	Merry Bros. D5, Baker Wood Preservative	1951

Brownsville: USX

1	SD38-2	EMD		1976

Burnham: Kovalchick Salvage Co.

614	25-ton	GE		
	S12	BLW		

Burnham: Standard Steel

368	SW8	EMD	CR 8697, EL 368	1952
811	SW8	EMD	CRI&P 811	1950
8590	SW1	EMD	CR 8590	1948
8897	SW7	EMD	CR 8897	1950

Butler: Armco Steel Co.

E77	100-ton	GE		1944

Birdsboro: Warner Corp.

1	80-ton	Whit		1953

Cambria: Cambria Slope Prep Plant

7	SW9	EMD	C&I 34	1951

Canonsburg: McGraw-Edison

1	65-ton	Whit	Westinghouse Air Brake	1948
2	45-ton	Porter	U. S. Navy 65-00024	1942

Carnegie: Columbia Steel & Shafting

	45-ton	GE	U. S. Army 4505	1941

Carnegie: Teledyne Columbia-Summerhill

4401	25-ton	GE	Columbia Steel, Panther Valley Coal	1955

Cementon: Whitehall Cement

1	10-ton	Vulcan		1937
2	35-ton	GE		1957

Clairton: USX

7502	SW1	EMD	URR 455	1949

Coatesville: Lukens Steel

42	SL120	GE		1977
44	NW2	EMD		

Creighton: Pittsburgh Plate Glass

104	50-ton	GE		

Curry Hollow: U. S. Bureau of Mines

1	6-ton	Plym		1972	42" gauge

Darlington: Garrett Railroad Car Parts

19	10-ton	Plym	Sharon Steel 19	1953

Darlington: Industrial & Rail Scrap Inc.

	20-ton	Plym	France Stone, Erie Stone Co.	1928

Dubois: Rescar Inc.

2	45-ton	GE	Donner-Hanna	1945
3	25-ton	Dav		1945

Duquesne: USX

11	70-ton	GE		1953
1	80-ton	GE		1957
2	SW1	EMD	TC&I 1000	1948
3	SW8	EMD	Donora Southern 810	1951

East Pittsburgh: Westinghouse Electric

2	40-ton	Plym		1000
11	80-ton	Whit		1944

Ebensburg: Bethlehem Mines

109	SC	1	EMC	1935

Ebensburg: Tanoma Mines

113	SW7	EMD	C&BL 113	1949

Erie: Erie Coke Co.

1028	NW2	EMD	Southern Coke 102, Chattanooga Coke & Chemical, Southern 1028	1946

Erie: General Electric

3	SL136	GE	C&NW 1198, BN 1100	1978

Erie: Hammermill Paper Co.

16	SL110	GE		1980

Export: Dura-Bond

	44-ton	GE		
	45-ton	Whit	MCP Corp., U. S. Navy 65-00165	1941

PENNSYLVANIA (continued)
Fairless Hills: USX

1	SW9r	EMD	ATSF 1238, ATSF 2438	1953
2	S2	Alco	BN 106	1938
3, 4, 6-8	S12d	BLW		1951-1952
5A	TR04Ar	EMD	ATSF 1242	1951
5B	TR04Br	EMD	ATSF 1243	1950
13A	S8A	BLH	Oliver Iron Mining 1205A	1951
13B	S8B	BLH	Oliver Iron Mining 1205B	1951
14	SW9u	EMD	ATSF 1234, ATSF 2434	1953
15	SW9u	EMD	ATSF 1236, ATSF 2436	1953
17	NW2	EMD	EJ&E 437	1949
	Remote control			
18	NW2u	EMD	EJ&E 458	1949
	Remote control			
20	20-ton	Plym		1952
23	SW1200	EMD	MP 1101	1963
23A	S8A	BLH	Oliver Iron Mining 1204A	1951
23B	S8B	BLH	Oliver Iron Mining 1205B	1951
24	H12-44	FM	PC 8330, PRR 8714	1952
25	H12-44	FM	PC 8309, NYC 9120	1951
26A	S8A	BLH	Oliver Iron Mining 1201A	1951
26B	S8B	BLH	Oliver Iron Mining 1206B	1951
27	H12-44	FM	U. S. Army 1845	1953
28	B40/40	GE		1973
30	SW1200	EMD	MP 1189	1964
31	NW2	EMD	SOO 301	1939
32	DS44-10001	BLW	Oliver Iron Mining 932	1949
106, 108, 112, 113, 115	SW1200	EMD	Farmrail, N&PB, same numbers	1956
155	SW1200	EMD	BN 155, GN 155	1961
243	SW1	EMD	Erie 609	1949
928	DS44-1000	BLW		1949
929	DS44-1000,	BLW	Oliver Iron Mining 929	1949
930	DS44-1000	BLW		1949
930, 931	DS44-1000	BLW	Oliver Iron Mining 930, 931	1949
1201	S8	BLH		1951
1206	S8B	BLH		1951
A	Slug		AW&W 6, rebuilt Alco S2	1952

GE3, GE4	S12	BLH		1951
	Remote control			
GE6-GE8	S12	BLH		1952
GE15, GE16	H12-44	FM		1952
GE18	S12	BLH		1952

Farrell: Sharon Steel

7, 8	SW900	EMD	B&O	1955

Fox Chapel: J.F. Casey Co.

29	20-ton	Plym	Kanawha Central 1	1941

Glassport: Glassport Transport

	SW7	EMD	YS	1950

Glen Mills: Koppers Co.

1	40-ton	Plym	General Crushed Stone	1957
2	20-ton	Plym	Bradford Hills Quarry, U. S. Army	1941

Hazelton:
Jeddo Highland Coal Co.

2	35-ton	GE		1955
694	50-ton	Atlas	U. S. Navy 65-00223	1939

Homestead: USX

4, 5, 11	100-ton	GE		1958

Indiana: Kovalchick Salvage

1	5-ton	Plym		1953
	36" gauge			

Landisville: Amherst Industries

91	25-ton	Plym		1941
92	25-ton	Plym		1928
8256	45-ton	GE	Ontario Midland 5, U. S. Army 8256	1944

Leetsdale: Bethlehem Steel

4411	25-ton	Whit		1941
	35-ton	GE		

McKees Rocks: USX

202	100-ton	GE		1956
	Remote control			

McKeesport: USX

7	50-ton	GE		
	36" gauge			
701	S1	Alco	TC&I 701	1942
	Stored unserviceable			
1003	S2	Alco	McKeesport Connecting 1003	1948

PENNSYLVANIA (continued)
Midland: Crucible Steel

72-82	95-ton	GE		1949-1956
83	70-ton	GE	Georgia Pacific 24	1957
	Last 70-tonner built			

Monaca: St. Joe's Minerals

1	50-ton	GE		
2	50-ton	GE		1949
3	95-ton	GE	Chicago Mill & Lumber	1949
4	25-ton	GE		

Monaca: Teledyne-Vasco

1	DL	Plym		36" gauge
2	DLHT	Plym	1953	36" gauge
3	FLB	Plym	1945	36" gauge

Monessen: Monessen-Southwestern

5	SW900	EMD		1953
19	80-ton	GE	Pittsburgh Steel 19	1949
21	SW7	EMD		1950
22	SW8	EMD		1950
23-25	SW9	EMD		1951-1952
26	SW8	EMD		1952
27	SW9	EMD		1953
28	SW8	EMD	Donora Southern 805	1951
29, 30	SW900			1964, 1965

Moon Township: Briggs & Turivas

1	65-ton	Whit	J&L 4, U. S. Army 8130	1941

Morrisville: Morrisville Scrap Metals

1	45-ton	GE		1947
2	45-ton	GE	A. E. Staley Co.	
89	S2	Alco	SB 89	1950	Stored
unserviceable				
8564	SW1	EMD	CR 8564	1950

Morrisville: Warner Sand & Gravel

14	DS44-750	BLW		1950
15	SW1	EMD		
8381	DS44-750	BLW	PC 7886, PRR 5569	1950

Milton: ACF Industries

1, 2	45-ton	GE		1942, 1944
3	44-ton	GE		
4	25-ton	GE		1942

Neville Island: Vulcan Detinning

16	45-ton	Porter	PA&M 16	1941

Neville Island: Dravo Corp.

SW24	40-ton	Whit		1952
SW50	44-ton	GE	B&O 8301, 8801, 19	1950

New Bethlehem: Terry Coal Sales

8653	SW9	EMD	CR 8653	1951

New Eagle: USX

61	80-ton	GE		1959

Oreland: Tank Car Co. of America

	44-ton	Dav	U. S. Army 1235	1953

Palmer: N.J. SALES

51	VO660m	BLW		1941
101	S6	Alco	Keystone Coke 101, Alabama By-products 101, UM&P	1956

Parkhill: Luria Brothers

1006	S2	Alco	Relco 1006, KCT 57	1945

Philadelphia: Farmers Export Inc.

1001	GP7	EMD	SLSF 594	1951
2001	GP9	EMD	CR7353, PC 7353	1956

Philadelphia: Gulf Oil Co.

7	65-ton	Whit	U. S. Army 8467	1944

Philadelphia: Philadelphia Electric

20	45-ton	Whit	U. S. Gypsum 103	1952
26	44-ton	GE		1945

Philadelphia: Philadelphia Thermal Energy Co.

1	35-ton	GE	Atlantic Electric	1962

Philadelphia: Tidewater Grain Co.

1, 2	GP7	EMD	Precision National Corp.	

Philadelphia: Union Tank Car Co.

6	SW1	EMD	Monon 6	1949

Pittsburgh: Allegheny Contracting Industries

	35-ton	Whit	American Steel Foundries	1942

Pittsburgh: Allegheny Ludlum Steel

21-23	SL85	GE		1975-1978
88	80-ton	GE	Dow Chemical 88	1945

Pittsburgh: Babcock & Wilcox

9	NW2	EMD		1941
8705	NW2	EMD		1947
8743	NW2	EMD		1949

PENNSYLVANIA (continued)

Pittsburgh: Damien Heintz

65	65-ton	GE	Federal Steel & Wire		

Pittsburgh: General Electric

4	45-ton	GE	Taylor-Piedmont	1953	Used as a loaner

Pittsburgh: Heppenstall Steel

	10-ton	Plym		1950	30" gauge

Pittsburgh: Jones & Laughlin Steel

10-15	80-ton	Whit		1951	

Pittsburgh: Neville Avenue Steam Plant

	25-ton	Plym	U. S. Navy 65-00279	1942	

Pittsburgh: USX

4	SW900	EMD		1958
14	SW8	EMD		1951
15	NW2	EMD		1947
15	SW8	EMD		1951
16-18	NW2	EMD		1941-1949
18	SW8	EMD		1951
19	NW2	EMD		1941
80, 81	SW8	EMD		1950
82	NW2	EMD		1947
82-85	SW8	EMD		1950-1951
88-90	SW1001	EMD		1970-1971
109	SW8	EMD		1951
110	65-ton	Dav		1951
114, 115	SW7	EMD		1950
128	SW8	EMD		1951
157	SW900	EMD		1957
170	MP15	EMD		1976
224	SW1	EMD		1940
240, 300, 301	NW2	EMD		1938-1947
705, 706	SW9	EMD		1952
720	F7A	EMD		1952
803, 810, 837	SW8	EMD		1951, 1953
934-939	SW9	EMD		1951, 1953
940-948	SW1200	EMD		1957
949-954	SW1500	EMD		1972
955-968	MP15	EMD		1976
1000	SW1	EMD		1948

1002	NW2	EMD		1947
1102	TR6A	EMD		1951
1116	SW8	EMD		1953
1207-1213	TR6A and B	EMD		1951-1953
1216	TR6A	EMD		1953
2203	SW7	EMD		1950
4906, 5206	NW2	EMD		1949
6367	25-ton	GE	U. S. Army	1939
LM1-LM5	NW2	EMD		1947-1948
SX1, SX2	SW1200	EMD		1957

Portland: Metropolitan Edison

1	40-ton	Plym		1955
2	45-ton	Plym		1957

Reesedale: West Penn Power

1	35-ton	GE	Sanderson & Porte	1956

Rimersburg: C&K Coal Co.

1052	RS1	Alco		

Saxonburg: USX

SX03	SW1	EMD	URR 459	1949

Sayre: Anabel Corp.

300	50-ton	Whit		

Sayre: Quality Coal Co.

503	SW1	EMD	Relco 503, Commonwealth Edison	1942

Sharpsburg: Deitch Corp.

1	65-ton	Whit	C&BL 51, U. S. Army 8138	1944
			Stored unserviceable	
1	65-ton	Whit	Babcock & Wilcox U. S. Army 8457	1944
2	65-ton	Whit	Etna & Montrose 1353, U. S. Army 8436	1944
			Stored unserviceable	
2	65-ton	Whit	Babcock & Wilcox U. S. Army 8456	1944
1339	65-ton	Whit	Greenville Steel 1339 U. S. Army 8418	1944
5539	65-ton	Whit	ACF 4439, U. S. Army 8438	1944
BP02	65-ton	Whit	J&L BP29, U. S. Army	1944

Spring Cove: Gladfelters Paper Co.

81	NW2	EMD	Maryland & Pennsylvania 81	1946

Springdale: West Penn Power

40		EMD	U. S. Army 7954	1942

Swissvale: Union Switch & Signal

	45-ton	Porter		1946

PENNSYLVANIA (continued)
Sharon: Sharon Steel

6	SW8	EMD	Donora Southern 806	1951	
7	SW8	EMD	EL 361	1951	
11-14	80-ton	GE			
15	65-ton	Whit	Pittsburgh Steel 4, U. S. Army 8415	1944	
15	80-ton	GE			
16-18, 20	80-ton	GE	Pittsburgh Steel	1948-1949	
31, 32	SW8	EMD	Cincinnati Union Terminal 31, 30	1964	

Tarentum: Grello Brothers

6492	12-ton	Plym	Scaife Co.	1947	Stored

Tarentum: Pittsburgh Plate Glass

101	50-ton	GE		1946
105	50-ton	GE		1948

Trafford: Westinghouse Electric

1	DDT	Plym		1962
2	16-ton	Plym		1977

Vandergrift: USX

	50-ton	GE	

Vandergrift: Wean-United

	45-ton	GE	UE&F	1940

Verona: Edgewater Steel

1	45-ton	GE		1942
2	45-ton	GE		1956

West Mifflin: Duquesne Slag Co.

103	65-ton	Whit	U. S. Army 8131	1944
105	50-ton	GE		1959

West Mifflin: Tube City Iron & Metal

1, 2	45-ton	GE	U. S. Army	
3	S2	Alco	URR 527	1945
469	SW1	EMD	J&SC 469, URR 469	1950

West Winfield: West Penn Power

116	40	EMD		1942

RHODE ISLAND

Davisville: Seaview Transportation

1	80-ton	GE		
2	80-ton	GE	U. S. Navy 65-000467	1943
3	S3	Alco	B&M 1175	1950
62	S4	Alco	PH&D 62, GNWR 36	1953

SOUTH CAROLINA

Allen: Duke Power Co.

83	SL82	GE		1980

Anderson Quarry: Winnsboro Granite Co.

	35-ton	Plym	Duval Corp. 4	1952

Bath: Dixie Clay Co.

	4-ton	Plym		1967	36" gauge

Bennetsville: Becker Sand & Gravel

1	45-ton	Dav	U. S. Army 7218	1942
3	S2	Alco	A&StAB 1001	1942
6	S2	Alco	NKP 33	1947
7	S2	Alco	SLSF 291	1948
11	80-ton	GE	Alcoa Terminal 6	1947
22	80-ton	GE	BS&G 2, U. S. Navy 65-00453	1942
44	80-ton	GE	Columbia Sand, U. S. Army 7008	1942
55	80-ton	GE	BS&G 5, U. S. Army 7388	1942
77	S2	Alco	Gifford-Hill	1946
88	S2	Alco	BRC 407	1948
99	S4	Alco	SD 108, P&LE 8654	1953

Blair: Lone Star Industries

1	16-ton	Plym	Budd Manufacturing.	1942
2	30-ton	Plym		
3	25-ton	GE		1946

Brownsville: Becker Sand & Gravel

2	44-ton	GE	SOO 300	1941
3	S1	Alco	TCT 31	1947
4	S1	Alco	Des Moines Union4	1946
6	S1	Alco	PC 9408	1941

Canadys: South Carolina Electric & Gas

1	80-ton	GE		1961

Cash: Becker Sand & Gravel

10	S2	Alco	N&SS 1003	1941
20	S2	Alco	NKP 29	1947
30	S2	Alco	NKP 35	1947
40	44-ton	GE	U. S. Navy 65-00241	1943

Cayce: Guignard Brick Co.

5	50-ton	Porter	U. S. Navy	1941

Charleston: Airco Alloys & Carbide

2	44-ton	GE	South Carolina Port Authority 104	1947

SOUTH CAROLINA (continued)

Charleston: Airco Alloys & Carbide (continued)

3	44-ton	GE	South Carolina Port Authority 101	1949	
4	44-ton	GE	South Carolina Port Authority 102	1951	

Charleston: Amoco Oil Co.

1047	NW2	EMD	Southern 1047	1947

Charleston: Koppers Co.

1	25-ton	GE	Pittsburgh Metallurgical	1942	36" gauge
6	25-ton	GE	Smith & Sons Lumber	1946	

Charleston: Massey Coal

100	NW2u	EMD	CR 9174, PRR 8672	1948
101	NW2u	EMD	CR 9179, PRR 9179	1949

Charleston: South Carolina Electric & Gas

4	SL110	GE	1984

Charleston: South Carolina Port Authority

729	S2	Alco	CRI&P 729	1945

Columbia: Carolina Eastman Co.

2103	SW9	EMD	SCL 183	1951

Columbia: Forster-Dixiana Sand Co.

	44-ton	GE	PRR 9339	1948

Columbia: Martin-Marietta

7588	RS11	Alco	CR 7588	1957

Conway: Central Electric Power Corp.

	40-ton	GE	1965

East Darlington: Fiber Industries

1	40-ton	Plym	1974

Florence: Koppers Co.

2	45-ton	GE	1945
7	25-ton	GE	1948

Floyd: Nucor Steel

3	S12	BLW	Gettysburg 407	1953

Gaston: Nassau Metals

1077	50-ton	GE	Western Electric	1955

Gaston: Nassau Recycle

	45-ton	GE	Georgetown

Gaston: Georgetown Steel

705	110-ton	GE	Port of Tacoma 705	1980
5049	65-ton	GE	Heckett Engineering 1, BR&L, U. S. Army 7346	1940
5050	65-ton	GE	Heckett Engineering 2, BR&L, U. S. Army 7284	1942

Gaston: South Carolina Public Service

2	S4	Alco	N&W 2076	1953

Greenwood: Monsanto Textiles

	12-ton	Plym	Chemstrand Corp.	1965

Hagood: Becker Sand & Gravel

20	45-ton	GE	Kiewit & Son, Pearl River. Valley 300	1945
21	80-ton	GE	U. S. Navy 65-00418	1943
22	80-ton	GE	U. S. Navy 65-00417	1944
364	80-ton	GE	Lavion 364	1942
2	80-ton	GE	U. S. Air Force 7890	1942
	45-ton	GE	Bailessey Construction, U. S. Army 7422	1942

Harleyville: Carolina Giant Cement

P140	25-ton	Plym	Muskogee Electric	1930

Harleyville: Gifford-Hill Inc.

930	45-ton	GE	TVA 3	1942

Hartsville: Carolina Power & Light

105	70-ton	GE	LRS 105	1948

Irmo: South Carolina Electric & Gas

2	80-ton	GE	1957

Jackson: B. F. Shaw

6	45-ton	GE	U. S. Army 7929	1944

Jackson: E. I. Du Pont

101	80-ton	GE	U. S. Navy 65-00524	1951
102	80-ton	GE		1951
103	80-ton	GE	U. S. Army 1692	1951
105-108	RS1	Alco		1951

Jackson: J. M. Huber Co.

1	50-ton	GE	1956

Laurens: B. F. Shaw

	8-ton	Whit	Fellsmere Sugar

Laurens: Vulcan Materials

6010	45-ton	GE	Olin-Mathieson	1949

Newry: Courteney Mills

1001	SW1	EMD	Southern 1001, CofG 3	1941

North Augusta: Hamburg Industries

1	44-ton	GE	Georgia Marble	1951

NORTH CAROLINA (continued)
Pacolet: Vulcan Materials
| 2264 | 45-ton | GE | Birmingham Slag, U. S. Army 7431 | 1941 |

Pelzer: Duke Power Co.
| 80 | 25-ton | GE | | 1950 |

Pickens: Singer Manufacturing Co.
| 1 | 10-ton | Whit | Poinsette Lumber | 1933 |

Rion: Superior Stone Co.
| 1 | SW1 | EMD | EL 351 | 1940 |
| 203 | 70-ton | GE | HPT&D | 1948 |

Robinson: Carolina Power & Light
| | 25-ton | GE | | 1950 |

Rock Hill: Celanese Corp.
| 2009 | SW1 | EMD | Southern 2009 | 1947 |

Russellville: Georgia-Pacific Corp.
| 20 | 45-ton | Plym | PA&M 3, U. S. Army 7484 | 1942 |
| 4 | 18-ton | Plym | Russellville Lumber, U. S. Treasury | 1936 |

South Congaree: Columbia Silicate
2378	45-ton	Dav	Ohio Lime 2378, U. S. Army 7442	1942
1	25-ton	GE	U. S. Navy 65-00379	1942
2	45-ton	Porter	U. S. Navy 65-00187	1942

South Congaree: Pennsylvania Glass Sand
| | 65-ton | GE | U. S. Navy 65-00275 | 1943 |

Spartanburg: Southern Wood-Piedmont
| 5 | 45-ton | GE | Taylor-Piedmont | 1949 |

Spartanburg: Spartan Iron & Metal
| 3 | 25-ton | GE | Pennsylvania Electric 806 | 1946 |

Sumpter: B. L. Montague Co.
| | 25-ton | GE | Pennsylvania Electric | 1945 |

Sumpter: Palmetto Brick Co.
| 2 | 10-ton | Plym | Richtex Brick | 1948 |

Wallace: Palmetto Brick
| | 10-ton | Plym | Du Pont | 1939 |

Wateree: South Carolina Electric & Gas
| 3 | 110-ton | GE | | 1969 |

SOUTH DAKOTA

Britton: Jarrett Ranches Inc.
| 622 | SW9 | EMD | MILW 622 | 1951 |

Corson: Benson-Quinn Terminals
| 606 | DS44-1000 | BLW | SCL 62, SAL 1450 | 1950 |

TENNESSEE

Alcoa: Todco
| 8 | S3 | Alco | Alcoa Terminal | |

Ashland City: Southern Iron & Equipment
| 251 | 25-ton | GE | Koppers Tie, Gulf Oil | 1952 |

Beech Grove: Beech Grove Processing
| 256, 259 | U28B | GE | TTI 256, 259 | 1966 |

Boyce: Central Soya
| 1 | SW8 | EMD | | |

Chattanooga: Chattanooga Industrial
| 118 | RS1 | Alco | A&StAB 911 | 1945 |

Chattanooga: Siskin Steel & Supply Co.
| 365 | NW2 | EMD | Southern 1042 | 1947 |

Chattanooga: Velsicol Chemical
| 102 | 80-ton | GE | Southern Coke 102, Chattanooga Coke & Chemical | 1951 |

Cleveland: World Equipment Sales
| | 8-ton | Plym | Peabody Coal Co. | 1978 |

Copperhill: Cities Service Co.
107	S6	Alco	Tennessee Copper	1957
108	SW1500	EMD		1970
109	MP15	EMD		1974

Copperhill: Tennessee Copper Co.
| 105 | S1 | Alco | | 1947 |
| 106 | S4 | Alco | | 1953 |

Cowan: Marquette Cement
| 1 | 44-ton | Dav | U. S. Army | 1944 |

Gallatin: Tennessee Valley Authority
| 24 | H16-66 | FM | | 1958 |

Grahamville: Tennessee Valley Authority
| 44 | S2 | Alco | U. S. Army 7124 | 1942 |

Harriman: Southern Salvage
| 1133 | SW9 | EMD | Southern 1133 | 1951 |

Jackson: American Creosoting
| 8 | 45-ton | GE | | 1952 |

Jackson: Tennessee Mill Division, Florida Steel
| 1 | 65-ton | GE | W. R. Grace | 1943 |

TENNESSEE (continued)

Kingsport: Kingsport Iron & Metal

6	VO1000	BLW	Tennessee Eastman	1943

Kingsport: Tennessee Eastman

2	MP15	EMD		1975
1549	SW1500	EMD	P&LE 1549	1971
1556	SW1500	EMD	P&LE 1556	1971

Kingston: Tennessee Valley Authority

13	80-ton	GE	U. S. Army	
14, 16	VO1000	BLW	U. S. Navy	1944, 1945
46	S2	Alco	U. S. Army	
8032	RSD1	Alco	U. S. Army 8032	1942
8679	RSD1	Alco	U. S. Army 8679	

Knoxville: Alcoa Aluminum

7	MP15DC	EMD		1980

Knoxville: Dixie Cement

	35-ton	Dav		

Memphis: Memphis Dock-Rivergate Terminal

	45-ton	GE	Philadelphia Electric	

Nashville: W. G. Bush Brick

2	18-ton	Plym		
93	25-ton	GE		1953

Nashville: Steiner-Liff Iron & Metal

6250	GP9	EMD	C&O 6250	1956

Rockwood: Direct Reduction Co.

112	S1	Alco	LV 112	1950

Watts Bar: Tennessee Valley Authority

26	65-ton	GE		1945
5958	RSD1	Alco	U. S. Army 8658	1945

TEXAS

Alvin: Monsanto Chemical

129	SW7	EMD	Inman Service 129, SCL 129	1950
135	SW9	EMD	Inman Service 135, SCL 135	1951

Amarillo: Grain Producers

	23-ton	GE	Du Pont Chemical	1941

Arlington: Gifford-Hill Inc.

1082	70-ton	GE	Graham County 102, Savannah Dock 102	1953

Avinger: Southwest Electric Power

1	35-ton	Plym	Midland-Ross	

Arenal: Thorstenburg Materials Inc.

55	S4	Alco		1951

Baytown: Gulf Oil Co.

3	SW1	EMD	CRI&P 4803, IC 609	1946

Baytown: Inman Service

2	65-ton	GE	U. S. Navy 65-00603	1942

Baytown: Intra-Plant Switching Corp.

40	SW7	EMD		
1227	SW9	EMD	SCL 134	1951

Beaumont: Gulf Oil Co.

101	MP15DC	EMD		1979

Beaumont: Houston Chemical

1	RS1	Alco	SR&N 1001	1951

Beaumont: Mobil Oil

117	SL110	GE		1981

Blewett: White Mines

249	35-ton	Plym		
	65-ton	GE	Uvalde Asphalt	1947

Brownwood: White Mines

2	65-ton	Plym		

Burnett: Texas Construction Materials Inc.

508	25-ton	Plym		

Cambello: San Miguel Electric Co-op

	115-ton	GE		1978

Cason: Southwest Electric Power

2	35-ton	GE	Du Pont	

Channelview: Cargill Grain Inc.

499	NW2	EMD	BN	
590	S2	Alco		Named Granny

Camp Bowie: FMC Inc.

1	45-ton	GE	U. S. Army 7251	1942

Dabney: Uvalde Rock Asphalt Co.

1-4	44-ton	GE		

Dallas: Kraft Foods

2608	CF7	ATSF	ATSF 2608	1972

Dallas: Lone Star Industries

1	70-ton	GE	Nipak Industries, Frankfort & Cincinnati 102	1947

Dallas: Texas Crushed Stone

1202	SW9	Alco	C&NW 1202	1952

TEXAS (continued)

Dallas: Trinity River Authority

	45-ton	GE	San Angelo Tank Car Co.	

Daingerfield: Texas Utilities

3307	B23-7	GE		1981

Deer Park: Shell Oil

1153	SW1200	EMD	MP 1153	1965

Denison: W. J. Smith Wood Preserving Co.

400	25-ton	GE		1950	
1655	12-ton	Whit			30"gauge
1656	25-ton	Whit			30"gauge

Devonia: Industrial Processing Co.

1007	SW1	EMD	NS 1007, Southern 1007	1947

El Paso: Asarco

4	CF7	ATSF	ATSF 2441	1979
5	CF7	ATSF	ATSF 2533	1978
6	CF7	ATSF	ATSF 2606	1975
7	CF7	ATSF	ATSF 2633	1975

Fannin: Central Power & Light

	35-ton	GE	Carborundum Co., Buffalo Slag 4	1956

Freeport: Dow Chemical Co.

1001-1003	SW1001	EMD		1980

Fort Worth: Freightmaster Inc.

3	50-ton	Whit	Quick Car Inc., Ventura County	1943

Fort Worth: TG Railway Service

1	Homebuilt		Gates Rubber	1951 Chain drive

Fort Worth: Tank Lining & Rail Car Repair

7	44-ton	GE	Pullman-Standard 7	1947

Galena Park: Agri-Industries

600	S2	Alco	
602	S4	Alco	
603	S4	Alco	
604	RS3	Alco	

Galveston: Cook Industries

1012	S2	Alco	Relco 1012	
1023	S2	Alco	Relco 1012, Far Mar Co.	
203	SW1	EMD	GW 203	1949
205	SW1	EMD	GW 205	1951

Grand Prairie: Gifford Hill Inc.

82	70-ton	GE		1940

Gregory: Reynolds Metals

	SW1	EMD		

Houston: Armco Steel Co.

1811	S6	Alco	
1823	S3	Alco	
1827-1829	S2	Alco	
1832-1835	S4	Alco	
1837	S2	Alco	
1839, 1840	SW1001	EMD	

Houston: Bauer Barge Co.

1	80-ton	Whit		
2	45-ton	GE	U. S. Navy, Byer Terminal	
	50-ton	Plym	Cargill Inc.	
122	VO660	BLW	MN&S 600	1940

Houston: Equity Export

97, 98	S2	Alco	Houston Belt & Terminal 14, 15	1945

Houston: General Electric

1253	S6	Alco	SP 1253	1956
2	45-ton	GE		1949

Houston: Houston Light & Power

1953	25-ton	GE	Kaiser Agricultural Chemicals	1953
1	SL65	GE		1982

Houston: Houston Truck & Equipment

7017	65-ton	GE	Port of Seattle, U. S. Navy 65-00354	1945

Houston: International Terminal

475	GP8	ICG	Prairie Trunk 475, ICG	
2355	S2	Alco	ATSF 2355	1945

Houston: Lubrizol Corp.

2297	SW1200	EMD	SP 2297	1965

Houston: Mobil Oil

116	MP15AC	EMD		1981

Houston: Proeler Steel

1243	S6	Alco	SP 1243	1956
1844	S4	Alco	SP 1844, 1566	1955

Houston: Shell Oil

2265	SW1200	EMD	SP 2265	1964

Houston: W. A. Smith

1	35-ton	Plym	U. S. Army 7615	1943

Houston: Southern Pacific Tie Plant

9020	S6	Alco	SP 1248	

220

TEXAS (continued)

Iola: Texas Municipal Power Authority

	SL125	GE		1979

Irving: Railroad Maintenance & Construction

7427	45-ton	GE	U. S. Army 7427	1941

Jasper: Texas Utilities Co-op

27	25-ton	Whit	Philadelphia Electric	1951

Jefferson: Gifford-Hill Inc.

101	S2M	Alco	

Jewett: Nucor Steel

51-53	DS44-1000	BLW	SCL 51, 52, 57; SAL 1439, 1440, 1449	1950
88	VO1000	BLW	SCL 88	1941

Knoxville: Mississippi Delta Equipment Co.

413	NW2	EMD	1948
415	NW2	EMD	1946

Lone Star: Lone Star Steel

26-28, 31, 32	S2r	Alco	ATSF 2332, 2364, 2390, 2336, 2326	1942-1949
33, 38	S4r	Alco	ATSF 1508, 1503	1951
53-56	S4d	Alco	ATSF 1533, 1514, 1500, 1526	1953

Longview: Rescar Inc.

1	45-ton	GE	Shippers Car Line	1942
2050	GP7r	EMD	ATSF 2050	1951
3688	25-ton	GE		

Lubbock: Continental Grain Co.

1	23-ton	Plym		1939
18	SW9	EMD	MKT 18	1952
2301	HH600	Alco	Palo Duro Grain 2301, ATSF 2301	1937

Lubbock: Goodpasture Grain

254	45-ton	GE		
2352	S2	Alco	ATSF 2352	1945

Lubbock: Producers Grain Co.

	25-ton	GE	Du Pont	1942

Lufkin: Lufkin Metals

3	25-ton	Plym	Lufkin 6ATSF 6	1930
4	40-ton	BLH	Standard Oil 2	1952

Midlothian: Mid Texas International

2576, 2615	CF7	ATSF	ATSF 2576, 2615	1979, 1977

Midlothian: Texas Cement Industries

2428, 2447	CF7	ATSF	ATSF 2428, 2447	1979

Mount Pleasant: Texas Utilities

1, 2	SL65	GE	1976, 1977
101, 102	U18B	GE	1974
3303, 3304	E25B	GE	1978, 1979

Nash: Texana Tank Car Co.

101	S3	Alco

Nolanville: Bandas Industries

1227, 1236	S6	Alco	SP 1227, 1236	1955, 1956

Nolanville: Franklin Lime Co.

	50-ton	Whit	Belton 1, Dresser Industries	1942

Odessa: El Paso Products Co.

2	45-ton	Vulcan	U. S. Air Force 7223, U. S. Army 7223	1941
3	65-ton	GE		
4	65-ton	GE	Southwestern Portland Cement	
	H12-44	FM	SP 2379, 1581	

Pasadena: Ethyl Corp.

1	SW1	EMD	MP 9021	1939
2	100-ton	GE		
3	60-ton	GE		

Pasadena: International Terminal

	S2	Alco	Lubrizol Corp.	
	S1	Alco	U. S. Army 7136	1943

Pasadena: Lubrizol Corp.

10	S1	Alco

Pasadena: Olin Mathieson Chemical

4	NW5	EMD	Fort Street Union Depot	1947

Pasadena: Pak-Tank Terminal Co.

	S2	Alco	A&NR 11	1949

Pasadena: Shell Oil

5	50-ton	Whit		1948

Pasadena: Union Equity Grain

1, 2	RS3	Alco	Bauxite & Northern 12, 13	1951
3	RS1	Alco	WT 41	1944

Plainview: Agri-Industries

536	SW1	EMD	CRI&P 536	1942

Plainview: Harvest Queen Elevator

100	35-ton	Plym	Cummins Diesel	1936

TEXAS (continued)

Plainview: Midwest Grain Co.

6	80-ton	GE	Eveready Machinery, U. S. Navy 65-00509, U. S. Army 7857	1943
716	65-ton	GE		
765	80-ton	GE		
2302	HH600	Alco	ATSF 2302	1937

Plainview: PLB Grain Co.

6	65-ton	GE	Midwest Storage	
536	SW1	EMD	CRI&P 536	1942
765	80-ton	GE	Producers Grain Co.	

Port Arthur: Econorail Inc.
2452, 2473, 2478, 2510, 2516, 2526, 2534, 2541, 2560, 2569, 2575, 2580, 2581, 2582, 2593, 2602, 2604, 2608, 2624, 2631, 2642

	CF7	ATSF	ATSF, same numbers	1974-1980

For lease

Port Arthur: Gulf Oil Co.

9	SW600	EMD		1955

Ranger: Quality Service Railcar Repair

1005	45-ton	GE	Texas Railcar, U. S. Air Force 8564, U. S. Army 8564	1944

Ranger: Texas Railway Car Corp.

200	SW8	EMD	RS&P 200	1953

Roscoe: Texas Railway Car Corp.

100	SW1	EMD	RS&P 100	1949

Scottsville: Berwind Railway Service

2	25-ton	GE	MK 86	1956
2020	45-ton	GE		

Seadrift: Union Carbide

1251	SW9	EMD	Relco 1251	
1256	SW7	EMD	Relco 1256	
1606	GP9	EMD	Relco 1606	

Seguin: Structural Metals Inc.

500	80-ton	GE	St. Mary's 500	1944

San Augustine: Champion International

1	SL110	GE	MC&SA 548	1978
6056	SW1001	EMD		1979
6210	NW2	EMD		1947

Sherman: Anderson Clayton Foods

7225	SW900	EMD	GTW 7225	1956

Texas City: Texas City Terminal

35-37	MP15DC	EMD		1982

Tulia: Continental Grain Co.

1043	S2	Alco	Relco 1043	

Tyler: Camp Ford Industrial Park

9417	SW900	EMD	B&O 9417	1955

Victoria: Safety Railway Service

201	23-ton	GE	U. S. Air Force 7072, U. S. Army 7072	1941

Waco: Brazos Electric Co-op

	SL115	GE	San Miguel Electric	1978

Williams: Arco Polymers Inc.

104	RS1	Alco	SR&N 104	1951

UTAH

Castle Gate: Utah Power & Light

5923	GP9	EMD	D&RGW 5923	1955	

Geneva: USX

21-26	DS44-1000	BLW	Columbia Steel	1947	EMD engine
27-32	DS44-1000	BLW	Columbia Steel	1948	
80, 81	Slug		Carbon County	1944	
	Rebuilt from VO1000				

Plymouth: Nucor Steel

341	DS44-1000	BLW	P&BR 341	1948	

Provo: Pacific State Pipe & Foundry

1043	NW2	EMD	UP 1043	1946	

Salt Lake City: Argee Inc.

71	SW1200	EMD	KCT 71	1964	

Wellington: USX

1217	S6A, S6B	Alco		1956	Cow and calf

VERMONT

Barre: Bombardier

3	25-ton	Whit	Jenkins Valve	1944
305	S4	Alco	GMRC 305, D&H 3050	1950

Graniteville: Rock of Ages

	18-ton	Plym		

VIRGINIA

Alexandria: Fruit Growers Express
2	45-ton	GE		1953

Alexandria: Newton Asphalt Co.
2432	CF7	ATSF	ATSF 2432	1977

Amonate: Consolidation Coal Co.
97	65-ton	GE	Hanna Coal, Truax-Traer Coal	1942

Appalachia: General Coal Co.
9	S2	Alco	Southern 2208	1944
11, 12	S2	Alco	Southern 6059, 6057	1941

Appalachia: Pocahontas Fuel Co.
1, 2	50-ton			1958

Arlington: Howat Concrete Co.
125	S2	Alco	WM 125	1943

Big Rock: United Coal Co.
7, 8	S2	Alco		
117	S12	BLW	USS 117	1950

Big Rock: Wellmore Coal Co.
1	S4d	Alco		1957

Bishop: Pocahontas Fuel Co.
3	50-ton			1957

Blue Ridge: Blue Ridge Stone Co.
	30-ton	Whit	Houston Shipbuilding	1940

Bremo Bluff: Virginia Electric Power
1117	S1	Alco	B&M 1171	1949
6350	S3	Alco	B&M 1173	1950

Carbo: Appalachian Power Co.
1	80-ton	GE		1957
2	100-ton	GE		1969

Carbo: Clinchfield Coal Co.
100	SW600	EMD		1958

Chesapeake: Cargill Grain Inc.
3	S6	Alco	WM 152	1956
5	65-ton	GE	U. S. Army 7180, U. S. Navy 65-00057	1943
111	SW1200	EMD	N&PB 111	1956

Chesapeake: Eppinger & Russell
1	12-ton	Plym	Foley-Beardsley	1946
	36" gauge			
2	14-ton	Plym	Owens Contractors	1949

3	25-ton	GE	Rangaire Corp., Lone Star Cement 9	1958

Chesapeake: Lone Star Industries
2	45-ton	GE		1944

Chesapeake: Virginia Electric Power
1	110-ton	GE		1982

Chester: Virginia Electric Power
4	80-ton	GE		1963

Chesterfield:
Virginia Electric Power
3	SL110	GE		1981

College Park: General Electric
126	50-ton	GE		

Dozier Corner: Tidewater Construction Co.
288	8-ton	Vulcan	U. S. Army Quartermasters Corps 106	1941

Edgarton: Vulcan Materials
12001	SW900	EMD	B&O 9424	1955
12002	SW900	EMD	B&O 9409	1955

Fishersville: Agmark Intermodal Systems
85	NW2	EMD	MPA 85, RDG 92	1941

Fordwick: Lehigh Portland Cement
1	23-ton	GE	Indiana Gas & Chemical	1939
4	65-ton	Whit	U. S. Army 8433, U. S. Army 1344	1944

Front Royal: Avtex Fibers
15	T61	Alco	M&D 15, PRR 8424	1958

Front Royal: FMC Corp.
48	65-ton	GE	American Viscose, U. S. Army 7188	1943
279	65-ton	GE	U. S. Army 7189	1943

Glen Lyn: Appalachian Power Co.
1	SW1000	EMD		1969
102	80-ton	GE		1955

Graham Station: Vulcan Materials
302	SW1	EMD	P&BR 157	1940

Henry: Blue Ridge Talc Co.
1, 2	3-ton	Bkvl		1942, 1943
	36" gauge			

Hopewell: Allied Chemical Co.
3	80-ton	GE	Lukens Steel	1953
7, 8	65-ton	GE		1947, 1950

Hopewell: Virginia Electric Power
2	110-ton	GE		1982

VIRGINIA (continued)

Jewell Valley: Jewell Ridge Coal Co.

1	S2r	Alco	BRC 408	1950

Lawrenceville: Vulcan Materials

2001	SW9	EMD	B&O	1952
2002	SW9	EMD	B&O 9601	1952
2003	SW900	EMD	B&O 9401	1955

Lynchburg: Buncher Rail Car Service

1	50-ton	GE	Pittsburgh Steel	1948
	20-ton	Whit	Mason Lumber 7	1952

Lynchburg: Lynchburg Foundry

1	65-ton	GE	Tennessee Chemical Corp.	1953
4	S4	Alco	C&O 5113	1953
5	65-ton	GE		
9030	S2	Alco	C&O 9030	1943

Mineral: Virginia Electric Power

61	RS1	Alco	WT 61	1950

Manassas: Vulcan Materials

13250	NW2	EMD	KCS 4217	1948

Newport News: Massey Coal

SL1-SL3	SL110	GE		1982

Newport News: Newport News Shipbuilding & Drydock

502, 503	50-ton	Whit		1954
504	43-ton	GE		1940
505-507	50-ton	GE		1962-1963

Newport News: Union Carbide

8401	SW1	EMD	C&O 8401	1942

Nora: Clinchfield Coal

14	F7A	EMD	CRR 821, L&N 1822	1949

Norfolk: Cargill Grain Inc.

	44-ton	Plym	U. S. Army	1958

Norfolk: Continental Grain Co.

123	70-ton	GE	Georgia Pacific, Meadow River Lumber	1957
589	S2	Alco	TRRA 589	1949
			Stored unserviceable	
1055	S2	Alco	GTW 8095	1944

Norfolk: Norfolk Steel

1	25-ton	GE	Atlantic Electric, Deepwater Operating	1948

North Springfield: Vulcan Materials

5	65-ton	GE	U. S. Army 1405	1942

Oakwood: Virginia Pocahontas

1	6-ton	Bkvl		30" gauge

Petersburg: Rail Link Leasing

8	NW2u	EMD	MKT 8	1947	1200 h.p.
22, 26, 27, 30	DS44-1000	BLW	MKT, same numbers	1946-1947	
	EMD engine				
208, 212	SW1200	EMD	N&W, IT, same numbers	1955	

Pilgrim Knob: Virginia Pocahontas

2	4-ton	Bkvl		30" gauge
3	S2r	Alco		

Portsmouth: Aluminum Co. of America

5	S1	Alco	Alcoa Terminal 5	1945

Portsmouth: Atlantic Creosoting Co.

1	25-ton	GE		1950
2	5-ton	Vulcan	Whitehall Cement	1939

Portsmouth: Bird International

	44-ton	GE	U. S. Navy 65-000200	1943

Portsmouth: Eveready Machinery

2100	NW5	EMD	Southern 2100	1947	Stored

Radford: Lynchburg Foundry

2	65-ton	GE	Tennessee Chemicals	1953
2	5-ton	Plym		1960	36" gauge
3	5-ton	Plym		1956	36" gauge

Richmond: Koppers Co.

3	25-ton	GE		1949
8	25-ton	GE		1951

Richmond: Old Dominion Iron & Steel

	20-ton	Whit	U. S. Army 7723	1942

Richmond: Peck Iron & Metal

1	65-ton	GE	Maryland Slag 8	1949
94	VO1000	BLW	SCL 94, CBQ 9372	1944
1167	S1	Alco	Vepco 1167, B&M 1167	1948
EQ08	80-ton	GE	Phillips Chemical	1952

Richmond: Richmond Marine Terminal

1	25-ton	Dav		1939
2	50-ton	Dav		1939

224

Ripplemead: Virginia Limestone
| | 45-ton | GE | | 1951 |

Roanoke: Citadel Cement Co.
| | 50-ton | Plym | | |

Roanoke: Cycle Systems Inc.
| 2072 | MRS1 | Alco | U. S. Army B2072 | 1950 |

Salem: Koppers Co.
| 12 | 25-ton | GE | | 1948 |

Saltville: Town of Saltville
| 1 | SW1 | EMD | Olin-Mathieson Chemicals | 1947 |
| 3 | 25-ton | GE | Olin-Mathieson Chemicals | 1950 |

Sealston: Mid Atlantic Materials
| 126 | S2 | Alco | WM 126 | 1943 |

Sealston: Sloite Corp.
| 50 | RS1 | Alco | WT 50 | 1947 |

Skippers: Trego Stone Co.
| 1 | 50-ton | Whit | Maryland Slag 10 | 1951 |

Stony River: Virginia Electric Power
| 6101 | 110-ton | GE | | 1964 |

Vansant: Island Creek Coal
1-4	U33C	GE	Southern	1970
3	S2m	Alco	Red Ash Coal	Cabless
1493	GP18	EMD	N&W 2703, NKP 703	1960

Vansant: Jewel Smokeless Coal Co.
| 1 | 35-ton | Porter | Russell Fork Coal | 1945 |

Vansant: Virginia Pocahontas
| 1, 2 | GP7 | EMD | N&W 3457, 3463 | 1951 |
| 5 | S2r | Alco | | |

WASHINGTON

Hanford: Rockwell-Hanford
| 3727 | MP15AC | EMD | | 1983 |

Kennewick: Lamson Equipment Co.
| | 23-ton | GE | G. F. Atkinson, U. S. Air Force 7791, U. S. Army 7791 | 1941 |

North Richland: Hennisingen Cold Storage
| 55 | SW1 | EMD | Walla Walla Valley 104 | 1939 |

Plymouth: U&I Grain Terminal
| 504 | S1 | Alco | Relco 504 | |

Port Gamble: Pope & Talbot Lumber Co.
| | 65-ton | GE | | 1943 |

Seattle: Cargill Grain Inc.
| 101 | NW2 | EMD | PTS 101, BN 594 | 1948 |
| | NW2 | EMD | BN 481 | 1945 |

Seattle: Purdy Co.
| | 10-ton | Plym | | 1930 |

Seattle: Salmon Bay Steel
1	44-ton	GE		
1	25-ton	GE	Schnitzer Steel, ARR, U. S. Army 7769	1944
15125	44-ton	GE	U. S. Navy 65-00190	1942
27606	45-ton	GE	U. S. Army 8528	1944

Tacoma: Continental Grain Co.
11	70-ton	Whit	Port Townsend Railroad	
104	NW2	EMD		
1235	S6	Alco	SP 1235	1958

Tacoma: General Metals
| 11 | SW1 | EMD | BN 102 | 1942 |

Wallula: Boise Cascade
| 100 | SW1 | EMD | Portland Traction 100 | 1952 |

WEST VIRGINIA

Appalachia: Westmoreland Coal
| S1-S5 | S4r | Alco | | 450 h.p. |

Ashford: Kanawha Coal Co.
| 7245 | S12 | BLW | Erie Mining 7245, Monongahela 419 | 1954 |

Barracksville: Barracksville Mine
| 143 | S2 | Alco | WM 143 | 1946 |

Belle: Du Pont Chemical
| 2469, 2496, 2572, 2575, 2579 | | | | |
| | CF7 | ATSF | ATSF, same numbers | 1974-1978 |

Boldair: Spring Ridge Coal
| 1005 | S2 | Alco | | |

Buckhannon: Bethlehem Mines
| 80 | S2 | Alco | SB 80 | 1942 |

WEST VIRGINIA (continued)
Cabin Creek: Appalachian Power Co.

1	14-ton	Vulcan		1942
2	40-ton	BLH		1956
3	45-ton	GE	Ohio Power 3	1950

Ceredo: Oglebay-Norton Co.

287	GP9	EMD	C&O 6147	1956
485	GP18	EMD	CRI&P 1346	1960
511	S2	Alco	TRRA 578	1942
586	GP9	EMD	C&O 6153	1956
1248	GP18	EMD	N&W 2700	1960

Elk Run: Elk Run Coal

1981	C415	Alco	Precision National 1851, UP 421, CRI&P 421	1966

Erbacon: Juliana Coal Co.

2500	CF7	ATSF	ATSF 2500	1978

Farrell: Westmoreland Coal

10	S2	Alco	Southern 6058	1941

Glasgow: Appalachian Power Co.

1	SL144	GE		1976
C6	80-ton	GE		

Harrison: Allegheny Power Systems

1	RS11	Alco	SP 2917	1959

Haywood: Allegheny Power Systems

1B	Slug	MK		Built from RS2

Huntington: American Car & Foundry

1309	10-ton	Plym		1959
4864	45-ton	GE		1942
6500	65-ton	GE		1957

Huntington: Amherst Industries

	25-ton	GE		1950

Huntington: Connor Steel

89	80-ton	GE	Dow Chemical 89	1949

Huntington: Ohio River Co.

21	RS1	Alco	LEF&C 21	1950

Huntington: Steel of West Virginia

9	80-ton	GE	Eveready Machinery, U. S. Navy 65-00260	1943

Kenova: Oglebay-Norton Co.

287	GP18	EMD		1960
485	GP18	EMD	CRI&P 1336	1960

Kenova: Ohio River Transfer

56	S6	Alco		1956

Kenova: Transfer Terminal Co.

5083	SW9	EMD	C&O 5083	1951

Moundsville: 7Ohio Power Co.

3	110-ton	GE		1975
	45-ton	GE	West.Palm Beach Terminal, U. S. Navy 65-00381	1943

Matewan: Sprouse Creek Mine — Rawl Sales

1499	GP18	EMD	NS 1499	1960
2108	SW8	EMD	N&W 2108	1952

Orgas: Cedar Coal Co.

2151	S2	Alco		1944

Pine Creek: Island Coal

1	S2	Alco	M&NF 16	Cabless

Port Amherst: Amherst Industries

97	45-ton	GE		1947

Rhodell: Maben Energy Corp.

3051	SD35	EMD	Southern 3051	1965

Ricard: Beckley Lick Run Coal

10	S2m	Alco		Remote control

Sutton: Oneida Coal

1750	GP9	EMD		
8060	GP9u	EMD	ICG 8060	1968

Uneeda: Mae West Prep Plant

1522	SD35	EMD	N&W 1522	1965

Volga: Rawhide Tipple Co.

12	VO1000	BLW	Middle Fork 12, U. S. Army 7143	1943

Wharton: Eastern Associated Coal

1, 2	SL110	GE		1977, 1976

Willow Island: Allegheny Power Systems

1	SL144	GE		1978

WISCONSIN

Algoma: Trans-Northern Inc.

105	60-ton	GE	Miller Compressing 105	1937
818	S4	Alco	MILW 818,811,1884	1951

Algoma: U. S. Plywood

2	12-ton	Plym	Hines Co. 434	1953

WISCONSIN (continued)

Ashland: Continental Forest Products

| 1175 | S3 | Alco | Virginia Electric Power, B&M 1175 | 1950 |

Beloit: Fairbanks-Morse Engine Division

| | 50-ton | GE | Colt Industries | 1956 |

Beloit: Wisconsin Power & Light

| 1 | 45-ton | GE | | 1941 |
| 3 | 30-ton | Plym | | 1952 |

Brokaw: Wausau Papers

| 1026 | S2 | Alco | Relco 1026, TRRA 582 | 1944 |

Cudahy: Crane Manufacturing & Service Corp.

| 15 | 20-ton | Plym | Veterans Administration | 1940 |

Cudahy: George J. Meyer Manufacturing Co.

| 20 | 9-ton | Plym | | 1952 |

Eau Claire: Wissota Sand & Gravel Co.

| | 30-ton | Plym | Bucyrus Erie, Phoenix Utility | 1930 |

Glendale: Fort Howard Paper Co.

| 180 | S2 | Alco | FP&E 103 | |

Green Bay: American Can Co.

1	SW600	EMD		1960
1201	S6	Alco	SP 1201	1955
1280	SW600	EMD	C&NW 1280	1954
	44-ton	GE	St. Paul Union Depot	1941

Green Bay: Fort Howard Paper Co.

| 138 | S2 | Alco | PT 38 | 1949 |

Kohler: Kohler Co.

| | 80-ton | GE | U. S. Army 7149 | 1941 |

Manitowoc: Medusa Portland Cement

| | 4-ton | Bkvl | | 1950 | 36" gauge |

Manitowoc: Rahr Malting Co.

| | 20-ton | Plym | Busch Brewing, McKenzie-Simmons | 1947 |

Manitowoc: Busch Agricultural Resources Inc.

| 1 | 50-ton | GE | Harnischfeger, Nat'l Cash Register | 1961 |

Milwaukee: Cargill Grain Inc.

| | 12-ton | Mack | | 1930 |

Milwaukee: Milwaukee Metropolitan Sewerage District

| 1 | 20-ton | Plym | Duluth Stone | 1927 |
| 2 | 35-ton | Plym | | 1954 |

Milwaukee: A. O. Smith

| 100 | SW9 | EMD | U. S. Plywood | 1953 |

Milwaukee: Wisconsin Public Service

| 8754 | S2 | Alco | | 1944 |

Necedah: Wisconsin River Power Co.

| | 12-ton | Plym | Algoma Plywood, Cooper-Bessemer | 1929 |

Nekoosa: Nekoosa Paper

| 22 | S6 | Alco | CM&N 1205, SP 1205 | 1955 |

Oak Creek: Wisconsin Electric Power

| | SW1000 | EMD | | 1968 |
| | Remote control | | | |

Park Falls: Flambeau Paper Co.

| 46 | 50-ton | GE | | 1946 |

Pleasant Prairie: Wisconsin Electric Power

| 1 | SL110 | GE | | 1979 |

Port Edwards: Nekoosa-Edwards Paper

14	S1	Alco	South Omaha Terminal 5	1947
15	S2	Alco	Portland Terminal 35	1943
17	S2	Alco	C&NW 1094, 1085, 1015, CGW 10	1947
18	RS2	Alco	GB&W 301	1950
19	S6	Alco	SP 1244	1956
21	C415	Alco	SP 2409	1968

Rothschild: Weyerhaeuser Paper

| 1 | SW600 | EMD | Marathon Southern 1 | 1968 |
| 106 | 44-ton | Whit | U. S. Army 1209 | 1941 |

Sheboygan: Wisconsin Power & Light

201	S2	Alco	KGB&W 201	1948
	Remote control			
2	25-ton	Whit	L. E. Meyer Construction	1927

South Milwaukee: Bucyrus-Erie Co.

1	S2	Alco	Proctor & Gamble1	1946
2	S3	Alco	BLH (Eddystone switcher)	1956
8007	50-ton	GE	U. S. Army 2	1938

Superior: Archer Daniels Midland Grain

| 1 | 25-ton | GE | | 1949 |

Superior: Continental Grain Co.

| 1 | 25-ton | Plym | | 1966 |
| 2 | 35-ton | Plym | | 1967 |

Superior: Farmers Union Grain Terminal Association

| | 12-ton | Mack | Spencer Kellogg | 1934 |

Superior: Osborn McMillan Grain Elevator

| 101 | 30-ton | Whit | GATX 21-1 | 1955 |

WISCONSIN (continued)

Wausau: Mosinee Paper Mills

1	65-ton	GE		1955

Wisconsin Rapids: Consolidated Papers

321	40-ton	Plym		1957

WYOMING

Bill: North American Car Co.

50	NW2	EMD	C&O 5210	1949

Casper: Western Railroad Builders

248	44-ton	GE	Kansas City Power & Light 6	1949

Casper: Casper Iron & Metal

29	S2	Alco	Neosho Construction, Illinois Northern	1946

Casper: Exxon Oil

2001	GP20	EMD	Relco 2001, WP 2002	1959

Chisholm: Chisholm Mine

1	SW900	EMD	B&O 9415	1955

Evanston: Wyoming Car Co.

8568	45-ton	GE	U. S. Army 8568	1944

Laramie: Monolith Portland Cement

3	80-ton	GE	Amoco Chemical, U. S. Army 7852	1943

Laramie: Mountain Cement

1	70-ton	GE	Monolith Portland Cement	

Torrington: Holly Sugar Co.

8	25-ton	GE		1948

Worland: Holly Sugar Co.

1	45-ton	GE	U. S. Army 7423	1942

Wright: Kerr-McGee

4	30-ton	Whit	

CANADA

ALBERTA

Calgary: Western Canada Steel

202	C420	Alco	Wabash Valley 202, Vermont Northern 202, LIRR 202	1963

Edmonton: Canada Cement LaFarge

1	50-ton	GE		1956

Edmonton: Dow Chemicals

1001-1004	S6	Alco	SP 1246, 1254, 1223, 1237	1956
1005-1007	RS11	Alco	BN 4197, 4188, 4195; NP 917, 908, 915	1960

Edmonton: M-4 Holdings

6619	S11	MLW	CP 6619	1959
7249	SW900	EMD	CN 7249	1958

Grande Prairie: Proctor & Gamble

136	SW9	EMD	SCL 136, ACL 654	1951

Hanna: Skibsted

DL10	44-ton	CLC	CP 13	1958

Joffre: Novacor

7434	GP9	GMD	CR 7534, PC 7434, NYC 6034	1957
W111	NW2	EMD		1948
W115	SW1500	EMD	SP 2647	1972

Kaybob: Dome Petroleum

3	GMDH1	GMD	Raritan River Steel, Paikin Steel Products 85, GMD demonstrator 800	1959

Medicine Hat: Canadian Fertilizer

6706	SW8	GMD	CP 6706	1951

Medicine Hat: Stay Sales

6531	S3	MLW	Canadian Fertilizer, CP 6531	1955

Ram River: Canterra Energy

4010, 4011	C415	Alco	BN 4010, 4011; SP&S 100, 101	1968

Redwater: Esso Agricultural Chemicals

901	S2r	Alco	Newburgh & South Shore 1006	1946
902	S2	Alco	BN 902, NP 702	1942
903	S2r	Alco	TRRA 590	1949

Taber: Alberta Sugar

1501	25-ton	GE		1956

West Carseland: ICI

1	35-ton	Plym		1985

BRITISH COLUMBIA

Beaver Cove: Fletcher Challenge Canada

4097	RS3	Alco	D&H 4097	1952

Crestbrook: Crestbrook Forest Industries

195	44-ton	CLC	CP 14	1958

Crofton: Fletcher Challenge Canada

9	65-ton	Whit	U. S. Navy 65-00342	1944
1500	SW1500	EMD	SP 2663	1972

BRITISH COLUMBIA (continued)

Fraser Flats: Northwood Pulp & Timber

101	65-ton	GE	U. S. Navy 65-00407	1943
	Stored unserviceable			
102	SW900u	EMD	CRI&P 780	1959
103	SW1200	EMD	MILW 630	1954

Harmac: MacMillan Bloedel

184	65-ton	GE	Sidbec Dosco 7, Fundy Gypsum 12, Canadian Gypsum 12	1948

Kemano: Tonto Drilling

1, 2	25-ton	Plym		1982, 1977

Kitimat: Alcan Canada

1003	SW900	GMD		1957

Kitimat: Eurocan Pulp & Paper

43	70-ton	GE	Hawker Siddeley 43, Acadia Coal 43, CN 43	1950
502	S13	MLW	British Columbia 502	1959
941	70-ton	GE	BC Hydro 941	1949

Mitchell Island: Western Canada Steel

6574, 6583	S3	MLW	CP 6574, 6583	1957

Nimpkish Valley: Canadian Forest Products

301-303	SW1200RS	GMD		1956
	Dynamic brakes			
4804	SW1200	EMD	Coos Bay Lumber 1203	1954
	Dynamic brakes			

North Vancouver: Canadian Occidental Petroleum

149	SW9	EMD	BN 149, GN 17	1951

North Vancouver: Neptune Bulk Terminals Canada

46	T6	Alco	PT 46	1968
80, 81	S6	Alco	SP 1271, 1217	1955

North Vancouver: Pioneer Grain

1, 2	45-ton	Plym		1979

North Vancouver: Vancouver Wharves

21, 26, 29	S6	Alco	SP 1220, 1232, 1240	1956
102	NW2	EMD	BN 595, SP&S 42	1948
660	S3	MLW	United Grain Growers, CP 6503	1951
822	SW1200	EMD	SP 2280	1965
824	SW1200	EMD	MP 1279	1966

Port Alberni: MacMillan Bloedel

16	SW900r	EMD	SP 1197	1954

Port Mellon: Howe Sound Pulp & Paper

950001	SW900	GMD	CN 9735	1951

Powell River: MacMillan Bloedel

699	65-ton	GE	Domtar, Donnacona Paper	1956
931	SW900	GMD	BCH 931, Midland Ry of Manitoba	1956

Prince Rupert: Prince Rupert Grain Terminal

1135	SW1200	EMD	MP 1135	1966
1219	S6	Alco	SP 1219	1955

Vancouver: Pacific Elevators Ltd.

A, B, C		Hunslett		1968-1974

Vancouver: United Grain Growers Terminal

3	SW1200	EMD	Houston Belt & Terminal 35	1966

MANITOBA

Brandon: Manitoba Hydro

6146C1	65-ton	GE		1957	

Brandon: Simplot Chemicals

2504	S3	MLW	CP 6571	1957	
6521	S3	MLW	CP 6521	1953	

Churchill: Ports Canada

1, 2	35-ton	Plym		1930	

East Selkirk: Manitoba Hydro

97C1	35-ton	GE		1968	

Flin Flon: Hudson Bay Mining & Smelting

1	70-ton	GE		1956	
2	70-ton	GE	Kennecott Copper 64	1949	
	Stored unserviceable				
3	70-ton	GE	Kennecott Copper 65	1949	
5	70-ton	GE	LC 60	1950	
6	70-ton	GE		1955	
	Stored unserviceable				
7-9	70-ton	GE		1954-1955	
42	60-ton	GE	Asbestos & Danville	1929	
90	50-ton	GE		1950	Electric
93	50-ton	GE		1928	Electric
95		BLW	CWLE 107	1910	Electric
97	50-ton	GE	St. Louis & Belleville Electric 10	1901	Electric

Fort Garry: Manitoba Sugar

1	45-ton	GE		1958	

MANITOBA (continued)
Fort Rouge: BN Manitoba Ltd.

2	GP9	GMD		1957	

Fort Whyte: Canada Cement LaFarge

1	15-ton	GE		1955	
6	7-ton	Plym		1923	
13	50-ton	GE		1954	
641	45-ton	GE		1947	

Limestone: Bechtel-Kumagi

2702	25-ton	Plym	Rohm & Haas	1958	

Pine Falls: Abitibi Price

101	S3	MLW	Manitoba Paper 30, CN 8454	1952	
7158	SW8	GMD	CN 7158	1951	

Point du Bois: Winnipeg Hydro

4	20-ton	Dav		1927	

Steep Rock: Canada Cement LaFarge

3	50-ton	GE		1954	

Thompson: Inco

1, 2	6-ton	Bkvl		1960	36" gauge
3, 4	6-ton	Bkvl		1961	36" gauge
2081	G8	GMD	CN 851	1954	
2082-2084	RS18	MLW		1968	

NEW BRUNSWICK

Bathhurst: Stone Consolidated Inc.

3	S3	MLW		1954	

Belledune: Belledune Fertilizer

97	50-ton	GE	General Aniline 120	1963	

Dalhousie: New Brunswick International Paper

1	50-ton	Whit	Brown Corp.	1950	

Penobsquis: Potash Corp.

92-010	S4	MLW	Union Carbide 45, Asbestos & Danville 46	1949	
92-018	S12	MLW	CN 8241	1958	
92-019	S4	MLW	Quebec North Shore Paper, Asbestos & Danville 47	1950	

Saint John: Saint John Drydock Co.

1	44-ton	GE	Dexter Sulphite Pulp, U. S. Army 7042	1940	

NEWFOUNDLAND

Labrador City: Iron Ore Company of Canada

431501-431509	SW1200MG	GMD		1963-1971	Electric

NOVA SCOTIA

Brooklyn: Bowater Mersey Paper Co.

2	25-ton	Vulcan		1928	

Hantsport: Fundy Gypsum

14	45-ton	GE	Canadian Arsenals 2	1953	

Milford Station: National Gypsum

1, 2	45-ton	GE		1955	
3	45-ton	GE	Malagash Salt	1949	
5	100-ton	GE	Appalachian Power 2	1971	

Point Tupper: Georgia Pacific

1	50-ton	GE	Canada Cement 1	1956	

River Denys: Georgia Pacific

1	45-ton	GE		1962	

Sydney: Sysco

2-4	65-ton	GE		1947	
7	80-ton	GE	National Harbours Board 3	1953	
8-10	80-ton	GE		1951-1958	
11, 12	SW8	GMD		1952	
14	SW9	EMD		1951	
72	50-ton	GE		1949	36" gauge
73	50-ton	GE		1957	36" gauge
74	80-ton	GE		1956	36" gauge

Trenton: Lavalin

2701	45-ton	GE	Eastern Car 3	1948	
2702	45-ton	GE	Eastern Car 4	1954	

Wallace: Wallace Stone Quarry

1	20-ton	Plym	Canadian Dock & Dredge, U. S. Army	1943	

Windsor: Fundy Gypsum

640-642	25-ton	GE		1947-1956	
647	45-ton	GE	U. S. Gypsum 10	1949	

ONTARIO

Acton: United Aggregates

97	S3	MLW	CN 8497	1954	

ONTARIO (continued)

Amherstburg: Canadian Occidental Chemicals

1	25-ton	GE	BCM Ltd., Francon 25002, Highway Paving 2	1947

Amherstburg: General Chemical Co.

8	S1	Alco	Solvay Processing 2	1946
9	S2	Alco	CN 8137	1949
3A	S4	Alco	Allied Chemical, B&O 9113	1957
		Stored unserviceable		
4A	S4	Alco	Allied Chemical	1953
B12	S4	Alco	Allied Chemical, B&O 9001	1953

Amherstburg: The Hearn Group

200	S1	Alco	Essex Terminal 101	1941
300	S3	MLW	Essex Terminal 103	1952

Beachville: Beachville Lime

1	45-ton	GE		1957
5	45-ton	GE	North American Cyanamid	1950

Burlington: General Electric Canada

1	23-ton	GE	Canada Car & Munition	1941

Cambridge: Babcock & Wilcox

V70	40-ton	Plym	Ontario Hydro	1953
V90	44-ton	CLC	CP 17	1959

Cardinal: Canada Starch Corp.

9	SW8	GMD	CN 7162	1951

Clarkson: St. Lawrence Cement

1	45-ton	GE		1956

Copper Cliff: Inco

1-3	10-ton	Rogr	1957	30" gauge
4	10-ton	Rogr	1960	30" gauge
5	35-ton	GE	1955	
101, 102	50-ton	West	1919	Electric
108	50-ton	West	1918	Electric
110, 111	65-ton	GE	1936	Electric
112, 113	100-ton	GE	1938	Electric
114	100-ton	GE	1942	Electric
115	65-ton	GE	1942	Electric
116	100-ton	GE	1948	Electric
117	80-ton	GE	1926	Electric
118-120	100-ton	GE	1953	Electric
121	65-ton	GE	1955	Electric
122-126	85-ton	GE		1940-1950 Elec.
201	65-ton	GE		1957
202	80-ton	GE	CIL	1957
203	S2	Alco	CIL 144, WM 144	1946
204	S2	MLW	CP 7091	1949

Cornwall: ABB Canada

1	16-ton	Bkvl	Caloric Corp.	1973

Cornwall: Courtaulds

7096	S2	Alco	CP 7096	1949

Cornwall: ICI

1	25-ton	GE	Gaspe Copper Mines 1	1954

Falconbridge: Falconbridge Nickel Mines

101	S2	Alco	N&W 3321, WAB 321	1949
103	S1	Alco	EL 309, Erie 309	1946
106, 107	80-ton	GE		1951, 1953

Hamilton: Case International Ltd.

1	S1	Alco	EL 308, Erie 308	1946

Hamilton: Dofasco

411, 412	SW8	GMD		1952, 1953
414	SW1200	GMD		1964
415, 417	NW2	EMD	JT 30, 32	1947
418	SW9	EMD	FEC 223	1952
419, 420	NW2	EMD	JT 34, 35	1949
422, 423	S2	Alco	N&W 2050, 2070	1951, 1952
424-426	NW2	EMD	CN 7936, 7958, 7937	1946-1947
427	SW8	GMD	CGT 102	1951

Hamilton: National Steel Car Co.

11, 12	50-ton	GE		1954

Hamilton: S. G. Paikin & Co.

1	35-ton	Whit	Standard Slag 37	1951
7	25-ton	Plym	Canada Crushed Stone	1927
8	80-ton	GE	Canada Starch, Kimberly Clark	1959

Hamilton: Stelco

51	65-ton	GE		1950
54	80-ton	GE		1952
74-77	SW8	GMD		1953
78-85, 87-93	SW900	GMD		1956-1967

Hamilton: Westinghouse Canada Ltd.

1	50-ton	CLC		1950

ONTARIO (continued)

Huntsville: G. W. Martin Wood Products

1	25-ton	Dav	Weldwood Industries, Hay Inc.	1939

Ingersoll: Stelco

1	25-ton	GE		1950

Iroquois Falls: Abitibi Price

80	S4	MLW		1950
Stored unserviceable				
1203	S4	MLW	ONT 1203	1950
1310	RS3	MLW	ONT 1310	1951

Kapuskasing: Kruger Inc.

108	S13	MLW		1967
109	RS23	MLW	Devco 201, Sydney & Louisburg	1960

Kenora: Boise Cascade

409	80-ton	GE	MD&W 14	1951

L'Orignal: Ivaco

1	65-ton	GE		1951
3	25-ton	Plym	Canadian Refractories, Hopkins Co.	1940
1007	S2	Alco	General Electric, N&SS 1007	1946
7016	S2	Alco	CP 7016	1944
25060	S4	Alco	Francon, PC 9792	1950

Little Current: Dominion Mines & Quarries

1	30-ton	Plym	Wayne County Road Board	1925
2	20-ton	Whit		1929
3	10-ton	Plym		1941

London: General Motors Canada-Diesel Division

57	SW9	GMD	TH&B 57	1951
102	SW8	GMD	Essex Terminal 102	1951

Maitland: Du Pont Canada Ltd.

1	SW900	GMD	CN 7945	1958

Marathon: James River Marathon

17101	S13	MLW		1966
17102	S2	Alco	CP 7024	1944

Marathon: Davco East Dock Co.

620	SW9	EMD	MILW 620	1951

Mountain Chute: Ontario Hydro

1	16-ton	Whit		1930

Nanticoke: Stelco

70	SW9	GMD		1951
Stored unserviceable				

71, 72	SW8	GMD		1951
Stored unserviceable				
451	SW900	EMD	Canton 45	1956
452	SW900	EMD	B&O 9404	1955
453, 454	SW8	EMD	SP 1108, 1124	1953

Orillia: Nelson Aggregates

7010	25-ton	GE	U. S. Army 7768	1944
7050	44-ton	Whit		1945
7060	S3	MLW	CP 6534	1955
7080	S3	MLW	CP 6564	1957
1	25-ton	GE	Flintkote Co., St. Joe Minerals, U. S. Air Force	1951

Ottawa: National Research Council

102	S4	MLW	R&S 102	1951
6593	S3	MLW	CP 6593	1957

Peterborough: General Electric Canada

1	50-ton	GE	Western Electric 3	1956

Port Colborne: Inco

1	45-ton	GE		1956

Port Robinson: Cyanamid Canada

104	65-ton	GE		1950

Port Robinson: B. F. Goodrich

1	45-ton	GE	Falconbridge Nickel	1946

Red Rock: Domtar

68	RS3	Alco	R&S 30, RDG 485	1952
7961	GP8	ICG	LVRC 7961, ICG 7961	1953

Sarnia: Byers Corp.

404	SW900m	GMD	CN 7604	1958
902	SW900	EMD	CCR 902	1961

Sarnia: Du Pont Canada

40	SW1000	EMD	Houston Belt & Terminal 40	1968

Sarnia: ICI

915	SW900	EMD	CRI&P 560	1958

Sarnia: Novacor

1	45-ton	GE	Maple Leaf Milling	1948

Sault Ste. Marie: Algoma Steel

1, 2	50-ton	GE		1951
	36" gauge			
3	25-ton	GE	Canadian Distillers	1946
4	25-ton	GE	MISS 1	1945

ONTARIO (continued)
Sault Ste. Marie: Algoma Steel (continued)

4-6	50-ton	GE		1955-1964
	36" gauge			
30	80-ton	GE		1949-1952
50	SW8	1 GMD		1952
51	SW900	1 GMD		1964
60, 61	110-ton	GE		1974, 1975

Sault Ste. Marie: Algoma Steel, Tube Division

1	80-ton	GE		1951

Sault Ste. Marie: St. Mary's Paper Co.

7751	80-ton	GE		1952

Scarborough: AFG Glass Co.

10	65-ton	GE	BAR 10	1950

Scarborough: Dufferin Concrete Products

9114	25-ton	Plym	Dufferin Construction Co.	1928

Scarborough: Nelson Aggregates

36910	50-ton	CLC	Limestone Products	1950

St. Catharines: General Motors Canada

2128	GMDH3	GMD	GMD demonstrator 275	1960
47074	SW900	GMD		1966

Terrace Bay: Kimberly Clark

41	SW1000	EMD	Houston Belt & Terminal 41	1968
6539	S3	MLW	CP 6539	1955

Thorold South: Beaver Wood Fibre Co.

1	25-ton	GE		1947

Thorold South: Quebec & Ontario Paper

1	SW900	GMD	CN 7212	1954
2	SW900	GMD	CN 7943	1958

Thunder Bay: Canadian Pacific Forest Products

575	SW900	GMD	CN 7608	1958

Thunder Bay: Cargill Grain

14	SW8	EMD	SCL 14, ACL 54	1952
206	RS1	Alco	Devco 206, C&NW 200	1944

Thunder Bay: Pioneer Grain

1	65-ton	GE		1951
4	65-ton	GE	Johnson Co. 7	1952

Timmins: Falconbridge Nickel

51-53	RS23	MLW		1966
54, 55	GP38-2	GMD		1976

Toronto: Victory Soya Mills

1	25-ton	Whit	Canada Creosote	1950	30" gauge

Vaughan: Nelson Aggregates

7070	70-ton	GE	St. Lawrence 11	1956

Welland: Atlas Steel

4	25-ton	Porter	Atlas Explosives	1943
8	50-ton	Atlas	U. S. Navy 65-00326	1939
9	S3	MLW	CP	1955

Welland: Shaw Pipe Protection Co.

6	65-ton	GE	North American Cyanamid 6	1952

Welland: Stelpipe, Page Hershey Works

1	45-ton	GE		1947
7	80-ton	GE	Stelco 53	1950

Welland: Stelpipe, Welland Tube Works

3	50-ton	GE	Stelco 3	1956
5	44-ton	GE	CN 5	1956

Welland: Union Carbide Canada

9	65-ton	GE	U. S. Army 1404	1941
10	80-ton	GE		1953

Whitby: Lake Ontario Steel

1	70-ton	GE	BCOL (PGE) 556	1950
2	70-ton	GE	SLSF 12, Okmulgee Northern 7	1950
3	70-ton	GE	N&W 703, Southern 703	1948

Windsor: Zalev Brothers

1	45-ton	GE	Canada Cement 1	1946
7	45-ton	GE	Allied Chemical 7, Brunner Mond 7	1949
128	45-ton	GE	Great Lakes Pulp & Paper, CCF 110, U. S. Army 7430	1941

QUEBEC

Alma: Abitibi Price

7910	SW900	GMD	CN 7910	1953

Baie Comeau: Quebec & Ontario Paper

1	50-ton	GE		1948
2	45-ton	GE	Ontario Paper 2	1947
3	50-ton	GE	PA&M 7, U. S. Navy 65-00262	1941
7	RS3	MLW	RS 22	1955
45	S3	MLW	Papiers Cascades 106, Price Bros. 1, MLW demonstrator 7004	1950
9137	SW9r	EMD	CR 9137, PRR 8537	1952

QUEBEC (continued)
Beauharnois: PPG Standard Chemical

1	45-ton	GE	Carey Canadian Mines, Bell Asbestos, U. S. Navy 65-000136	1942

Beaupre: Abitibi Price

1	50-ton	GE	1949

Berthierville: GLC Canada

1	25-ton	GE	Singer Manufacturing Co.	1949

Bromptonville: Kruger Inc.

1	65-ton	GE	M&E 1, EJT 17	1950

Cedars: Soulanges Industries

111, 114	DL535E	BBD	WP&Y 111, 114 (never delivered)	1982

Chandler: Abitibi Price

5801	80-ton	GE	1952

Clermont: Donohue Charlevoix

1	25-ton	GE		1949
1079	S2	Alco	CP 7098	1949

Contrecoeur: Sidbeck Feruni

411	SW900	GMD	CN 7936	1957
412	SW8	GMD	CN 7163	1951
413	SW900	GMD	CN 7940	1957

Contrecoeur: Stelco

73	SW8	1 GMD		1951
D8	S3	MLW	National Harbours Board	1951

Delson: Domtar

1	25-ton	Whit		1940	30" gauge
2	20-ton	CLC	Canada Creosoting	1959	30" gauge
3	25-ton	GE		1948	30" gauge

Dolbeau: Domtar

1	50-ton	GE	Canada Cement 5	1948
2	25-ton	GE		1952

Donnacona: Domtar

6	SW900	GMD	CN 7952	1958

East Angus: Papiers Cascade

1	65-ton	GE		1952
2	65-ton	GE	Domtar 65, Brompton Pulp & Paper	1953

Havre St-Pierre: Quebec Fer et Titane

5110	MP15DC	EMD		1980
5127, 5138	GP9	GMD	QNS&L 139, 156	1955, 1956
5139, 5140	GP9	GMD	Cartier 57, 54	1960
5146	RS18	MLW		1968
5169	MP15DC	EMD		1977

Hull: E. B. Eddy

1	45-ton	GE	1946
2	25-ton	GE	1951

Joliette: Ciment St-Laurent

Z100	S11	MLW	CP 6622	1959

Jonquiere: Abitibi Price

1	S2	MLW	CN 8122	1949
50	S13	MLW	Asbestos & Danville 50	1962

Jonquiere: Jonquiere Mills

1	10-ton	Plym	Watson-Hopkins Co.	1957

Lachine: Dominion Bridge Quebec

1	25-ton	GE	Ivaco	1944
N15	44-ton	Whit	MP 806	1941

Lauzon: Davie Shipbuilding

1	50-ton	GE	1958

La Tuque: Canadian Pacific Forest Products

2Y64	S2	Alco	IT 1007	1950
			Stored unserviceable	
2Y65	S13	MLW		1963
7180	SW8	GMD	CN 7180	1951

Lebel sur Quevillion: Domtar

28-12	NW2	EMD	CN 7961	1947
28-137	SW1200	GMD	CN 7734	1959

Lime Ridge: Dominion Lime

1	35-ton	Whit	FGE, U. S. Navy	1937

Matagami: Societe d'Engergie de la Baie James

575-01	65-ton	GE	Hydro Quebec 130	1948

Montreal: Bombardier Inc.

7000	HR412W	BBD	1981

Montreal: Canadian Steel Foundries

D1	44-ton	GE	Donohue Charlevoix 2, Claremont & Concord 17, Greenville & Northern 78	1951

Montreal: Canadian Steel Wheel

2	50-ton	GE	Alcan 72-291	1941

Montreal: General Electric Canada

30009	80-ton	GE	1949

QUEBEC (continued)
Montreal: Ports Canada

7601, 7602	SW1001	EMD		1976
8403-8406	MP15AC	GMD		1984

Montreal: Sidbec Dosco

407	80-ton	GE	National Harbours Board 2	1952

Montreal East: Union Carbide Canada

176	SW9	EMD	SCL 176, ACL 694	1951
177	SW9	GMD	C&O 5244	1951

Nascouche: Merrilees Corp.

1	SW9	GMD	International Iron & Metal, TH&B 55	1950
2	45-ton	GE	Ciments LaFarge, Canada Cement	1950
10	70-ton	GE	Potasco 10, SN 201	1955
2005	SW9	EMD	SCL 165, ACL 683	1951
7062	S2	Alco	CP 7062	1947
35071	S2	Alco	CP 7097	1949
N16	S4	MLW	CN 8020	1952

Noranda: Noranda Mines

18, 19	80-ton	GE		1956, 1957
25	80-ton	GE	PA&M 180, U. S. Navy 65-00280	1951

Portage du Fort: Stone Consolidated Corp.

2	S1	Alco	L&N 24	1945
4	SW900	EMD	CRI&P 911	1959

Quebec City: Produits Forestieres Daishowa

1	50-ton	GE		1948

Riviere des Prairies: Canadian General Transit Corp.

1	25-ton	GE	Domte	1948

Shawinigan: B. F. Goodrich

1	25-ton	GE	Iroqouis Construction Co., Shawinigan Engineering	1948

Sherbrooke: Canadian Ingersoll-Rand

733	25-ton	GE	Canada Cement	1947

Atlas Steel

1	44-ton	GE	Lake Ontario Steel WAB 51	1943

Sorel: Marine Industries

390	44-ton	GE	National Harbours Board 1	1946
392	20-ton	Plym		1945
393	30-ton	Plym		1945
394	50-ton	GE	Canadian International Paper, Mattagami 103	1951

Sorel: Quebec Fer et Titane

1	45-ton	GE	Canada Cement	1946
2	44-ton	GE	Narraganset Pier 42, B&M 118	1947
624	45-ton	GE	NYS&W 200	1958
638	50-ton	GE		1955
640	50-ton	GE	Canada Cement 2	1954
5162	44-ton	GE	SEPTA, A&LM 34, LRS 100	1946

St-Felicien: Donohue St-Felicien

44	S3	MLW	CN 8482	1953

St-Hubert: Provincial Diesel Service

5	50-ton	GE	Abitibi Price 5	1948

Trois-Rivieres: Kruger Inc.

1	25-ton	GE		1955

Valleyfield: Expro

50, 51, 54	8-ton	Bkvl		1941
57	8-ton	Bkvl		1942
60	25-ton	GE		1953

Varennes: ABB Canada

1	45-ton	GE	Dosco 1	1946

Villeneuve: St. Lawrence Cement

1	45-ton	GE		1954

SASKATCHEWAN

Bienfait: Bienfait Coal Co.

182	RS1	Alco	GN 182	1944
5706	TR4A	EMD	MILW 694A	1951

Coronach: Prairie Coal Co.

84160	SW1001	EMD	Saskatchewan Power	1978
84161	SW1001	EMD	Saskatchewan Power	1978

Moose Jaw: Elders Grain

800010	23-ton	GE	Saskatchewan Power, U. S. Air Force 7792, U. S. Army 7792	1941

Rocanville: Potasco

604	70-ton	GE	Canmore Mines, JW&NW 600	1947
			Stored unserviceable	
35082	SW900	GMD	Union Carbide 2388, CN 7947	1958
35084	T6	Alco	St. Lawrence Railroad 17, N&W 17	1959

Saskatoon: Saskatchewan Power Corp.

1	80-ton	GE		1957
			Stored serviceable	

UPDATE

These four pages contain rosters which were received after most of the book had gone to press — several are those of the regional subsidiaries of the National Railways of Mexico.

ALABAMA RAILROAD

Reporting marks: ALIR   **Miles:** 61
Address: Address: 101 North Tenth Street, Fort Smith, AR 72901
Alabama Railroad operates a former Louisville & Nashvillle line from Flomaton to Corduroy, Alabama.

Nos.	Qty.	Model	Builder	Date	Notes
1779	1	GP9	EMD	1956	1
5730, 5732	2	GP7	GMD	1951	2
Total	3				

Notes:
1. Ex-Burlington Northern 1779, Great Northern 706
2. Ex-Chesapeake & Ohio 5730, 5732

ARKANSAS MIDLAND RAILROAD

Reporting marks: AKMD   **Miles:** 132
Address: P. O. Box 183, Lake Hamilton Br., Hot Springs, AR 71951
The Arkansas Midland operates four separate lines in Arkansas: Helena to Helena Junction (ex-Missouri Pacific), North Little Rock to Carlisle (ex-Rock Island), Malvern to Mountain Pine (ex-MP), and Gurdon to Birds Mill (ex-MP). Arkansas Midland is part of the Pinsly Railroad Company.

Nos.	Qty.	Model	Builder	Date	Notes
7700, 7703, 7704, 7707, 7722, 7726	6	GP8	ICG		
Total	6				

DALLAS, GARLAND & NORTHEASTERN RAILROAD

Reporting marks: DGNO   **Miles:** 62
Address: P. O. Box 460009, Garland, TX 75046-0009
The Dallas, Garland & Northeastern operates a former Missouri-Kansas-Texas route from Garland, Texas, northeast of Dallas, through Greenville to Trenton. It is part of the RailTex group of railroads.

Nos.	Qty.	Model	Builder	Date	Notes
173	1	GP9	EMD	1955	Ex-Southern Pacific
Total	1				

FORT SMITH RAILROAD

Reporting marks: FSRR   **Miles:** 49
Address: 101 North Tenth Street, Fort Smith, AR 72901
Fort Smith Railroad operates a former Missouri Pacific line between Fort Smith and Paris, Arkansas.

Nos.	Qty.	Model	Builder	Date	Notes
1791, 1902	2	GP9	EMD	1957	1
7802	1	RS3d	Alco	1955	2
Total	3				

Notes:
1. Ex-BN 1791, 1902
2. Ex-Wabash & Grand River 7802, Conrail 9931, Central of New Jersey 1549

MORRISTOWN & ERIE RAILWAY

Reporting marks: ME   **Miles:** 41
Address: P. O. Box 2206, Morristown, NJ 07962-2206
Morristown & Erie operates several short routes in northern New Jersey: Morristown to Roseland, Lake Junction to Randolph, Wharton to Rockaway, and Kenvil to Bartley.

Nos.	Qty.	Model	Builder	Date	Notes
16, 17	2	C430	Alco	1967	1
18, 19	2	C424	Alco	1964	2
Total	**4**				

Notes:
1. Ex-Conrail (New York Central) 2054, 2053
2. Ex-Toledo, Peoria & Western 800, 801

MOUNTAIN LAUREL RAILROAD

Reporting marks: MLRR   **Miles:** 128
Address: 1 Glade Park East, R. D. 3, Kittanning, PA 16201
Mountain Laurel operates former Conrail lines from Driftwood, Pennsylvania, through Du Bois and Brookville to Lawsonham (ex-Pennsylvania Railroad) and from Rose to Gretchen (ex-New York Central). It is affiliated with the Pittsburg & Shawmut Railroad and the Red Bank Railroad.

Nos.	Qty.	Model	Builder	Date	Notes
10-15	6	GP10	VMV		
Total	**6**				

NATIONAL RAILWAYS OF MEXICO

In 1986 National Railways of Mexico merged the four regional railroads of that country: Chihuahua Pacific, Pacific Railroad, Sonora-Baja California, and United South Eastern Railways. The first three are listed with NdeM in some railroad industry publications but not others; the South Eastern is listed separately (if at all). For some years prior to 1986, new power had been arriving painted NdeM colors and numbered in NdeM's numbering scheme, but the process of consolidating rosters is a slow one. The rosters of the four roads appear below in sequence by the new numbers; previous numbers are given too.

NATIONAL RAILWAYS OF MEXICO, NORTHERN REGION

The Northern Region is the former Chihuahua Pacific Railway, extending from Ojinaga, across the Rio Grande from Presidio, Texas, southwest through the city of Chihuahua to the Pacific at Topolobampo, with a branch north to Ciudad Juarez.

Reporting marks: CHP   **Miles:** 942
Address: P. O. Box 46, Chihuahua, Chih., Mexico

Nos.	Qty.	Model	Builder	Date	ChP Nos.	Notes
6300-6305	6	GP28	EMD	1964-1965	801-806	
6306, 6307	2	GP28	EMD	1965	808, 809	1
7000-7010	11	GP40-2	EMD	1975-1976	1008-1018	
7011-7014	4	GP40-2	EMD	1976	1019-1022	1
7015-7028	14	GP40-2	EMD	1981-1982	1023-1036	
8410-8417	8	GP40	EMD	1971	1000-1007	
9428-9437	10	GP38-2	EMD	1979	900-909	
9438, 9439	2	GP38-2P	EMD	1980	910, 911	1
Total	57					

Notes:
1. High short hood containing steam generator

NATIONAL RAILWAYS OF MEXICO, PACIFIC REGION

Reporting marks: FCP   **Miles:** 1,402
Address: Apartado Postal 15-M, Guadalajara, Jal., Mexico
The main line of the Pacific Region extends from the U. S. border at Nogales, Arizona, to Guadalajara. It is the former Pacific Railroad (Ferrocarril del Pacifico); previously the Southern Pacific of Mexico. There are branches east from Nogales and west from Guadalajara.

Nos.	Qty.	Model	Builder	Date	FCP Nos.	Notes
17, 19	2	PA1u	Alco	1948, 1947	17, 19	1
5310-5313	4	S6	Alco		Note 2	

Nos.	Qty.	Model	Builder	Date	FCP Nos.	Notes
7474-7485, 7487	13	RSD12	Alco	1959-1961	Note 3	
7600-7610	11	M420TR	MLW	1975	Note 4	
7700, 7702, 7703, 7705-7712	11	RSD5ru	Alco		Note 5	5
7800-7810	11	RSD5ru	Alco		576-586	6
7902-7904, 7906	4	U30C	GE	1969, 1971	403, 404, 406, 408	
8000-8002, 8004-8008, 8010-8013	12	M636	MLW	1972-1973	Note 7	
8332-8337	6	C628	Alco	1966, 1968	Note 8	
8989, 8991-8996	7	U36C	GE	1975	411, 413-418	
9182-9191	10	U23B	GE	1975	537-546	
9556-9567	12	M424W	MLW	1980-1981	Note 9	
9653-9660, 9662-9667, 9669-9680, 9682-9687	32	C30-7	GE	1979, 1981	Note 10	
11132-11135	4	C30-7	GE	1986	460-463	11
11149-11152	4	C30-7	GE	1986	456-459	12

Notes:
1. Ex-Delaware & Hudson 17, 19; Santa Fe 60L, 66L.
2. FCP 712, 719, 721, 722.
3. FCP 502-510, 512, 514, 515, 518.
4. FCP 522-525, 529-535.
5. FCP 547, 549, 550, 552-559; rebuilt by FCP 1979-1983; classed API620.
6. Rebuilt by FCP 1979-1983; classed BX620.
7. FCP 651-653, 655-659, 661, 663-666.
8. FCP 603, 605, 606, 608-610; ex-Delaware & Hudson.
9. FCP 561, 562, 564, 565, 567, 569-575.
10. FCP 419-426, 428-432, 434, 436-447, 450-455.
11. Delivered as NdeM 11132-11135.
12. Numbered into occupied slots; to be renumbered.

NATIONAL RAILWAYS OF MEXICO — BAJA CALIFORNIA DIVISION

Reporting marks: SBC   **Miles:** 379
Address: P. O. Box 3-182, Mexicali, B. Cfa., Mexico
The Baja California Division is the former Sonora-Baja California Railway. The line runs from Mexicali to Benjamin Hill, where it connects with National of Mexico's Pacific Region.

Nos.	Qty.	Model	Builder	Date	SBC Nos.
2203	1	FTA	EMD	1945	2300
7029-7034	6	GP40-2	EMD	1979-1982	2309-2314
7035-7043	9	GP40-2	EMD	1972-1976	2104-2112
7537-7539	3	GP18	EMD	1961	2304-2306
8257, 8258	2	GP35	EMD	1965	2307, 2308
Total	20				

SAN JOAQUIN VALLEY RAILROAD

Reporting marks: SJV   **Miles:** 128
Address: P. O. Box 937, Exeter, CA 93221
The San Joaquin Valley Railroad operates several former Southern Pacific routes in California's San Joaquin Valley: from Exeter west through Visalia and Hanford to Stratford and Huron, from Exeter north through Sanger to Fresno, and from Exeter south to Terra Bella.

Nos.	Qty.	Model	Builder	Date	Notes
101-103	3	GP9r	EMD	1954	Ex-UP 224, 498, 308
104	1	GP9r	EMD	1955	
Total	**4**				

SOUTH ORIENT RAILROAD

Reporting marks: SO　　　　　　**Miles:** 394
Address: P. O. Box 232, San Angelo, TX 76902
The South Orient Railroad operates former Santa Fe track (long ago the Kansas City, Mexico & Orient) from San Angelo Junction, between Brownwood and Coleman, Texas, through San Angelo, Fort Stockton, and Alpine to Presidio, Texas

Nos.	Qty.	Model	Builder	Date	Notes
200-204	5	GP9	EMD		
Total:	5				

TWIN CITIES & WESTERN RAILROAD

Reporting marks:　　　　　　**Miles:** 218
Address: 723 11th Street E., Glencoe, MN 55336
The Twin Cities & Western owns the former Milwaukee Road main line between Hopkins and Appleton, Minnesota. Trackage rights extend operations east to the Twin Cities and west to Milbank, South Dakota.

Nos.	Qty.	Model	Builder	Date	Notes
401, 402	2	GP10	ICG	1974	1
403-408	6	GP10	ICG	1971-1977	2
Total	8				

Notes:
1. Ex-BN 1420, 1401
2. Ex-ICG 8146, 8270, 8251, 8118, 8091, 8028

UNITED SOUTH EASTERN RAILWAYS (Ferrocarriles Unidos del Sureste)

Reporting marks: FUS　　　　　　**Miles:** 843
Address: Calles 43 y 44 No. 429-C, Mérida, Yucatán, Mexico
United South Eastern Railways extends from a connection with National of Mexico at Coatzacoalcos, in the state of Vera Cruz, east to Mérida. Several branches, some narrow gauge, fan out from Mérida.

Nos.	Qty.	Model	Builder	Date	FUS Nos.
5301-5306	6	S6	Alco	1955 1960	104-107, 109, 110
5418-5420	3	GA8	EMD	1964	83-85
5421, 5427	2	GA8	EMD	1967	86, 92
5700, 5701	2	C420	Alco	1965	510, 511
7295	1	RS11	Alco	1956	401
8255, 8256	2	GP38	EMD	1971	512, 513
9181	2	B23-7	GE	1979	523
9415-9421	7	GP38-2	EMD	1972, 1975	515-521
9422-9427	6	GP38-2	EMD	1981, 1982	528-533
9553-9555	3	M424W	MLW	1981	525-527
13005-13008	4	SD40-2	EMD	1972	601-604